Biochemistry and Molecular Biology in the Post Genomic Era

nova
science publishers

www.novapublishers.com

Biochemistry and Molecular Biology in the Post Genomic Era

Omics for Animal Sciences: Principles and Approaches
Asif Nadeem (Author)
Maryam Javed (Author)
2022. ISBN: 978-1-68507-722-8 (Hardcover)
2022. ISBN: 978-1-68507-901-7 (eBook)

The Future of Metabolic Engineering
Abhishek Sharma, PhD (Editor)
Dhruti Amin, PhD (Editor)
2022. ISBN: 978-1-68507-362-6 (Hardcover)
2022. ISBN: 978-1-68507-464-7 (eBook)

Recent Advances in Computer Aided Drug Designing
Akhil Varshney, PhD, Ashutosh Mani, PhD (Editors)
2021. ISBN: 978-1-53619-739-6 (Hardcover)
2021. ISBN: 978-1-53619-904-8 (eBook)

A Closer Look at Proteolysis
Jelena Radosavljević (Editor)
2020. ISBN: 978-1-53618-677-2 (Hardcover)
2020. ISBN: 978-1-53618-743-4 (eBook)

Caspase-3: Structure, Functions and Interactions
Lunawati L. Bennett, PhD (Editor)
2020. ISBN: 978-1-53618-610-9 (Softcover)
2020. ISBN: 978-1-53618-686-4 (eBook)

More information about this series can be found at
https://novapublishers.com/product-category/series/biochemistry-and-molecular-biology-in-the-post-genomic-era/

Elaine J. Mathis
Editor

Structure and Function of Cytochromes

nova
science publishers
www.novapublishers.com

NOTICE TO THE READER

Library of Congress Cataloging-in-Publication Data

ISBN: 979-8-88697-102-6

Published by Nova Science Publishers, Inc. † New York

Contents

Preface

This book contains five chapters exploring the structure and function of cytochromes, which are redox-active proteins that contain a heme with a central Fe atom at its core as a cofactor. Chapter One explores the functional relevance of cytochrome c-oxidase subunit IV isoform expression for cellular energy sensing, ATP production, and oxidative stress levels and its role in neurodegeneration. Chapter Two deals with the factors determining the orientation of ligated histadines in hemoproteins, cytochromes, and model systems. Chapters Three and Four include comparative studies of native and synthetic cytochrome b. Lastly, Chapter Five discusses some examples of synthetic and native cytochromes used as central components of biosensors in bioelectronic/biocatalytic applications.

Chapter 1 - Cytochrome c-oxidase (CCO) is the terminal enzyme complex of the respiratory chain, known as complex IV, in all mitochondria. Mitochondria are responsible for cellular energy and, therefore, the biochemical adaptation and evolution of this enzyme is key to understand its role in humans. CCO's main role is to transduce energy from the catalysed redox reaction into an electrochemical proton gradient which, apart from ATP formation, ultimately, drives the electrical membrane potential and the pH gradient. In neurodegenerative diseases, there is an impairment of mitochondrial energy production and increased ROS production; however, the involvement of CCO as a consequence and/or cause of pathological processes is still unknown. In this review, the authors will explore the functional relevance of CCO subunit IV isoform expression for cellular energy sensing, ATP production, and oxidative stress levels and its role in neurodegeneration. The authors will discuss its crucial role in the progression and prevalence of numerous human neurodegenerative diseases with special emphasis on the importance of CCO as a diagnostic and therapeutic tool for these diseases. Finally, the authors will approach brain photobiomodulation as a promising strategy to enhance the metabolic capacity of neurons and stimulate anti-inflammatory, anti-apoptotic, and antioxidant responses, as well as neurogenesis and synaptogenesis.

Chapter 2 - Native heme proteins cover a broad range of biological functions, including those which show redox-active ligand complexes. Therefore, it is of great interest to understand the role of the protein environment in modulating the heme cofactor's properties. Different orientations of the imidazole rings relative to heme can shift the heme redox potential and control the coordination of substrates to heme proteins and cytochromes. The molecular model systems built with the minimalistic structural requirements can help further to sort out the different factors and control mechanisms. The entire RCSB PDB database was searched to reveal these factors, and the heme proteins with at least one histidine ligand were analyzed. The authors analyzed a total of 693 hemes in 432 different crystal structures of heme proteins. The exhaustive analysis of available crystal structures of heme proteins has shown that two major factors determine the orientation of axial histidine ligands: (1) the interaction of the imidazole ring with the heme propionic acids and (2) the interaction with its histidine backbone. The results of the PDB study have been interpreted by a molecular force field and quantum-chemical computations, confirming that the emphasized interactions are mainly of electrostatic origin. The protein environment offers a sufficient number and different oriented hydrogen-bond acceptor groups, as was found in several detailed case studies. Surprisingly, based on the same analysis, the orientation of the NδH group of imidazole axially coordinated to heme does not often depend on a specific hydrogen-bond pattern with acceptor groups of the protein. Thus, the hydrogen-bond pattern is not often decisive, and it is probably the natural pathway to fine-tune the orientation of the ligated imidazoles. DFT computations revealed two main factors determining the imidazole orientation: (1) the direct intramolecular electrostatic interactions of propionic groups with the polar NδH groups of imidazole, and (2) the electrostatic interaction of the total dipole moment of the imidazole-heme complex with the reaction field. The theoretical and computational outstanding challenges in this area were also discussed. One of the purposes of the presented study is to shed some light on the histidine orientation in proteins and find a set of rules that control or govern histidine orientation.

Chapter 3 - In the previous chapter, the authors have generated atomic coordinates of a synthetic protein recently synthesized (SynCytb) to model the central part of the native cytochrome b (Cytb) subunit consisting of a four-helix bundle with two hemes as cofactors. Since the X-ray structure is not solved, the structural elements of SynCytb were assembled from scratch using all known chemical and structural information available, thereby avoiding strain and atomic clashes. The molecular mechanics was subsequently employed to fully relax the model structure of SynCytb by applying the stepwise constraint energy minimization. Molecular dynamics simulations are applied to the computer-generated synthetic protein structure and the crystal structure of native Cytb, revealing very similar root-mean-square deviation (RMSD) profiles over time and, thus, demonstrating a relative rigidness, stability, and strain-free of the finally modeled structure.

In this chapter, to further test and validate the computer-generated structure of SynCytb, characteristic observables have to be computed and compared with measured quantities. One of these is the redox potential, a key parameter of redox-active proteins. The results of the redox potentials calculation are discussed here. For comparison, protonation patterns and heme redox potentials in the SynCytb model structure and native Cytb from the respiratory cytochrome bc1 (Cytbc1) complex are calculated from the electrostatic energies by solving the Poisson-Boltzmann equation. The computed redox potentials generally agree within 20 mV with the experimental measurements. The factors of protein environment and different energy terms determining a shift in the heme redox potentials are elucidated and analyzed in detail. In addition, the pH dependence of the redox potential of all hemes is evaluated in the physiologically relevant pH range. This chapter also rationalizes the effects of the applied ionic strength, external redox potential of bulk solvent (aqueous solution), the coupling of protonation and reduction reactions, and the distribution and population of the microscopic redox states in the studied biomolecular systems. The redox-Bohr effect in native Cytb is modeled successfully, whereby the essential protonatable groups responsible for this phenomenon are identified. A good agreement between calculated and measured redox potentials offers the possibility to analyze the factors of the protein environment determining the shift in the redox potentials of hemes, which are cofactors in the synthetic and native cytochromes. That opens new avenues to understand how nature tunes the redox potentials of cofactors in proteins to perform their biological functions and how one can construct synthetic proteins whose cofactors have desired redox potentials. The applied approach is valuable in studying structure–function correlation and understanding protein–cofactor interactions in atomic detail.

Chapter 4 - The authors defined a procedure to generate the atomic coordinates of an artificial protein (SynCytb), recently synthesized using the template-assembled synthetic strategy. A four-helix bundle, binding two hemes, constitutes the catalytical heart of the photosynthetic cytochrome b6f (Cytb6f) complex and the respiratory cytochrome bc1 (Cytbc1) membrane-bound protein complex. The model structure mimics the central part of the Cytb subunit and reproduces a native-like function. It consists of a four-helix bundle, two hemes, and a cyclic decapeptide as a template. Since the X-ray structure is not yet solved, the structural elements of SynCytb were assembled from scratch using all known chemical and structural information available, thereby avoiding strain and atomic clashes. A great deal of work was done to generate the coordinates of this protein by a sophisticated stepwise modeling technique. The molecular mechanics was subsequently employed to fully relax the model structure by applying the stepwise constraint energy minimization. To validate the stability and robustness of the computer-generated model structure, SynCytb is tested by monitoring the conformational stability and fluctuations during a long-term molecular dynamics (MD) simulation and comparing the results with values obtained from the crystal structure of native Cytb. The RMSD (root-mean-square deviation) and RMSF (root-mean-square fluctuation) data from the native and artificial proteins MD simulations reveal very similar profiles over time, demonstrating a relative rigidness, stability, and strain-free of the final model structure. The results of the theoretical work presented in this chapter offer an extensive cheminformatics/ bioinformatics analysis to elucidate and compare the structural details and heme conformations of both proteins. Specifically, the details related to redox cofactors and their close surroundings have been examined. The authors analyzed the relative orientation of axial histidines ligated to heme, the heme conformations, propionates orientations, and orientation of the tilted helical axes in native and SynCytb and compared them with available experimental data. The implemented analysis provides valuable information for further research and facilitates protein design and molecular modeling since the heme redox potential intrinsically and internally depends on the axial ligands' type, orientation, alignment, and basicity. The protein design presented here can be used to generate novel proteins with tailored structural and functional properties.

Chapter 5 - The field of bionanotechnology has recently demonstrated a broad spectrum of applications with nano–bio materials, and these novel hybrid systems have found extensive use for sensing, conductivity, catalysis, energy harvesting, storage or transduction, and the design of new "smart" devices. Specifically, the integrated biomolecule–nanoparticle systems assembled on highly conductive electrodes have been successfully exploited in designing enzyme-based bioelectrochemical devices. The de novo protein systems studied here can find use in bionanotechnology and bioelectronics as the central components of biosensors. Hence, understanding their redox properties is essential for fine-tuning the function of such biosensors. In this chapter, a molecular modeling approach has been employed by combining molecular mechanics, molecular dynamics, and quantum mechanics to generate the atomic coordinates of 25 small synthetic mono-heme proteins with a four-helix bundle motif that resembles the catalytic site of native cytochrome b. The protonation pattern of proteins and the redox potentials of the heme cofactor are evaluated from the electrostatic energies by solving the Poisson-Boltzmann equation. The factors of protein environment and effects of mutations determining the shift in the redox potential are analyzed and elucidated. The computed redox potentials generally agree well with available experimental measurements (RMSD $\leq$ 15 mV). The predictions of redox potential are made for the structures with Em values outside of the range used for a redox indicator. The factors responsible for the shift in the redox potential are analyzed in great detail, and the concept presented here may pave the way for the design of proteins with desired properties customized to specific functions while providing new valuable information in both fundamental and applied sciences. In this chapter, some examples of synthetic and native cytochromes used as central components of biosensors in bioelectronic/ biocatalytic applications are also discussed.

Chapter 1

Cytochrome C-Oxidase as a Neuroprotective Target

Joseph Allison[1], Ana Jimenez[2], Victor Echeverry[2], Sara Uceda[2] and Natalia Arias[2,3,4,*]

[1]Department of Basic and Clinical Neuroscience, Institute of Psychiatry, Psychology and Neuroscience, King's College London, London, UK
[2]Department of Psychology, Faculty of Life and Natural Sciences, BRABE Group, University of Nebrija, Madrid, Spain
[3]Institute of Neurosciences of the Principality of Asturias (INEUROPA), Spain
[4]Health Research Institute of the Principality of Asturias, Spain

Abstract

Cytochrome c-oxidase (CCO) is the terminal enzyme complex of the respiratory chain, known as complex IV, in all mitochondria. Mitochondria are responsible for cellular energy and, therefore, the biochemical adaptation and evolution of this enzyme is key to understand its role in humans. CCO's main role is to transduce energy from the catalysed redox reaction into an electrochemical proton gradient which, apart from ATP formation, ultimately, drives the electrical membrane potential and the pH gradient. In neurodegenerative diseases, there is an impairment of mitochondrial energy production and increased ROS production; however, the involvement of CCO as a consequence and/or cause of pathological processes is still unknown. In this review, we will explore the functional relevance of CCO subunit IV isoform expression for cellular energy sensing, ATP production, and oxidative stress levels

[*] Corresponding Author's Email: narias@nebrija.es.

In: Structure and Function of Cytochromes
Editor: Elaine J. Mathis
ISBN: 979-8-88697-102-6

and its role in neurodegeneration. We will discuss its crucial role in the progression and prevalence of numerous human neurodegenerative diseases with special emphasis on the importance of CCO as a diagnostic and therapeutic tool for these diseases. Finally, we will approach brain photobiomodulation as a promising strategy to enhance the metabolic capacity of neurons and stimulate anti-inflammatory, anti-apoptotic, and antioxidant responses, as well as neurogenesis and synaptogenesis.

Keywords: cytochrome c-oxidase, ATP, oxidative stress, neurodegeneration, photobiomodulation

CCO Structure, Evolution and Biochemistry across Species

The Evolution of CCO

Cytochrome c-oxidase (CCO) is the terminal enzyme of the electron transport chain (ETC) and utilizes most of the oxygen delivered to cells. As a result, CCO is a key element in the energetic production of organisms, although some species of anaerobic protists have lost key CCO genes (Gawryluk, R. M. R. et al. 2016).

This enzyme is ubiquitous with its origin having been attributed to prokaryotes, including the α-proteobacterium that gave rise to mitochondria as an endosymbiotic event (Sagan, L. 1967, Zimmer, C. 2009, Richards, T. A. et al. 2011). This event facilitated a 15-fold increase in the capacity of oxidative phosphorylation to generate adenosine triphosphate (ATP) through this new cellular component. The oxidative phosphorylation reactions in the eukaryotic cell were evolutionarily conserved insomuch as almost all the eukaryotic lineages maintain this ability. However, this process has undergone many changes along the years specifically focusing on preventing apoptosis of the cell due to the production of reactive oxygen species parallel to the oxidative phosphorylation process (Avery, S. V. 2011, Pierron, D. et al. 2012).

Within the oxidative phosphorylation reaction, CCO has a critical role. In this sense, it has been described that a reduction in CCO activity leads to decreased ATP production (Villani, G. et al. 1997, Pacelli, C. et al. 2011, Huttemann, M. et al. 2012). Specifically, this enzyme oxidizes cytochrome c along with the reduction of an oxygen molecule to water. The energy liberated in this exogenic reaction is used to pump protons from the mitochondrial matrix into the intermembrane space (IMS) generating a proton gradient that allows ATP synthase to generate ATP (Brischigliaro, M. et al. 2021).

Like oxidative phosphorylation, evolution has changed the CCO enzyme, which was initially conformed of three or four subunits but now currently consists of fourteen units in mammals (Raitio, M. et al. 1987, Steinrucke, P. et al. 1987, Balsa, E. et al. 2012, Pitceathly, R. D. S. et al. 2018, Zong, S. et al. 2018). However, the subunits I and II, present in prokaryotes, are enough to replicate CCO function. Thus, it is believed that CCO was functional in the pre-eukaryotic cell just after the endosymbiotic event, and that selective pressure adapts CCO to eukaryotic necessities (Pierron, D. et al. 2012).

It is also important to highlight that the location of the genes encoding the subunits are not a reflection of their origin because some genes have been transferred in the endosymbiotic process to the nuclear genome (Szklarczyk, R. et al. 2010). Specifically, three subunits are synthesized by mitochondrial DNA (COX1, COX2 and COX3) and 11 by the cell nucleus (COX4, COX5A, COX5B, COX6A, COX6B, COX6C, COX7A, COX7B, COX7C, COX8 AND NDUFA4/COXFA4). These nuclear subunits were thought to be accessory genes, however, mutations in these subunits are proved to have a great negative impact on the activity of CCO (Arnold, S. et al. 1997, Fornuskova, D. et al. 2010). There are homologs to these nuclear proteins in many invertebrates (*Drosophila melanogaster*), fungi (*Neurospora crassa, Saccharomyces cerevisiae*) or plants (*Arabidopsis thaliana*) suggesting that these nuclear genes appeared before the evolution of the eukaryote lineage. Lastly, another interesting aspect in the evolution of the CCO is the absence of some subunits in differing species, suggesting that secondary losses occured during evolution (Das, J. et al. 2004).

CCO Isoforms

As previously suggested, the different tissues in a pluricellular organism have different energetic requirements which have made the CCO enzyme to acquire different tissue-specific isoforms for numerous subunits such as COX4, COX6A, COX6B, COX7A, COX8, SCAF1 or SCAFI (Lobo-Jarne, T. et al. 2018, Brischigliaro, M. et al. 2021). Moreover, the subunits responsible for the catalytic action seem to be the same in all tissues, whereas the nuclear encoded subunits could be linked to the differences between isoforms (Pierron, D. et al. 2012). In other words, most cells have the same number of COX1, COX2 and COX3 genes but can have different number of copies of the different nuclear subunits (Pandit, S. K. et al. 2013). These phenomena could be explained by phylogenetic studies which indicate that isoforms originate by a duplication of the same gene at timepoints during evolution which are called paralogs. In this regard, in vertebrates there are two paralogs of COX4 and COX6a, three for COX7a and no paralogs for the isoforms COX6c and COX7C. Some variability has been demonstrated in other species; in teleosts an extra copy for COX5a, such as COX5b have been found. In primates, paralogs have been found in COX6b, COX7b, COX8. Thus, it is clear that CCO subunit paralogs vary in the vertebrate lineage (Little, A. G. et al. 2010).

Furthermore, the distribution of the different paralogs changes depending on the tissues. That is the case for COX6a-1, which is expressed in most mammals' tissues whereas COX6a-2 is only expressed in striated muscle. This discrepancy in COX6a-2 expression was associated to a double efficiency in responses to low cellular energy levels in the tissue (Linder, D. et al. 1995, Huttemann, M. et al. 1999). By contrast, in fish, COX6a1 is dominant in the brain and the gill, while COX6a2 is abundant in skeletal muscle. Regarding COX4, in mammals COX4-1 is present in all tissues, whereas COX4-2 expression is restricted to specfic cells and tissues. Indeed, the expression of COX4-2 has been linked to hypoxia and cancer (Huttemann, M. et al. 2001, Sinkler, C. A. et al. 2017). However, in fish amphibians or reptiles, COX4-2 does not respond to hypoxia, being predominant in the heart and skeletal muscle, whereas COX4-1 is dominant in the brain.

Regarding the three isoforms of COX7a (COX7a-1, COX7a-2 and COX7a-3), COX7a-1 is dominant in the heart and the skeletal muscle in mammals, but not present in fish where a pseudogenization has been proposed as an explanation for this discrepancy. COX7a-2 was found in all tissues in both mammals and fish, whereas COX7a-3 is dominant in the liver and kidney (Sinkler, C. A. et al. 2017).

Moreover, the same variability applies to other subunits such as COX5a in fish which have an extra copy, COX5a-1 and COX5a-2. Both are expressed in all tissues although COX5a-1 is more common. There is also two copies for COX5b (COX5b-1, COX5b-2) with COX5b-1 being dominantly expressed in the brain whilst COX5b-2 is the more ubiquitous transcript in fish (Little, A. G. et al. 2010). Also, additional copies for COX6b, COX7b, COX8 in other species such as primates, have been found.

Relevant Mutations in CCO

There is a subclass of oxidative phosphorylation (OXPHOS) disorders, characterized by defects in CCO biogenesis and/or function. One of the most relevant is the human mitochondrial-CCO deficiency (OMIM #220110) which is a mitochondrial condition, fatal in childhood, but mildly affected individuals can survive through adolescence or even adulthood. A vast spectrum of partial CCO deficiency conditions, are often caused by gene defects in mitochondrial DNA-encoded (mtDNA) tRNAs or assembly factors of CCO which secondarily affect the enzyme. In addition, the degree of CCO deficiency in mutations of mtDNA genes, including CCO-encoding subunits but also the RNA genes often associated with this biochemical defect, are substantially influenced by the degree of heteroplasmy in different tissues of the affected subject, which further expands the biochemical and clinical heterogeneity of these conditions (Brischigliaro, M. et al. 2021).

In line with this, nowadays, there is evidence of over 30 nuclear and mitochondrial genes causing CCO deficiency. An example of this is the gene SURF1, which typically triggers CCO deficiency through complete loss of functional variants, likely due to redundancy. However, in both human and non-human organisms, the complete lack of the gene and its corresponding function is still compatible with the individual's development. This is due to the fact that CCO is still fully assembled and functional even when SURF1 is missing or severely impaired (Brischigliaro, M. et al. 2021). This finding also points out the existence of compensatory mechanisms which could replace the function of the lost protein. Indeed, similar results were found when other CCO genes were affected such as COX10 and COX15, which did not completely abolish the function of the gene product but were sufficient in causing the disease.

Overall, the number of mutations in mtDNA CCO genes are higher than the number of mutations in the genes encoded by the nuclear DNA. Also, more mutations are reported in mtDNA genes and nuclear-encoded genes than in the structural subunits of CCO. Finally, the vast majority of mutations associated with CCO deficiency have been observed in assembly or accessory factors, such as SURF1 previously described, or tRNA genes which impact CCO activity.

Beside point mutations, there are large-scale rearrangements of mtDNA, such as deletions which have been linked to human mitochondrial disorders with respiratory chain deficiencies and, more specifically, partial CCO deficiency. In line with this, a mouse model of mtDNA rearrangement was generated (Inoue, K. et al. 2000) showing the lethality of CCO deficient activity in muscle fibers.

Furthermore, mutations affecting mt-tRNAs underlie some CCO activity defiencies such as mutations in mt-tRNALys (Shoffner, J. M. et al. 1990, Yoneda, M. et al. 1990, Boulet, L. et al. 1992, Silvestri, G. et al. 1992, Zeviani, M. et al. 1993, Santorelli, F. M. et al. 1996, Tiranti, V. et al. 1999, Rossmanith, W. et al. 2003), mt-tRNAAla (Spagnolo, M. et al. 2001, Horvath, R. et al. 2003, Swalwell, H. et al. 2006, Lehmann, D. et al. 2015), mt-tRNAPhe (Mancuso, M. et al. 2004, Deschauer, M. et al. 2006, Zsurka, G. et al. 2010), mt-tRNALeu (Goto, Y.-i. et al. 1990, Kobayashi, Y. et al. 1990, Zeviani, M. et al. 1991, Moraes, C. T. et al. 1992, Silvestri, G. et al. 1994), mt-tRNATrp (Nelson, I. et al. 1995, Santorelli, F. M. et al. 1997, Silvestri, G. et al. 1998, Tulinius, M. et al. 2003, Maniura-Weber, K. et al. 2004, Sacconi, S. et al. 2008, Smits, P. et al. 2010, Baric, I. et al. 2013) and mt-tRNAAsp (Seneca, S. et al. 2005) which cause respiratory defective phenotypes with CCO deficiency. One reversible mutation has been found in the nucleotide of tRNAGlu (DiMauro, S. et al. 1983, Roodhooft, A. M. et al. 1986, Zeviani, M. et al. 1987, Salo, M. K. et al. 1992) where the patients present mitochondrial myopathy at birth which recovers at 2-3 years of age. To date, only two mouse models have been modelled for mt-tRNALys and mt-tRNAAla mutations (Shimizu, A. et al. 2014, Kauppila, J. H. K. et al. 2016) with phenotypes less severe than in humans. Moreover, yeast models of mt-tRNALeu (Moraes, C. T. et al. 1993, Hao, H. et al. 1996), mt-tRNAPhe (Francisci, S. et al. 1998), mt-tRNAGly (Rohou, H. et al. 2001) and mt-tRNASer (Toompuu, M. et al. 2002) have shown growth failure in aerobic conditions but further mechanisms have not been explored.

Finally, it is important to note that post-translational modification of CCO subunits could happen and may alter the structural features of proteins, the interactions with their partners and their functions. In Table 1 we summarize the current relevant mutations in CCO and their main effects on different species.

Table 1. Mutations in CCO structural subunits and their effects on different species

Gene	Type of mutation	Species	Effects
MT-CO1 (OMIM #516030)	m.7444G>A	Human	Leber's Hereditary Optic Neuropathy (LHON) (Brown, M. D. et al. 1992)
	m.7444G>A	Human	Converts the AGA termination codon of MT-CO1 to a lysine codon (AAA) (Mkaouar-Rebai, E. et al. 2013)
	m.1555A>G	Human	Aminoglycoside-induced hearing loss (Pandya, A. et al. 1999, Yuan, H. et al. 2005)
	m.7444G>A	Not reported	Not reported
	m.6480G>A	Human	Encephalopathy with sensorineural hearing loss, myoclonic epilepsy and mental retardation (Jaksch, M. et al. 1998)
	m.6742T>C	Human	Acquired idiopathic sideroblastic anaemia (AISA) by changing p.Ile280Thr (Gattermann, N. et al. 1997)
	m.6721T>C	Human	Acquired idiopathic sideroblastic anaemia by changing p. Met273Thr (Gattermann, N. et al. 1997)
	m.6930G>A	Human	Premature stop codon at Gly343 resulting in the loss of 33% percent of the peptide (Bruno, C. et al. 1999)
	m.6489C>A	Human	Epilepsia partialis continua by the conversion of Leu196 to Ile via substitution (Varlamov, D. A. et al. 2002)
	m.6328C>T (p.Ser142Phe)	Human	Muscle weakness and cardiac hypertrophy (Lucioli, S. et al. 2006)
COX1	p. Ser157Leu	*Saccharomyces cerevisiae*	Severe CCO deficiency (Lemaire, C. et al. 1998)
	Two nonsense mutations and eight different missense mutations in invariant or conserved residues	*Saccharomyces cerevisiae*	Typical pet (from 'petite', the French word for 'small') colony phenotype, caused by abolished respiratory growth on non-fermentable carbon sources (Lemaire, C. et al. 1998)
	mt:CoIA302T	*Drosophila melanogaster* (mt:CoI)	Mild CCO deficiency, flies exhibit wild-type behavior (Xu, H. et al. 2008)
	mt:CoIR301L	*Drosophila melanogaster* (mt:CoI)	Mild CCO deficiency but are healthy except for male sterility (Xu, H. et al. 2008)
	mt:CoIR301S	*Drosophila melanogaster* (mt:CoI)	Severe phenotypes associated with $\leq$50% COX activity compared to controls. Also, flies exhibit age-related neurodegeneration, with impaired motor skills and reduced lifespan (Xu, H. et al. 2008)
	m.6589T>C	*Mus musculus* (mt-Co1)	Substitution of Val421 to Ala, associated with $\leq$50% CCO activity (Kasahara, A. et al. 2006)

Table 1. (Continued)

Gene	Type of mutation	Species	Effects
MT-CO2 (OMIM #516040)	m.7587T>C	Human	Severe ataxia, cognitive impairment, optical atrophy with retinopathy and muscle atrophy (Clark, K. M. et al. 1999)
	m.7671T>A	Human	Affects the stability of the binding of heme a3 by MT-CO1 (Rahman, S. et al. 1999)
	m.8042delAT	Human	Causes a frameshift at Met153 that generates a premature stop codon at 155 amino acids (68% of the wild-type protein) (Wong, L. J. et al. 2001)
	m.7896G>A	Human	Psychomotor-delay, failure to thrive and mild cardiomyopathy (Campos, Y. et al. 2001)
	Mutations and ablation	*Saccharomyces cerevisiae* (COX2)	Absent respiration and no growth in non-fermentable carbon sources (Fox, T. D. 1979, Weiss-Brummer, B. et al. 1979, Khalimonchuk, O. et al. 2007)
	mt:CoIIG177S	*Drosophila melanogaster* (mt:CoII)	20% reduction of COX activity in adult individuals, it also affects male fertility (Patel, M. R. et al. 2016)
MT-CO3 (OMIM #516050)	m.9804G>A	Human	Alters a highly conserved site (Ala200Thr) and removes five amino acids from the third hydrophobic domain of MT-CO3, including two evolutionarily conserved Phe residues (p.Phe94_Phe98del) (Keightley, J. A. et al. 1996)
	m.9952G>A	Human	Converts Trp249 codon to a stop codon, resulting in the loss of the last, highly conserved, 13 amino acids at the C-terminal region of MT-CO3 (Hanna, M. G. et al. 1998)
	m.9537insC	Human	Causes a frameshift truncating MT-CO3 after 110 amino acids (42% of the wild-type) (Tiranti, V. et al. 2000)
	m.9379G>A	Human	Results in a stop codon at Thr58 (Horvath, R. et al. 2002)
	cox3Δ	*Saccharomyces cerevisiae* (COX3)	Displays increased sensitivity to exogenous oxidative challenges (Khalimonchuk, O. et al. 2007)

Gene	Type of mutation	Species	Effects
COX4I1 (OMIM #123864)	c.303_304delinsTT	Human	Affecting two amino acids in COX4I1 peptide, a highly conserved Lys(number) and a non-conserved Thr(number). CCO activity was approximately 50% (Abu-Libdeh, B. et al. 2017)
	c.454C>A	Human	Results in p.Pro152Thr, affecting a conserved residue and leading to approximately 15% of the norm in skeletal muscle (Pillai, N. R. et al. 2019)
	COX5B	*Saccharomyces cerevisiae*	Wild-type growth on non-fermentable carbon sources (Trueblood, C. E. et al. 1987)
	COX5A	*Saccharomyces cerevisiae*	Petite phenotype (Trueblood, C. E. et al. 1987)
	COX4L	*Drosophila melanogaster*	No genetic models, but post-transcriptional silencing show strong behavioural effects under normoxic conditions (Klichko, V. et al. 2014)
COX4I2 (OMIM #607976)	c.412G>A (p.Glu138Lys)	Human	Role for COX4I2 as a hypoxia-inducible CCO subunit (Shteyer, E. et al. 2009)
	c.253C>T (p.Arg85Trp)	Human	CCO activity in muscle mitochondria was 50% of the lowest normal value (range 285–1080 nmol/min/mg) (Vondrackova, A. et al. 2012)
		Saccharomyces cerevisiae	Null mutants show normal growth (Trueblood, C. E. et al. 1987) and COX5B itself is strictly regulated by oxygen levels (Hodge, M. R. et al. 1989, Burke, P. V. et al. 1997)
		Mus musculus (Cox4i2)	Triggers CCO deficiency and degenerating lung pathology, with airway hyporeactivity likely caused by ATP depletion (Huttemann, M. et al. 2012)
COX5A (OMIM #603773)	c.212G>A	Human	Results in p. Arg71His missense, altering a conserved residue, associated with failure to thrive, psychomotor delay, cortical atrophy, hypotonia, anaemia and enlarged liver and spleen (Vondrackova, A. et al. 2012)
	c.319C>T	Human	Results in p.Arg107Cys, affecting a conserved residue at the interface between COX5A and COX4 (Baertling, F. et al. 2017)
		Saccharomyces cerevisiae (COX6)	COX6 null mutant displays reduced growth rate under strict respiratory conditions (Glerum, D. M. et al. 1997, Glerum, D. M. et al. 1998)
	COX5Atend	*Drosophila melanogaster* (COX5A)	Despite having ATP levels sufficient to maintain cell survival and growth, COX5A larvae displayed lethality due to cell cycle arrest; eye-specific adult mutants lose lens secreting cones and pigment cells (Mandal, S. et al. 2005)

Table 1. (Continued)

Gene	Type of mutation	Species	Effects
COX5A (OMIM #603773)	COX5A KD	*Drosophila melanogaster* (COX5A)	Lethality at the third-instar larval stage (Kemppainen, K. K. et al. 2014)
	Knockdown	*Danio rerio* (cox5aa, cox5ab)	Severe developmental defects in the endodermal tissues, altered cardiac function, muscle weakness and loss of motor neurons (Baden, K. N. et al. 2007)
COX6A1 (OMIM #602072)	Same 5-bp deletion c.247-7_247-3del5 near a splicing acceptor site of exon 3.	Human	Linked to a form of Charcot-Marie-Tooth disease (lately named CMT recessive intermediate type D, CMTRID) (Tamiya, G. et al. 2014)
	Null mutants	*Saccharomyces cerevisiae* (COX13)	Decreased growth rate on non-fermentable carbon sources (Taanman, J. W. et al. 1993) but CCO was fully assembled
	Mutant in levy1	*Drosophila melanogaster*	Leads to encephalomyopathic effects in adults such as neurodegeneration, motor dysfunction and bang-induced paralysis plus premature death (Liu, W. et al. 2007)
	Levy knockdown	*Drosophila melanogaster*	Developmental arrest of most of the flies at the pupal stage (Kemppainen, K. K. et al. 2014)
	Absence of COX6A1	*Mus musculus* (COX6A1)	Displays partial CCO-deficiency in muscle with sciatic nerve thinning and neurogenic muscle atrophy with reduced motor nerve conduction (Tamiya, G. et al. 2014)
COX6A2 (OMIM #602009)	c.117A>C	Human	Causes p.Ser39Arg substitution. Profound CCO deficiency in skeletal muscle (Inoue, M. et al. 2019)
	c.127T>C	Human	Causes p.Cys43Ala substitution. Profound COX deficiency in skeletal muscle (Inoue, M. et al. 2019)
	COXVIaH-/-	*Mus musculus* (Cox6a2)	20% residual CCO activity in the heart (Radford, N. B. et al. 2002)
COX6B1 (OMIM #124089)	c.221G>A	Human	Leads to the transition p.Arg19His and severe encephalomyopathy linked to abnormal MRI (magnetic resonance imaging) patterns, muscle weakness and pain, growth retardation and metabolic acidosis (Massa, V. et al. 2008)
	c.58C>T	Human	Leads to p.Arg20Cys substitution of a conserved amino acid involved in the stabilization of the subunit in the holocomplex. Clinical manifestations are encephalomyopathy, hydrocephalus and hypertrophic cardiomyopathy (Abdulhag, U. N. et al. 2015)

Gene	Type of mutation	Species	Effects
	cox12Δ	*Saccharomyces cerevisiae* (COX12)	Poor growth at room temperature and no growth at 37 ◦C (LaMarche, A. E. et al. 1992)
	Knockdown	*Drosophila melanogaster* (COX6B)	Severe developmental arrest and lethality (Kemppainen, K. K. et al. 2014)
COX7A1 (OMIM #123995)	3-bp deletion c.91_93delAAG	Human	Loss of a conserved lysine (p.Lys31del) and CCO deficiency (Vondrackova, A. et al. 2012)
	cox7aΔ	*Saccharomyces cerevisiae* (COX7)	The activity of CCO is absent and the holoenzyme fails to assemble (Aggeler, R. et al. 1990)
	partial loss of COX7A	*Drosophila melanogaster* (COX7A)	CCO deficiency with lethality at the pupal stage. Development of the CNS is not affected but results in locomotor dysfunction and seizure sensitivity (Kemppainen, K. K. et al. 2014)
	Lack of the isoform	*Mus musculus* (Cox7a1)	CCO deficiency (Huttemann, M. et al. 2012, Lee, I. et al. 2012)
COX7B		Human	Mutations in its sequence cause an unusual X-linked disease named MLS (microphthalmia with linear skin lesions) or MIDAS (microphthalmia, dermal aplasia and sclerocornea) (Morleo, M. et al. 1993, Indrieri, A. et al. 2012)
COX8A (OMIM #123870)	c.115-1G>C	Human	Loss of 31C-terminal amino acids out of the 44 composing the wild-type peptide. COX activity was 5–10% of the norm in muscle and fibroblasts respectively (Hallmann, K. et al. 2016)
COXFA4 (OMIM #603933)	c.42+1G>C	Human	Causes a frameshift in exon 3 that gives a protein of 49 amino acids, 14 of which are in frame, whereas the wild-type protein consists of 81 amino acids (Pitceathly, R. D. et al. 2013)

The Proton Pumping of the Cytochrome C-Oxidase

The enzyme CCO, or complex IV, is the last step in oxidative phosphorylation of the inner mitochondrial membrane, which is essential in the aerobic metabolism of animals and numerous aerobic bacteria (Tsukihara, T. et al. 1996, Little, A. G. et al. 2018, Wikström, M. et al. 2018). The Electron Transport Chain (ETC) is formed by four important structural proteins - Complexes I to IV- and the enzyme ATP synthase -Complex V- that jointly produce free energy from the electrochemical proton gradient across the inner mitochondrial (Wikström, M. et al. 2018).

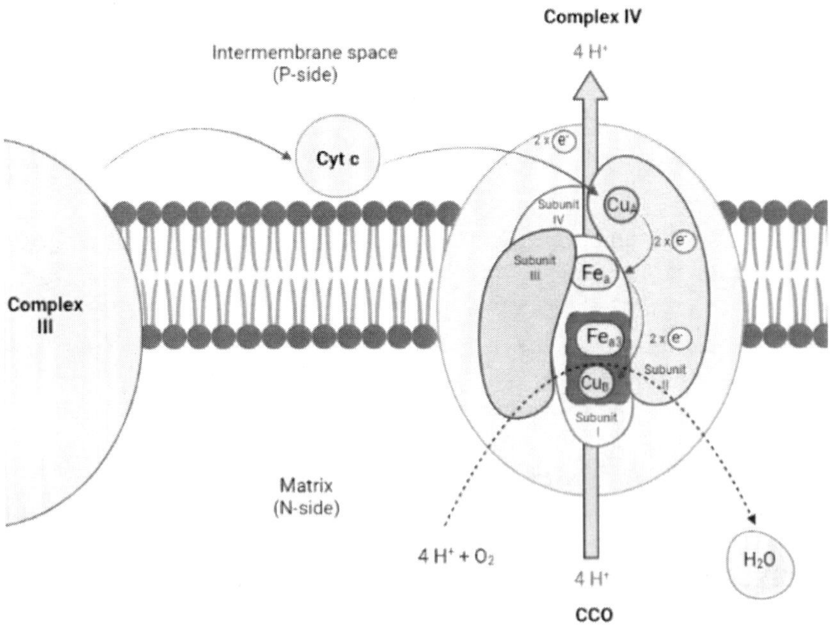

Figure 1. The proton pumping mechanism of cytochrome c-oxidase. CCO is made up of several protein subunits that contain the prosthetic groups involved in electron transport: two heme groups (Fe_a and Fe_{a3}) and two nuclei containing Cu^{2+} ions (Cu_A and Cu_B centers). Electrons are transported one by one until four accumulate in the last redox center (formed by Fe a3, CuB center and Tyr244); being donated to an oxygen molecule. During proton pumping the passage of protons to the intermembrane space (P- side) is accompanied by the reduction of the final acceptor, converting O_2 to to H_2O. This figure has been created using BioRender.

The number of subunits varies depending on the entity under study, such as two to three in the bacterium *Paracoccus denitrificans* (Iwata, S. et al. 1995), eleven in *Arabidopsis* (Millar, A. H. et al. 2004) and fourteen in the *Bos taurus* (Wikström, M. et al. 2018). In mammals, CCO consists of ten nuclear-coded subunits (COX4, COX5a, COX5b, COX5a-1, COX5b-1, COX5c-1, COX7a, COX7b, COX7c, and COX8), along with three catalytic mitochondrial-coded subunits (COX1, COX2 and COX3) (Tsukihara, T. et al. 1996, Kadenbach, B. et al. 2000). Recent studies have found another subunit, COXFA4, which was traditionally assigned to Complex I but is now known to be a subunit of CCO (Pitceathly, R. D. S. et al. 2018, Zong, S. et al. 2018). The catalytic subunits of this complex, CCO, are highly hydrophobic and are typically encoded by the mitochondrial genome. Moreover, the structure presented by the complex is a heme-copper oxidase with two heme fractions (a and a_3) in addition to a mononuclear copper center (Cu_B), bound to COX1 and another binuclear copper center (Cu_A) bound to COX2, where it also contains the binding site for the cytochrome c (Tsukihara, T. et al. 1996, Wikström, M. et al. 2018). Fe_{a3} heme closely associated with a Cu_B-$(His)_3$ unit form the binuclear center (BNC).

The Proton Pumping Stoichiometry

CCO reduces molecular O_2 to water, coupled to proton pumping, and produces ATP at Complex V due to the proton motive force (PMF), the membrane potential electrical force, and the pH gradient (Wikström, M. et al. 2018). The reduction of oxygen to two water molecules requires four electrons and four H^+, and, in addition, four residual protons are translocated that contribute to the synthesis of ATP. The translocation of the two H^+ molecules releases energy and contributes to the generation of the PMF.

In vivo, the CCO enzyme is presented as a dimer which is stably bound to the inner mitochondrial membrane and operates as a transmembrane system associated with both cytochrome c and the mitochondrial matrix. The cytochrome oxidase dimer exists in two conformations, such as oxidized and reduced. To explore the cycle represented in Figure 2, we can start with the BNC in the reduced state or R form, where Fe_{a3} and Cu_B have respective oxidation states, Fe^{2+} and Cu^+. For the entire CCO protein to be in the fully reduced state, the other metal centers also must be reduced. For CuA, the reduced copper dimer state is then Cu+–Cu+.

The reaction cycle commences by coordination of O_2 to Fe_{a3} to form the adduct state A. The cycle then progresses quickly through the proposed Peroxo and Hydroperoxo states to the experimentally identified states P_M/P_R and further to state F. In this process (A to P_R form), three electrons from Fe $_{a3}^{2+}$ and Cu $_B^{1+}$, along with one proton and one electron from Tyr^{244} -OH are transferred to the bound O_2, forming O^{2-} and O_{H-}. The subsequent reduction of the heme a3 ferryl-oxo complex via another electron from Fe/CuA and uptake of another proton is significantly slower and leads to active cycling oxidized state OH via the postulated intermediate state F_H (Shimada, A. et al. 2021). In summary, at this point, consumption of the 4 available electrons to reduce O_2 results in the active cycling oxidized State O_H. To continue cycling, two electrons must be delivered from CCO to the DNC to reduce Fea3 and CuB transiting via CuA/Fe, and soon after two more electrons total from CCO are needed at Fe/CuA to regenerate the fully reduced State R.

Studies have indicated that the translocation of protons from the mitochondrial matrix (N-side) takes place in $P_R \rightarrow F$ and $F \rightarrow O_H$ -intermediate oxidase state- transitions. This process promotes the positivity of the intermembrane space (P-side) during the proton pumping (Wikström, M. et al. 2018, Ishigami, I. et al. 2019, Shimada, A. et al. 2021). Furthermore, when the O_H reduction occurs immediately after the O_2 reduction reaction, intermediate states (E form) that promote proton translocation ($O_H \rightarrow E$ and $E \rightarrow R$) may appear. However, if the reduction does not occur immediately, the O_H state relaxes back to the fully oxidized resting state (O form), preventing that the protons are pumping to the P side (Ishigami, I. et al. 2019).

Irreversible electron transfer is one of the necessary conditions that have been established for energy transduction to be effective in the redox-coupled proton pump system. In this case, the neutral radical state of Tyr244 and ferryl of Fe_{a3} are sites that provide irreversible transitions (P→F and F→O, respectively) and high electronic affinity (Shimada, A. et al. 2021). For each pair of electrons that is transported from complex IV, two protons are translocated from the mitochondrial matrix (N-side) into the intermembrane space (P-side), thus contributing to the efflux of protons into the intermediate space and becoming CCO a real proton pump (Wikström, M. et al. 2018). The presence of water in the matrix seems to have high dielectric constants (up to $\varepsilon_W \sim 30$ approximately) and its presence is essential to provide the means for the transfer of proton pumping in the enzyme and the electrostatics in it (Farahvash, A. et al. 2018).

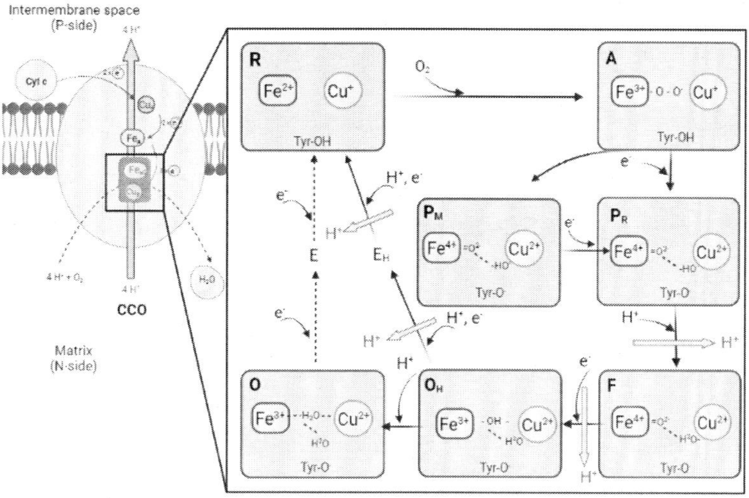

Figure 2. The change from the reduced form (R Form) to the oxidized form (O), which allows the translocation of electrons and protons from one side to the other to the membrane, is represented. The proton pumping from the mitochondrial matrix (N-side) to the intermembrane space takes place in $P_R \rightarrow F$ and $F \rightarrow O_H$ transitions as it has been represented in red arrow. When the OH reduction occurs immediately after the O_2 reduction reaction, intermediate states (E and E_H form) also promote the proton translocation. This figure has been created using BioRender.

The Proton Pumping Function

As discussed, the proton pump is regulated by structural changes within the protein molecule during the reaction cycle. Research on changes in pH has found that CCO has high stability given its wide pH range (4.6–8.6). For this reason, changes between this range will not affect the efficiency of the enzyme allowing the proton pumping to remain normal (Kagi, J. H. et al. 1968, Lepp, H. et al. 2009). On the other hand, and as mentioned above, the elimination of different subunits in CCO reduces the membrane potential. This leads to a lower production of ATP during the redox cycle which causes a compensatory process in the mitochondria by upregulating the oxidative metabolism. As a consequence, there is an accumulation of lactate that reduces the pH levels below the stability range and, with it, more than 80% of the enzymatic activity is reduced as in the Leigh's disease (DiMauro, S. et al. 1987).

Some heavy metals, such as Zinc and Thorium, have been shown to inhibit CCO by increasing thorium-induced cellular toxicity by altering membrane potential and production of reactive oxygen species (Yu, L. et al. 2022) or by inhibiting proton motility in the pathway output and reducing the efficiency of energy transduction after zinc administration (Dineley, K. E. et al. 2003).

While most of the studies have focused on the catalytic nucleus of CCO due to their involvement in the catalytic cycle, the function of the rest of the subunits is certainty unknown. Nevertheless, these isoforms seem to play an important role in the stabilization of the enzymatic core and in the regulation of the activity of the CCO, being able to reduce it, or even disassemble the enzyme, when these subunits are affected by a mutation.

The Role of CCO in Neurodegeneration

Neurodegenerative disorders are considered as a set of pathological conditions featured by slow progressive and irreversible dysfunction and loss of neurons and synapses in selected areas of the nervous system, determining its clinical presentation and course (Jellinger, K. A. 2009). The major basic processes associated to neurodegeneration are considered multifactorial (genetic, environmental and endogenous factors related to aging), but the basic molecular mechanisms are not fully understood. In this sense, a hallmark of neurodegenerative diseases is a dysfunction of mitochondria including CCO, the terminal enzyme complex of the respiratory chain (Arnold, S. 2012).

Mitochondria are key cytoplasmatic organelles with a well-known role in energy metabolism as the major cellular ATP producers. Also, mitochondria regulate several cellular processes, such as Ca^{2+} homeostasis, reactive oxygen species (ROS) signaling, ion channel regulation, and exocytosis (Trigo, D. et al. 2022). Alterations in mitochondrial metabolism influence all cellular activities and may be related with the onset and progression of many neurodegenerative disorders (Jellinger, K. A. 2010). The dynamic properties of mitochondria include mitochondrial fission, fusion, transport, biogenesis and degradation, and these processes are critical for neuronal function and health. Thus, dysregulation of mitochondrial dynamics has been strongly implicated in the pathogenesis of neurodegenerative diseases such as Alzheimer's disease (AD) (Flannery, P. J. et al. 2019), Parkinson's disease (PD) (Van Laar, V. S. et al. 2013), and Huntington's disease (HD) (Guedes-Dias, P. et al. 2016).

Both aging and neurodegenerative diseases are associated with mitochondrial dysfunction. With aging, respiratory chain efficiency decreases, with electron leakage and reduced ATP generation, increasing ROS levels and eventually exacerbating age associated damage (Trigo, D. et al. 2022). Thus,

ROS have both a physiological role and pathological effects, depending on their type and rate of production. ROS are mainly generated from the mitochondrial electron transport chain (ETC) through one electron reduction of molecular oxygen (O_2). Superoxide is then converted to hydrogen peroxide (H_2O_2) by superoxide dismutases. Complexes I (NADH dehydrogenase) and III (ubiquinone cytochrome c reductase) of the ETC are the most important sources of ROS production. Complex IV (CCO) itself is not a source of ROS, but this enzyme exerts effects on the respiratory chain modulating the ROS production at complexes I and III (Srinivasan, S. et al. 2012, Jurcau, A. 2021).

Although mitochondrial dysfunction is commonly seen to involve failures of more than one complex of the ETC, several studies have suggested that CCO dysfunction is associated with increased mitochondrial ROS production and cellular toxicity (Srinivasan, S. et al. 2012). Here, we will refer to some hallmarks in the most prevalent neurodegenerative diseases, with focus in the role of CCO.

Alzheimer's Disease

The main pathophysiological characteristics of this form of dementia are the aggregation of the amyloid beta (Aβ) protein into plaques, the formation of neurofibrillary tangles by the tau protein, and mitochondrial dysfunction (Selkoe, D. J. et al. 2016, Gao, Y. L. et al. 2018, Swerdlow, R. H. 2018).

Regarding CCO, deficient activity of this enzyme in AD has been reported in different brain regions (i.e., frontal cortex, temporal cortex, and hippocampus), in platelets, and fibroblasts, wich could increase tissue vulnerability to excitotoxicity or reduced oxygen viability (Arnold, S. 2012). In addition, this impaired activity of CCO may be accompanied by structural degradation and induction of apoptosis (Cardoso, S. M. et al. 2004).

On the other hand, neuronal loss has been described as a secondary cause for CCO defects, and decreased expression of CCO or its mtDNA encoded subunits as a result of decreased activity of neurons and free radical mediated cell damage, respectively. Also, the abnormal interactions between OXPHOS polypeptides and compounds 2012 #289 such as amyloid β has been discovered (Shoffner, J. M. 1997). Some data suggest that amyloid precursor protein and amyloid β are targeted to mitochondria impairing CCO activity in AD brains (Reddy, P. H. et al. 2012, Cadonic, C. et al. 2016).

Parkinson's Disease

PD is the second most prevalent neurodegenerative disease, and characterized by resting tremor and bradykinesia, largely due to the degeneration of the dopaminergic neurons of the substantia nigra (Ebanks, B. et al. 2022). Genetic risk factors for PD include mutations in the SNCA gene (formation of Lewy bodies), and PINK1 and parkin genes (failed mitophagy, neuronal defects and apoptosis) (Wood-Kaczmar, A. et al. 2008, Mahul-Mellier, A. L. et al. 2020, Ebanks, B. et al. 2022).

Increasing evidence suggests that mitochondrial dysfunction may be involved in the etiology of PD, but the role of the respiratory chain remains unclear. Impaired complex I activity is central to the pathogenesis of dopaminergic neuronal death in PD and is evidenced by the fact that α-synuclein can interact with the ETC, decreasing complex I activity, ATP production, and respiratory capacity, which further aggravates pathology (Dauer, W. et al. 2003, Malpartida, A. B. et al. 2021).

With aging, it has been shown that a large proportion of neurons in the substantia nigra lose CCO, which could be associated with the very high levels of mtDNA deletions in CCO-deficient neurons observed (Itoh, K. et al. 1996, Kraytsberg, Y. et al. 2006). Interestingly, the activities of the mitochondrial complexes I and IV were reduced in leukocytes from patients with idiopathic PD compared with age-matched controls (Muftuoglu, M. et al. 2004).

The application of MPTP (1-methyl-4-phenyl-1,2,3,6-tetrahydropyridine) in vivo and its toxic derivative 1-methyl-4-phenylpyridinium (MPP+) in vitro represent a well-accepted experimental model of Parkinson's disease. Data obtained by Boyalla et al. (Sundar Boyalla, S. et al. 2011) show a specific increase of COX4-2 transcript and protein levels in male mesencephalic astrocytes caused by MPP+, involving CCO isoform expression in an impairment of energy production and elevated oxidative stress levels.

Besides, several findings, both in animal models of PD and fibroblasts of patients that carry pathogenic PINK1 mutations, support that mitochondrial dysfunction in PD may require the structural remodeling of respiratory complexes. It has been observed, for example, decreased activity of CCO during the degeneration process, reduced CCO in PD fibroblasts, or defective assembly of complex I and complex IV (Novack, G. V. et al. 2020).

Huntington's Disease

An autosomal-dominant inherited disorder caused by an excessive number of CAG repeats in the HTT gene, resulting in PolyQ repeats near the amino terminal. In addition, severe deficiencies in the activities of respiratory chain complexes have been associated with abnormal energy metabolism in the striatum of these patients. The major clinical presentations in HD include involuntary movements and bradykinesia, with atrophy of the medium spiny GABAergic neurons of the striatum (Rikani, A. A. et al. 2014, Bates, G. P. et al. 2015).

Biochemical studies of mitochondria in striatal neurons from late-stage HD patients revealed reduced activity of the complexes II, III, and IV of the ETC (Reddy, P. H. 2008). Furthermore, a study of moderate-to-severe grade HD patients showed a reduction in COX2 protein levels that corresponded with disease severity and it was consistent with the observed mitochondrial loss (Kim, J. et al. 2010). Using a toxin-mediated in vitro model of HD (3-nitropropionic acid, NPA), they observed an upregulation of COX4-2 mRNA transcription and protein expression in neural cells. Also, in striatal astrocytes, the authors found that upregulation of COX4-2 produced an increased enzyme activity at the expense of elevated mitochondrial peroxide production, causing increased cell death (Singh, S. et al. 2009, Arnold, S. 2012).

Amyotrophic Lateral Sclerosis

ALS Is a fatal adult-onset disease characterized by progressive degeneration of motor neurons. ALS is a highly heterogeneous disease, with both familial and sporadic forms (more than 90%). Although many of the disease mechanisms remain elusive, abnormal mitochondrial structure, impaired ATP production, dysregulation of energy metabolism and calcium homeostasis, as well as apoptosis, have been implicated in the neurodegenerative process (Benson, B. C. et al. 2021). Several missense mutations in the gene encoding Cu/Zn superoxide dismutase (SOD1) have been identified in approximately one-fifth of autosomal-dominant cases of ALS (Reddy, P. H. 2008).

In several animal and cell culture models they have observed a reduction in mitochondrial respiratory chain activity, mainly in complexes I and IV, among other effects, such as reduced ATP generation and mitochondrial membrane potential (Arnold, S. 2012). In addition, decreased CCO activity in ventral horn motor neurons from autopsied ALS patients has been described

(Borthwick, G. M. et al. 1999), as well as in gray matter in both cervical and lumbar spinal cords from ALS patients (Delic, V. et al. 2018). Moreover, both complex I and IV activities were decreased in skeletal muscle from patients (Vielhaber, S. et al. 2000). In some ALS brains from patients, it has been shown an increased complex I activity, perhaps as a compensatory response to CCO dysfunction (Browne, S. E. et al. 1998), (Delic, V. et al. 2018).

The Therapeutic Potential of Targeting CCO

CCO is of major significance in the development of neurodegenerative diseases, implicated in oxidative stress and production of ROS as well as a major gatekeeper to the activation of caspase-mediated apoptotic cascades (Arnold, S. 2012). Surprisingly, little research has been conducted into modulators that selectively influence CCO. That being said, a large amount of data and research indicates that modulating CCO contributes towards the overarching mechanisms through which some neuroprotective therapies work. These treatments include estrogen therapy and similar derivatives, resveratrol, and vitamin D3; and whilst they may have differing targets, common mechanistic links between all of them include modulation of CCO activity and downstream events.

One could presume that in many conditions where there is oxidative stress, a CCO modulator could prove efficacious even if, for example, the root cause does not lie directly with impairments in CCO. For instance, in the event that complex I is impaired, resulting ROS levels and oxidative stress would increase. This may ultimately lead to cell death through mitochondrial-dependent apoptotic pathways, with CCO release often being the primary trigger (Kushnareva, Y. et al. 2002). As such, targeting and preventing this release may therefore be therapeutic even if complex I remains inhibited. In line with this, studies where release of CCO was prevented also proved to be neuroprotective. Indeed, the administration of methazolamide, an FDA-approved drug in treating glaucoma, reduced the onset of disease in the R6/2 Huntington disease (HD) mouse model (a transgenic line expressing an exon 1-containing expanded CAG repeat) (Carter, R. J. et al. 1999, Wang, X. et al. 2008). Similarly, minocycline - a tetracyline antibiotic – was capable of preventing CCO release and the subsequent recruitment of apoptotic initiators in the same R6/2 HD mouse model, as well as mutant SOD1 (superoxide dismutase) mice with ALS (Zhu, S. et al. 2002, Wang, X. et al. 2003), abnormal human tau mouse lines as models for Alzheimer´s disease (AD)

(Noble, W. et al. 2009), and zinc-induced Parkinson's in mice (Kumar, V. et al. 2016).

However, whilst CCO may have broad therapeutic potential, it cannot prevent all cases of apoptosis and oxidative stress. In fact, age-related neuronal loss in a knockout-mouse model of *Dicer* – a helicase with RNAse motif-failed to show neuronal survival upon treatment of minocycline (Cheng, S. et al. 2015). Nevertheless, the study still showed reductions in inflammation which still lends support that CCO modulation may be beneficial even if subclinical.

Thus, in this section, we will be analysing the evidence of therapeutic interventions which impact CCO dynamics across wide ranging systems which are implicated in, and integral to the aetiology and epidemiology of aging and neurodegenerative diseases. Therefore, paving way for future research into CCO modulators in the development of therapies for neurodegenerative conditions.

Estrogen Is Neuroprotective through Modulation of CCO Dynamics

Epidemiologically, the neurodegenerative diseases tend to show sex-dependent pathogenesis. Parkinson´s disease (PD) is 1.5 times more likely to occur in men than women (Wooten, G. F. et al. 2004). Both PD and AD also show significant increases in incidence around fluctuations in sex hormones. For example, pre-menopausal women have much lower likelihood of being diagnosed with AD or PD compared to post-menopausal women (and men) as well as those taking estrogen blockers, such as tamoxifen for breast cancer, have increased risk of PD (Benedetti, M. D. et al. 2001, Davey, D. A. 2013, Lin, H. F. et al. 2018, Mosconi, L. et al. 2018). Moreover, pre-menopausal women diagnosed with PD also reported increased severity of symptoms in the lead up to and during menstruation (Quinn, N. P. et al. 1986).

Regarding amyotrophic lateral sclerosis (ALS) and frontotemporal dementia (FTD), the results are less convincing. In a systematic review by Curtis et al. (2017), the authors reported that only C9orf72 (Chromosome 9 open reading frame 72) and GRN (granulin) mutations had higher prevalence in females (Curtis, A. F. et al. 2017). On the converse, in HD, onset is not defined by sex; however, the severity of motor and cognitive symptoms is worse in women than men which may be in part to difference in brain structural changes such as cortical thickness (Lee, J. K. et al. 2017, Hentosh, S. et al. 2021). These epidemiological studies nevertheless highlight the possible involvement of sex hormones in the development of neurodegenerative diseases. For that reason, extensive research has been conducted to explain these sex differences with most of the research focusing on estrogen therapies.

One of the large attractions about investigating estrogen is that the estrogen receptors α and β (Erα, Erβ) are distributed throughout the brain – albeit in different patterns. ERβ is of particular interest as it is localised to the mitochondria thereby indicating a potential role in the modulation of mitochondrial dynamics, energy production and oxidative stress – all key components of neurodegenerative diseases (Yang, S. H. et al. 2004). Estrogens were, indeed, found to modulate these pathways through direct interactions as well as the triggering of long-term alterations in general transcription and the co-ordination of mitochondrial and nuclear gene transcribed subunits of the complexes in the electron transport chain (Arnold, S. 2012, Arnold, S. et al. 2012). Notably, estrogen and other derivatives were capable of upregulating different mitochondrially-encoded CCO subunits in cortical astrocytes, neurons and primary spinal cord neurons as well as increasing mRNA expression of the nuclear-encoded CCO subunits IV and VII (Watanabe, T. et al. 1998, Nilsen, J. et al. 2007, Araujo, G. W. et al. 2008, Irwin, R. W. et al. 2008, Johann, S. et al. 2010).

Nevertheless, the therapeutic potential of CCO may vary between the neurodegenerative diseases despite being regulated heavily by estrogen signalling. For instance, PD is usually considered a disease with deficits in complex 1 of the ETC - also evidenced by the fact that 1-methyl-4-phenyl-1,2,3,6-tetrahydropyridine (MPTP), a major toxin model used, impairs complex I activity (through MPTP's cleavage to 1-methyl-4-phenylpyridinium, MPP^+) (Dauer, W. et al. 2003, Sterky, F. H. et al. 2012). Moreover, the toxin models of HD which utilise 3-nitropropionic acid and malonate inhibit complex II (Greene, J. G. et al. 1993, Borlongan, C. V. et al. 1995). In spite of this, targeting CCO is not redundant in these diseases. Many of the disease models are caused by mitochondrial toxins targeting the ETC, therefore, boosting the efficiency of the ETC through CCO may influence the dynamics of the other complexes. For instance, in a rotenone model of PD administration of exogenous CCO was sufficient to reduce the elevated levels of ROS and oxidative stress seen caused by complex I inhibition (Kushnareva, Y. et al. 2002). The significant presence of mitochondrial dysfunction across sporadic and familial cases of PD, AD, HD, ALS and FTD therefore may indicate broad therapeutic potential of CCO modulation.

For example, in AD, administration of 17α-estradiol (which has less hormonal activity than endogenous estrogen) was found to potently prevent cell death through the reduction and termination of lipid peroxidation pathways (Dykens, J. A. et al. 2005). Under peroxidative conditions, cardiolipin – a major constituent of the inner mitochondrial membrane - is frequently peroxidised. Moreover, an environment which is high in peroxides can mobilise CCO to interact and create large pores in cardiolipin-containing lipid bilayers such as the mitochondrial membrane. This may result in membrane destruction and pemeabilisation thereby allowing the release of CCO into the cytosol, triggering apoptotic cascades (Puchkov, M. N. et al. 2013). Importantly, it is the change in ratios of peroxidised cardiolipin to regular cardiolipin that seem to have deleterious effects (Petrosillo, G. et al. 2007). The ability for 17α-estradiol to essentially block the release of CCO and apoptotic pathways is therefore testament to its powerful neuroprotective effects.

Nevertheless, there are considerable drawbacks in the use of estradiols. Long-term, these drugs are carcinogenic when administered preventing them from having any meaningful clinical effect (Russo, J. et al. 2003). Moreover, they can have deleterious effects when oxidised, increasing ROS levels as well as increasing the incidence of stroke (Wassertheil-Smoller, S. et al. 2003, Nilsen, J. 2008, Henderson, V. W. et al. 2012). Patients suffering from neurodegenerative disorders can already experience hypoperfusion in the CNS, therefore, combined with these side-effects may present too much of a risk (Zhao, Y. et al. 2015). When administered, they may not have any net effect due to increasing ROS and long-term side-effects. As such, extensive research and drug testing has gone into identifying suitable analogs which also do not have a hormonal impact. An emerging class are the phytoestrogens - common in foodstuffs such as soy and ginseng. These estrogens operate through similar mechanisms to achieve reductions in neuroinflammation as well as modulation of CCO, ROS and proteins of apoptotic caspase cascade.

Genistein – a phytoestrogen – upregulated Bcl-2 (B-cell lymphoma 2) in the MPTP mouse model. This helps to anchor CCO to the mitochondrial membrane thereby preventing release of CCO into the cytosol. This prevents activation of the mitochondrial-dependent apoptotic pathway through activation of caspase 3 and 9 (Liu, L. X. et al. 2008). In animal models of HD (3-nitropropionic acid injection), genistein was found to restore ATP production levels as well as increase expression of Bcl-2 and reduce caspase-3 (Menze, E. T. et al. 2016). Restoration of ATP production efficiency is important as this is commonly disrupted in neurodegeneration. Moreover,

ATP could even itself be neuroprotective as it binds and prevents fibrillization of FUS which aggregates and causes cytotoxicity in ALS and FTD (Kang, J. et al. 2019). Therefore, as CCO is also capable of increasing ATP efficiency, it stands to reason that specific modulation of CCO may be neuroprotective in this regard.

Genistein was also found to protect hippocampal cell cultures incubated with amyloid-β_{25-35}. Notably, beyond preventing caspase 3 activation and neuronal death, indicative that CCO remains mitochondrially bound, genistein also reduced intracellular Ca^{2+} concentrations as well as ROS levels which are primary mechanisms through which amyloid-β plaques cause neuronal death (Zeng, H. et al. 2004). When the phytoestrogens genistein, daidzein and equol were used in conjunction there was an additional increase in oxygen consumption, as well as a 63% increase in CCO activity accompanied by greater neuronal survivability than when genistein, or indeed estradiol (considered the gold standard) were administered alone. Both genistein alone, and when in combination, upregulated Bcl-2 similarly which helps CCO remain bound to the mitochondrial membrane and therefore prevent activation of the pro-apoptotic caspase-3 pathway. This suggests that combinations of phytoestrogens that further modulate and improve CCO functioning show greater neuroprotection via a two-pronged mechanism targeting bioenergetics and anti-apoptotic pathways (Zhao, L. et al. 2009). Moreover, this study used phytoestrogen doses in line with regular dietary phytoestrogen intake amongst the Asian population (due to high phytoestrogen content in soy-rich diets). The results illustrate that dietary intake alone (without supplementation) can exert powerful neuroprotective effects and is condordant with the Asian population having lower incidence of AD and cognitive deficits in aging (Zhao, L. et al. 2007).

However, different genetic causes or progression of pathology may greatly affect the efficacy of targeting CCO. Indeed, in a paper by Cottrell et al. (2002) (Cottrell, D. A. et al. 2002) there is evidence to suggest that reduction in CCO is independent of amyloid plaque formation and apoptotic cells. This stands in contrast to many studies which report the opposite and may be indicative of normal mitochondrial function in the presence of neurodegeneration.

Another phytoestrogen, ginsenoside RG1, a pharmacologically active constituent of ginseng, was found able to restore Bcl-2 levels in the 6-hydroxy-dopamine (6-OHDA) injected rat model (Xu, L. et al. 2009). This was concordant with other studies which found ginsenoside to prevent release of CCO from the mitochondrial membrane in MPTP and rotenone models, thereby halting apoptosis (Chen, X. C. et al. 2002, Leung, K. W. et al. 2007). In addition, Ginsenoside RG3 could reduce production of ROS, reducing Ca^{2+} levels and increasing ATP production (Tian, J. et al. 2009). Moroever, ginsenoside RG1 recapitulated the effects of genistein by protecting against amyloid-β_{25-35} by preventing apoptosis and reduced ROS production accompanied by decreased expression of CCO and increased expression of Bcl-2 thereby demonstrating efficacy in AD models (Liu, J. F. et al. 2015).

In a rotenone model of PD increases of CCO were seen (through release into the cytosol) as well as the caspases 3 and 9 indicative of apoptosis. Moreover, the anti-apoptotic regulator Bcl-2 was reduced which favours the release of CCO into the cytosol (Yang, J. et al. 1997, Tamilselvam, K. et al. 2013). A plant-based flavanol and phytoestrogen, hesperidin, was administered which was found to reverse the effects of the rotenone toxin by reducing ROS and increasing ATP production (Tamilselvam, K. et al. 2013). Interestingly, in a model of HD induced by striatal lesioning with quinolinic acid which had reduced mitochondrial complexes-I, II, III and IV activities, hesperidin was found to have a no-significant effect. Yet, when combined with minocycline, these alterations and resulting behavioural deficits of the lesion were reversed (Kumar, A. et al. 2013).

Selective agonists for both Erα and Erβ 1,3,5-tris(4-hydroxyphenyl)-4-propyl-1H-pyrazole (PPT) and 2,3-bis(4-hydroxyphenyl) propionitrile (DPN) respectively, were also successful in showing neuroprotective properties. Both agonists showed to increase the expression of Bcl-2 in a hippocampal glutamate-induced excitotoxicity neuronal model. As a result, neuron survival was increased in a dose dependent manner likely due to the prevention of CCO release from the mitochondrial membrane and initiation of apoptosis (Zhao, L. et al. 2004).

The extensive evidence of estrogenic influence in neuroprotection of the degenerative diseases particularly through modulation of CCO and prevention of release from the mitochondrial membrane supports the notion that direct CCO modulation may be beneficial. There is some evidence to suggest that the phytoestrogens are less potent and unable to elicit significant neuroprotective effects. Given that administration of estradiol is unsuitable long-term due to side effects, combinations of phytoestrogens represent a potential alternative as well as other mechanistic alternatives such as modulators of ROS production, CCO, Bcl-2 and the caspases.

Targeting CCO to Reduce Metal Accumulation

Transition metals are of increasing concern due to bioaccumulatory events in the food chain as a consequence of human impact from improper waste disposal and mining. Chronic exposure in wildlife has been shown to be toxic in part due to its modulatory effect on CCO (Olsson, P.-E. et al. 1998, Sabolic, I. 2006). For instance, copper reduces the activity of CCO, whilst nickel was associated with increased CCO activity (Pierron, F. et al. 2009). Importantly, in neurodegenerative diseases, accumulation of metals is commonly reported and, in some cases, believed to induce pathogenesis. Animal models infused with metals such as zinc recapitulate parkinsonian symptoms (Kumar, A. et al. 2010). Moreover, lysosomal disorders such as GRN deficiency, which causes FTD, Gaucher's disease and the neuronal ceroid lipofuscinoses (NCLs), show accumulations of iron in heterogenous aggregates known as lipofuscin (Ahmed, Z. et al. 2010, Regenboog, M. et al. 2016). Heterozygous GRN mutations show FTD, whilst homozygotes have lysosomal storage pathology (Heetveld, S. et al. 2015). Similarly in PD, heterozygosity for a mutant allele of GBA-1 (glucocerebrosidase-1) is associated with PD pathology and onset whereas, homozygosity is linked to the storage disorder knows as Gaucher's disease (Brockmann, K. et al. 2015). In homozygous patients, significant accumulations of metals and other aggregates occur. Nevertheless, the exact role of these aggregates is not entirely confirmed as lipofuscin has been correlated to cognitive decline in aging as well as disease (Sharma, D. et al. 1993).

As previously mentioned, estrogen shows neuroprotective effects through improvements in mitochondrial function inclusive of CCO modulation. Through modulation of the ETC, estrogen also reduces production of ROS limiting oxidative stress and thereby preventing the build-up and aggregation of lipofuscin (Bayati, S. et al. 2011, Kumar, P. et al. 2011, Hara, Y. et al. 2014). However, this remains a complicated therapeutic strategy because in neurodegenerative disease the cellular environment is pro-oxidising with high levels of metallic elements, particularly in GRN-FTD, GBA-1 mutations and NCLs. This is a result of iron partaking in the Fenton reaction resulting in the production of a highly toxic hydroxyl radical (Rivera-Mancia, S. et al. 2010). This is triggered by the hydroxylation of the 2- and 4- carbon positions of estradiol resulting in redox cycling producing hydrogen peroxide and a resulting hydroxyl radical (overviewed in Figure 3). Thus, exacerbating, rather than ameliorating, oxidative stress (Li, Y. et al. 1994). This can also lead to changes in the mitochondrial cardiolipin membrane bilayer, reducing mitochondrial membrane potential and impairing ETC activity (Sen, T. et al. 2006). Moreover, the formation of hydroxyestrogen quinones involved in the redox cycling also form the basis of carcinogenesis through formation of depurinating DNA adducts (Chandrasena, R. E. et al. 2008, Nilsen, J. 2008).

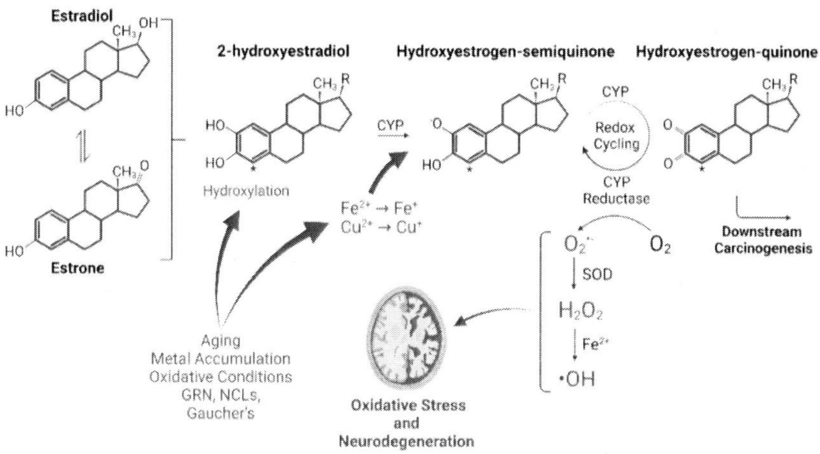

Figure 3. Estradiol toxicity in a highly oxidative and metallic environment. Under normal conditions the hydroxylation of estradiol and estrone naturally occurs. However, increased oxidative conditons from metal accumulation, aging and neurodegenerative disorders increases the likelihood of this conversion. This results in the hydroxylation of the 2-carbon or 4-carbon (denoted by the asterisk) position. Both hydroxyl sites undergo the same following reactions. Notably, metal ions – Fe (II) and Cu (II) - can be reduced, instead of enzymatic conversion via CYP, to produce the semiquinone and this is exacerbated under high metal ion concentrations. This the enters a redox cycle of the semiquinone and quinone resulting in production of superoxide radicals. This is dismutated by to hydrogen peroxide by superoxide dismutase (SOD) which then, with Fe (II), undergoes fenton catalysis to form a highly toxic hydroxyl radical. Superoxide, hydrogen peroxide and hydroxyl radicals all exert oxidative stress on the cell which can contribute to neurodegeneration. Hydroxyestrogen quionones also form the basis of downstream carcinogenesis through depurinating DNA adducts. Abbreviations: CYP, Cytochrome P450 enzymes; GRN, Granulin (mutations); NCLs, Neuronal Ceroid Lipofuscinoses. Figure was made in Biorender.

For instance, AD has been shown to have increased copper levels, as well as trace copper in drinking water has been shown to induce AD-like pathology (Squitti, R. et al. 2002, Sparks, D. L. et al. 2006). Indeed, copper seemingly modulates aggregation in many neurodegenerative disorders with copper chelators preventing accumulation of amyloid-β, toxic Huntington aggregates

(Cherny, R. A. et al. 2001, Fox, J. H. et al. 2007, Xiao, G. et al. 2013) as well as accumulations of copper and iron promoting prion-like spread of α-synuclein in PD (Li, Y. et al. 2020), as well as playing a role in FTD and ALS disease progression through dyshomeostasis (Andersen, P. M. 2006, Santillo, A. F. et al. 2009, Violi, F. et al. 2020). Therefore, any introduction of estrogen into these systems may result in neuroprotective and neurotoxic side effects (Nilsen, J. 2008). However, if we can recapitulate the therapeutic effects of estrogen (see above the limitations regarding redox imbalance and carcinogenicity), i.e., by targeting CCO independent of drug administration, then it remains a promising avenue of exploration.

In this regard, Sen et al. (Sen, T. et al. 2006) showed vitamin D3 to be neuroprotective in a mouse model injected with zinc. Administration of zinc increased lipid peroxidation, cytoplasmic CCO and terminal deoxynucleotidyl transferase dUTP nick end labelling (TUNEL) - indicative of apoptotic events. This precedes but leads to dopaminergic degeneration. However, vitamin D3 treatment reduced lipid peroxidation as measured by the levels of malondialdehyde and dihydropyridine polymers in a spectrofluorometer. This was accompanied by cytosolic release of CCO and subsequent neuron degeneration. It is likely that vitamin D3 was able to prevent release of CCO by scavenging ROS thus, preventing the peroxidation of lipids and subsequent decrease in mitochondrial membrane potential known to induce release of CCO (Sen, T. et al. 2006). Furthermore, transition metals such as zinc and iron also elicit peroxidative conditions (Jimenez Del Rio, M. et al. 2000, Lin, A. M. et al. 2003, Sen, T. et al. 2006). This could lead to the release of CCO into the cytosol and may allude to vitamin D3 being effective as well for these metals.

A green tea polyphenol, epigallocatechin gallate (EGCG), is a potent metal chelator as well as ROS scavenger. Jung et al. (Jung, J. Y. et al. 2007) found that EGCG was neuroprotective against a nitric oxide donor, sodium nitroprusside, by preventing ROS release and subsequent upregulation of Bax. Moreover, Bcl-2 levels were restored, and CCO was, as a result, tethered to the mitochondrial membrane preventing cytosolic release and activation of the caspases. Moroever, EGCG was found to be neuroprotective in the MPTP PD-induced murine model. It reduced ROS levels and increased levels of the iron exporter, ferroportin, in dopaminergic neurons improving survivability. Reduced functioning of ferroportin can lead to accumulations of iron deposits which can elicit ROS production and resulting CCO release and apoptosis (Dixon, S. J. et al. 2014, Chen, D. Z., Y.; Lyons, K.E.; Pahwa, R.; Reddy, M.B 2015, Pietrangelo, A. 2017, Xu, Q. et al. 2017). Therefore, it is likely that reduced iron levels via ferroportin upregulation, is a likely mechanism through which EGCG prevents release of CCO from the mitochondrial membrane (Weinreb, O. et al. 2009). EGCG has also been shown to be effective against many metal depositions such as cadmium and platinum by increasing CCO activity (Pan, H. et al. 2015). However, it suffers similar problems to estrogen whereby it can be oxidized by iron (Fe^{3+}) ions to impact glutathione and mitochondrial functions (reviewed by Zolak, 2021) (Zwolak, I. 2021).

Other studies have shown that the blockade of hydrogen peroxide, which may be caused by excess iron and copper, by astaxanthin, a supplemental carotenoid found in krill and microalgae, prevented the release of CCO. Astaxanthin also upregulated Bcl-2 - which helps to anchor CCO to the mitochondrial membrane further. This had a knock-on effect of restoring ATP production in peroxide-treated cells thus promoting cell survival (Kim, J. H. et al. 2009).

Taken together, these results are suggestive that maintaining proper function of CCO by anchoring it to the mitochondrial membrane and avoiding peroxidative conditions may limit the impact of aging and neurodegenerative disease influenced by depositions and accumulations of transition metals. This supports the notion that treatments such as vitamin D3 and astaxnthin, which can retain CCO in the mitochondrial membrane may have neuroprotective effects across the broad spectrum of neurodegenerative diseases and as an anti-aging drug (Yang, L. et al. 2019).

CCO as a Neuroprotector against Hypoxia

With aging comes declining circulatory functions, cardiovascular disease, and build-up of atherosclerotic plaques. Together, these can result in subclinical hypoperfusion in the brain which over time, may bring about significant impairment in ATP production and energy impairment (reviewed by de la Torre, 2012) (de la Torre, J. C. 2012). With energy dyshomeostasis comes increased ROS production, impairment of the ETC and it can precede neurodegenerative changes by promoting the accumulation of amyloid-β and α-synuclein through disruption of autophagy (Wu, J. J. et al. 2009, Pahnke, J. et al. 2013, Gan, M. et al. 2015).

Moroever, hypoperfusion can result in hypoxia causing increased levels of lipid peroxidation, which as mentioned before, can alter CCO dynamics and trigger apoptosis (Wilhelm, J. et al. 1999, Liu, H. et al. 2012, Zhao, Y. et al. 2015). Therefore, reducing hypoxic events, or mitigating the deleterious consequences in hypoxia represent an important intervention in prevention of cognitive dysfunction associated with aging, as well as the triggering, or exacerbation, of neurodegeneration (Sapin, E. et al. 2015, Macheda, T. et al. 2019, Korte, N. et al. 2020).

Hypoxic events often lead to necrosis where oxygen starvation directly induces neuronal mortality. However, over longer periods of time, apoptotic mechanisms take hold as neighbouring cells undergo programmed cell death (Northington, F. J. et al. 2001). Thereby suggesting that if apoptosis can be reduced, neuronal survival may be possible. Indeed, resveratrol, a polyphenol found mainly in grapes and red wine, was able to significantly reduce apoptosis and activation of caspase-3 as well as upregulating Bcl-2 and downregulating Bax (Bcl-2-associated X protein) (Keshtkar, S. et al. 2020). Furthermore, resveratrol was capable of reducing both lipid peroxidation and release of CCO into the cytosol in ischemic rats. This was accompanied by restored mitochondrial function and reduced hippocampal pathology of CA1, which is particularly prone to insult, as well as reductions in neuroinflammation and glial cell infiltration (Yousuf, S. et al. 2009).

It has also been discovered that hypoxic events result in the phosphorylation of CCO subunits I, IV and Vb which reduce activity and increase production of ROS (Prabu, S. K. et al. 2006). Phosphorylation was prevented by administration of a myristoylated peptide inhibitor and H89 which selectively block the PKA (protein kinase A) pathway and resulted in reductions in ischemic injury due to increased CCO activity. In addition, PI3K (phosphatidylinositol kinase-3) inhibition resulted in reductions of phosphorylated Akt (protein kinase B) but increased cytosolic release of CCO and phosphorylation of Bad (Bcl-2-associated death promotor) further supporting the role of phosphorylation in the regulation of CCO induced apoptosis (Hirai, K. et al. 2004). Upregulating the PI3K pathway and phosphorylated Akt may therefore present as an attractive therapuetic due to their downstream effects on mitochondrial CCO release and potential triggering of pro-apoptotic pathways when in short supply. Indeed, lovastatin was proved to be neuroprotective by increased phosphorylation of Akt resulting in a reduction of cytosolic CCO preventing activation of caspase-3 apoptotic pathways (Xu, R. et al. 2008). Moreover, EGCG, from green tea, was also capable of upregulating PI3K and phosphorylation of Akt in the G93A SOD1 mouse model of ALS as well as when given as a pretreatment prior to exogenous hydrogen peroxide insult. This prevented release of CCO and increased cell viability via reductions in activated caspase-3 (Koh, S. H. et al. 2003, Koh, S. H. et al. 2006).

Herbal remedies such as a trans-himalayan phytococktail (consisting *of H. rhamnoides* fruit pulp, *P. armeniaca* fruit pulp, and *R. imbricata* dry root extract; 100:50:1) have also recapitulated the effects of resveratrol in reducing hypobaric hypoxia as a result of high-altitude simulation in rats. Memory impairment by the hypoxic conditions was restored upon receiving the phytococktail and, at a cellular level, prevented the loss of hippocampal neurons. This was found to be a result of upregulation of the PI3K/AKT pathway reducing expression of activated caspase-3 and is likely a result of CCO remaining bound to the mitochondrial membrane as evidenced by lovastatin above which also operates through this pathway (Dhar, P. et al. 2019).

Modulation of CCO remains an important factor in hypoxic events which may influence cognitive function in aging and neurodegenerative processes. In many studies, triggering of apoptosis through the caspase-3/9 cascade resulted in neuronal loss. However, treatment with resveratrol and herbal remedies which may impact CCO activity and translocation have been seen to be effective. Moreover, these interventions also influence phosphorylation pathways which manipulation is sufficient for efficacious impact.

CCO, Insulin and Neuroprotection

In recent years, increasing interest in insulin signalling has risen due to several pathological crossovers of AD and diabetes. This includes impaired brain responsiveness to insulin which may aggravate inflammatory reactions as well as cause oxidative stress and mitochondrial dysfunction (de la Monte, S. M. 2014). Indeed, a higher prevalence of AD in type 1 and type 2 diabetes sufferers indicates a relationship between the two. This has led to some suggestion that sporadic AD may be viewed as a type of diabetes located in the CNS, so-called type 3 diabetes (reviewed by de la Monte, 2014) (de la Monte, S. M. 2014). In recent years, this research has expanded to other neurodegenerative disorders such as Parkinson's and Huntington's which also show relationships to diabetes (Farrer, L. A. 1985). Interestingly, in PD it has been shown that there may be no significant increase in incidence of diabetes, however, there remains high prevalence of insulin resistance in subclinical, non-diabetic PD patients (Becker, C. et al. 2008, Hogg, E. et al. 2018). There is some suggestion that insulin resistance may also influence the non-motor symptoms of PD due to patients suffering from concurrent dementia (PD with dementia) being more than two times likely to have insulin resistance (Bosco, D. et al. 2012).

Research into targeting this resistance to insulin has shown neuroprotective effects through modulation of the mitochondrial-dependent activation of caspase-3 via downregulation of ROS and alteration of CCO dynamics. For instance, insulin-like growth factor I (IGF-I) has been shown to have anti-apoptotic effects before insult. When combined with erythropoietin, this prevented caspase-3 activation and loss of neurons (Digicaylioglu, M. et al. 2004). Moreover, in MPTP-induced neurotoxic SH-SY5Y cell cultures, insulin was found to lower nitric oxide release, reduce ROS production and Ca^{2+} influx. Insulin also upregulated Bcl-2 levels and downregulated Bax levels reducing the likelihood of CCO being released into the cytosol thereby triggering apoptosis (Ramalingam, M. et al. 2015, Ramalingam, M. et al. 2016).

Sanderson et al. (2013) (Sanderson, T. H. et al. 2013) found that insulin was neuroprotective in a global brain ischemic model induced by bilateral carotid occlusion for 8 minutes. The effect of insulin prevented CCO release resulting in survival of hippocampal CA1 neurons, prone to neurodegeneration and insult (Pandya, V. A. et al. 2021). The mechanism through which this occurred was due to phosphorylation of the tyrosine-97 (Tyr97) residue in CCO which persisted long after ischemic insult. Interestingly, insulin phosphorylated Tyr97 even in the absence of any ischemic insult (Sanderson, T. H. et al. 2013). More research is needed, however, if phosphorylated-Tyr97 can retain CCO to the mitochondrial membrane then insulin and other therapies which can induce this phosphorylation may be neuroprotective in other models of disease which have CCO-release induced apoptosis.

The glycoside paeoniflorin, found in the roots of the peony plant, has also been found to be neuroprotective by improving dysfunctional insulin signalling. A mouse model of cognitive impairment and insulin resistance induced by intracerebroventricular administration of streptozotocin (STZ) was given paeoniflorin which dramatically increased activity levels of CCO as well as ATP production, mitochondrial membrane potential and oxidative stress in the hippocampus and the cortex of mice. Furthermore, insulin signalling was restored via upregulation of phospho-PI3K and phospho-Akt as well as reducing p-IRS-1 (insulin receptor substrate 1) proteins (Wang, D. et al. 2018). Increasing usage of the STZ-model in AD indicates the crossover and restoration of insulin signalling as a therapeutic option (Salkovic-Petrisic, M. et al. 2011, Grieb, P. 2016). Moreover, in PD models of 6-OHDA and MPTP, paeoniflorin was also found to be protective, specifically in the MPTP model, through upregulation of Bcl-2 and inhibition of caspases-3/-9 indicating the cross-disease role that CCO has to play through insulin dynamics and resulting neurodegeneration (Gu, X. S. et al. 2016, Zheng, M. et al. 2017).

Novel Therapeutics in CCO

CCO sits at a convergence in apoptotic mechanisms which underlie major pathways implicated in the progression of neurodegeneration. It is an influencing factor in hypoxia-induced insult, insulin-resistance and accumulations of metals – all of which can lead to neurodegeneration via activation of caspase-3 and resulting apoptosis. Many therapies such as estrogens and its analogues, polyphenols found in tea and wine, and herbal remedies have proven efficacious against these insults via the modulation of CCO, either directly or indirectly. As such, there is convincing evidence that CCO remains a viable target for anti-aging and anti-neurodegenerative therapeutics. Nevertheless, there remains significant hurdles – current treatments are not always suitable due to side effects (i.e., resveratrol and the toxicity of wine; estrogen and it's carcinogenic, pro-stroke properties). Specific modulation of CCO however, remains relatively unexplored. Bilirubin, which specifically targets CCO can be both toxic and protective, illustrating that CCO is indeed a valuable target (Vaz, A. R. et al. 2010, Mendez, N. V. et al. 2013). Nevertheless, pharmacological intervention may be difficult to balance and further research is required.

CCO Promising Therapies

CCO is complex IV of the mitochondrial respiratory chain and consists of 13 protein subunits containing two heme centers (heme a and heme a3) as well as two copper centers (CuA and CuB). This enzyme transfers electrons from cytochrome c to O_2 via CuA and from heme a to the heme a3/CuB binuclear center (Hill, B. C. 1994). The proton gradient so formed drives the activity of ATP synthase.

Cytochromes, as CCO, together with a range of chromophores such as water, oxyhemoglobin (HbO2), deoxyhemoglobin (Hb), myoglobin, melanin and flavins can be found in biological tissues. As chromophores, they can absorb light energy at different wavelengths. For example, water molecules absorb light energy at wavelengths greater than 970 nm, while wavelengths shorter than 600 nm are absorbed by flavins, hemoglobin, and melanin. Special interest has been taken in the excitation of mitochondrial cytochrome c-oxidase (CCO) which serves as a primary photoacceptor in the red to near infrared (NIR) region (Karu, T. I. 1986).

This feature of CCO has been corroborated by the absorption peaks in the red (heme a, 605 nm; CuA reduced, 620 nm; heme a3/CuB, 655 nm; CuB oxidized, 680 nm) and the NIR spectral regions (CuB reduced, 760 nm; CuA oxidized, 825 nm) (Karu, T. I. et al. 2005, Ball, K. A. et al. 2011). Moreover, it has been shown that the purified enzyme, cytochrome c-oxidase was shown to be activated in vitro by red laser (633 nm) (Pastore, D. et al. 2000).

The abovementioned results suggested that photon energy could be absorbed by the various metal centers of CCO, so the electrons are excited from the ground state to upper excited states (Santana-Blank, L. et al. 2010). Furthermore, since the principal chromophores are located inside the mitochondria, it follows that cells with a large number of mitochondria and a high metabolic activity are particularly responsive to light. This consideration applies to muscle cells (both skeletal and cardiac), liver, kidney and other internal organs, but also neurons (especially CNS neurons) shedding light into a new therapeutic field for neurodegeneration, as we will later describe.

The proposed hypothesis for the red or NIR light effects on CCO is that nitric oxide, which inhibits CCO by non-covalently binding between heme-a3 and CuB (Sarti, P. et al. 2012), can be photodissociated (Lane, N. 2006). Since it is known that NO inhibits electron transport in the mitochondrial respiratory chain, its dissociation can increase the mitochondrial membrane potential (MMP), increase oxygen consumption and hence the proton gradient ultimately leading to an increase in ATP production (Hamblin, M. 2008). This mechanism of action would also explain the potential of red or NIR light as a therapy in diseased conditions, because unhealthy or hypoxic cells are more likely to have inhibitory concentrations of NO. These consequences are also accompanied by long-lasting effects on cells, due to the production of reactive oxygen species (ROS) and release of Ca^{2+} as versatile second messengers, in turn leading to activation of transcription factors and signaling mediators such as NF-κB (Karu, T. 2000, Hamblin, M. et al. 2006, Gao, X. et al. 2009, de Freitas, L. F. et al. 2016).

It is important to point out that this hypothesis has been recently critiziced by Sommer (Sommer, A. P. 2019). The author suggested that all the changes previously described occurs via the interaction of photons with intracellular interfacial water layer (IWL), where the minimum dose is related to the minimum light intensity necessary to reduce IWL viscosity and the maximum dose is related to interference of the photons with CCO, either directly or via ROS generation. Although this could bring more light into the problem the lack of experimental evidence does not allow to draw a conclusion.

Identifying the Therapeutic Dosage

Wong-Riley et al. was a pioneer in testing the effect of different wavelengths on CCO activity. When irradiation using LED light (4 J/cm2) at wavelengths of 670 and 830 nm was applied in cultured rat visual cortical neurons, upregulation of CCO activity was observed (Wong-Riley, M. T. et al. 2001, Wong-Riley, M. T. et al. 2005). Another study in cultured cortical neurons showed that, with a constant irradiance (25 mW/cm^2), the peak efficiency was obtained at 3 J/cm^2 for ATP production, along with an increase in MMP as well as calcium levels. Whereas both low (0.03 and 0.3 J/cm^2) and high (10 J/cm^2) doses showed minor stimulatory effects and even inhibitory effects due to mitochondrial damage occurred at higher fluence (30 J/cm^2) (Sharma, S. K. et al. 2011).

Similar results were obtained when applied to naïve rats, where LED light irradiation at 633 nm resulted in augmentation of CCO activity in their prefrontal cortex (PFC) by 14% (10.9 J/cm^2) (Rojas, J. C. et al. 2012), in the superior colliculus by 26% and in the whole brain by 60% (3.6 J/cm^2) in a rat model of rotenone induced neurotoxicity (Rojas, J. C. et al. 2008). However, in another study on mice, the application of light at wavelengths of 627 and 660 nm did not have significant stimulatory effects (Ball, K. A. et al. 2011).

These results together suggest a biphasic or inverted U-shaped dose-response curve of CCO to light. According to this relationship, light at very low doses has no significant stimulatory effects (below threshold), whereas small (but still larger) doses above this threshold produce beneficial effects. On the other hand, much higher doses of PBM can have inhibitory and even harmful effects.

CCO as a Rescue Factor in Neurodegeneration

Under neurodegenerative conditions, mitochondria may undergo major changes including reduced respiratory chain complex activity which will lead to lower ATP synthesis, overproduction of ROS, and the loss of MMP, inner mitochondrial permeability transition, and the release of cytochrome c into the cytosol (Nunnari, J. et al. 2012). However, it has been shown that 3 J/cm^2 application was able to increase the activity of CCO in the hippocampus in a murine model of Alzheimer´s disease (AD) (Lu, Y. et al. 2017). In line with this, Purushothuman et al. (Purushothuman, S. et al. 2014) demonstrated a significant restoration of CCO expression patterns in the neocortex and hippocampus following 4 weeks of transcranial LED therapy (670 nm). Also, transcranial PBM therapy using 808 nm laser increased cerebral ATP levels in amyloid protein precursor (APP) transgenic mice (6 J/cm2) (De Taboada, L. et al. 2011), Aβ-induced AD mice (3 J/cm2) (Lu, Y. et al. 2017), as well as a mouse model of major depression (41 J/cm2) (Xu, Z. et al. 2017).

However, contradictory results have been shown in cultured cells, where a single-session of LED treatment (670 nm, 4 J/cm2) increased the ATP content of 1-methyl-4- phenylpyridinium (MPP+) exposed striatal neurons (Ying, R. et al. 2008), but a single session of laser irradiation did not enhance ATP levels either in Aβ treated PC12 cells (670 nm, 1 J/cm2) (Sommer, A. P. et al. 2012) or in the PD cybrid cell lines (at 810 nm, 2 J/cm2) (Trimmer, P. A. et al. 2009). This evidence is supporting the need to define the effect of irradiation time on brain or cultured neurons. Studies have proved that low fluence/low irradiance and the high fluence/high irradiance group (both being applied the same amount of time) together with the high fluence/low irradiance (taking 10 times more) were all equally effective. The only combination that was not effective was the low fluence at the high irradiance (only taking 1 min) (Salehpour, F. et al. 2018). All these results highlight how the modulation of CCO through a specific light wavelength could be used as a powerful tool to recover neuronal function in neurodegeneration.

It is important to note that NIR- lightemitting has been applied for years to humans to have live monitoring of the changes in the concentration of oxyHb and deoxyHb. Indeed, fNIRS which is a reliable tool relies on continuous-wave monochromatic near-infrared light. However, it was often overlooked that that when applied on the head, light in the infrared spectrum can also determine changes in neural activity. Indeed, Zomorrodi and colleagues (Zomorrodi, R. et al. 2019) showed that when applying NIR light pulse (810 nm wavelength, 40 Hz) on different regions (frontal, left/right temporal, precuneus cortex, and intranasal), frequency-dependent effects on endogenous brain activity were found. Similar findings were found on an Alzheimer's disease patient (Yue, L. et al. 2015), a low frequencies power reduction (i.e., delta and theta) and a higher frequencies power increase (i.e., alpha, beta, and gamma) were recorded following light stimulation. These findings might suggest that modulation through NIR light could also modulate cortical oscillations, and then brain connectivity, in a frequency-dependent manner in humans (Martini, M. et al. 2021).

Conclusion

In this chapter we have highlighted the importance of understanding the numerous mutations in structural and assembly genes identified in CCO which ultimately lead to deficient patients. Also, we have presented strategies to modulate CCO in order to maintain or recover ATP production, ultimately recovering the respiratory integrity of mitochondria.

The current advances in gene therapy where the development of vectors allows targeting of a specific organ, could be driven to treat mitochondrial disorders. Alternative approaches to gene therapy have also been explored to fight CCO deficiency and other respiratory chain defects such as dietary changes which enhances mitochondrial biogenesis, which in turn raises the cellular concentrations of ATP among other positive effects (Wang, Y. et al. 2015, Kaliszewska, A. et al. 2021). Also, light modulation of CCO has been presented; however, more studies are needed regarding the biological effects of the power density of red or NIR-light (mW/cm^2), and not on the total dose (J/cm^2).

Although more studies are needed to decipher the mechanisms responsible for the biogenesis of the respiratory chain complexes such as CCO, and the role of the factors involved in this process, especially when it comes to neurodegenerative diseases; we currently have some promising tools to slow down the course of neurodegeneration.

References

Abdulhag, U. N., Soiferman, D., Schueler-Furman, O., Miller, C., Shaag, A., Elpeleg, O., Edvardson, S. and Saada, A. (2015). "Mitochondrial complex IV deficiency, caused by mutated COX6B1, is associated with encephalomyopathy, hydrocephalus and cardiomyopathy." *Eur J Hum Genet* 23(2): 159-164.

Abu-Libdeh, B., Douiev, L., Amro, S., Shahrour, M., Ta-Shma, A., Miller, C., Elpeleg, O. and Saada, A. (2017). "Mutation in the COX4I1 gene is associated with short stature, poor weight gain and increased chromosomal breaks, simulating Fanconi anemia." *Eur J Hum Genet* 25(10): 1142-1146.

Aggeler, R. and Capaldi, R. A. (1990). "Yeast cytochrome c oxidase subunit VII is essential for assembly of an active enzyme. Cloning, sequencing, and characterization of the nuclear-encoded gene." *J Biol Chem* 265(27): 16389-16393.

Ahmed, Z., Sheng, H., Xu, Y. F., Lin, W. L., Innes, A. E., Gass, J., Yu, X., Wuertzer, C. A., Hou, H., Chiba, S., Yamanouchi, K., Leissring, M., Petrucelli, L., Nishihara, M., Hutton, M. L., McGowan, E., Dickson, D. W. and Lewis, J. (2010). "Accelerated lipofuscinosis and ubiquitination in granulin knockout mice suggest a role for progranulin in successful aging." *Am J Pathol* 177(1): 311-324.

Andersen, P. M. (2006). "Amyotrophic lateral sclerosis associated with mutations in the CuZn superoxide dismutase gene." *Curr Neurol Neurosci Rep* 6(1): 37-46.

Araujo, G. W., Beyer, C. and Arnold, S. (2008). "Oestrogen influences on mitochondrial gene expression and respiratory chain activity in cortical and mesencephalic astrocytes." *J Neuroendocrinol* 20(7): 930-941.

Arnold, S. (2012). "Cytochrome c oxidase and its role in neurodegeneration and neuroprotection." *Adv Exp Med Biol* 748: 305-339.

Arnold, S. and Kadenbach, B. (1997). "Cell respiration is controlled by ATP, an allosteric inhibitor of cytochrome-c oxidase." *Eur J Biochem* 249(1): 350-354.

Arnold, S., Victor, M. B. and Beyer, C. (2012). "Estrogen and the regulation of mitochondrial structure and function in the brain." *J Steroid Biochem Mol Biol* 131(1-2): 2-9.

Avery, S. V. (2011). "Molecular targets of oxidative stress." *Biochem J* 434(2): 201-210.

Baden, K. N., Murray, J., Capaldi, R. A. and Guillemin, K. (2007). "Early developmental pathology due to cytochrome c oxidase deficiency is revealed by a new zebrafish model." *J Biol Chem* 282(48): 34839-34849.

Baertling, F., Al-Murshedi, F., Sanchez-Caballero, L., Al-Senaidi, K., Joshi, N. P., Venselaar, H., van den Brand, M. A., Nijtmans, L. G. and Rodenburg, R. J. (2017). "Mutation in mitochondrial complex IV subunit COX5A causes pulmonary arterial hypertension, lactic acidemia, and failure to thrive." *Hum Mutat* 38(6): 692-703.

Ball, K. A., Castello, P. R. and Poyton, R. O. (2011). "Low intensity light stimulates nitrite-dependent nitric oxide synthesis but not oxygen consumption by cytochrome c oxidase: Implications for phototherapy." *J Photochem Photobiol B* 102(3): 182-191.

Balsa, E., Marco, R., Perales-Clemente, E., Szklarczyk, R., Calvo, E., Landazuri, M. O. and Enriquez, J. A. (2012). "NDUFA4 is a subunit of complex IV of the mammalian electron transport chain." *Cell Metab* 16(3): 378-386.

Baric, I., Fumic, K., Petkovic Ramadza, D., Sperl, W., Zimmermann, F. A., Muacevic-Katanec, D., Mitrovic, Z., Pazanin, L., Cvitanovic Sojat, L., Kekez, T., Reiner, Z. and Mayr, J. A. (2013). "Mitochondrial myopathy associated with a novel 5522G>A mutation in the mitochondrial tRNA(Trp) gene." *Eur J Hum Genet* 21(8): 871-875.

Bates, G. P., Dorsey, R., Gusella, J. F., Hayden, M. R., Kay, C., Leavitt, B. R., Nance, M., Ross, C. A., Scahill, R. I., Wetzel, R., Wild, E. J. and Tabrizi, S. J. (2015). "Huntington disease." *Nat Rev Dis Primers* 1: 15005.

Bayati, S. and Yazdanparast, R. (2011). "Antioxidant and free radical scavenging potential of yakuchinone B derivatives in reduction of lipofuscin formation using H2O2-treated neuroblastoma cells." *Iran Biomed J* 15(4): 134-142.

Becker, C., Brobert, G. P., Johansson, S., Jick, S. S. and Meier, C. R. (2008). "Diabetes in patients with idiopathic Parkinson's disease." *Diabetes Care* 31(9): 1808-1812.

Benedetti, M. D., Maraganore, D. M., Bower, J. H., McDonnell, S. K., Peterson, B. J., Ahlskog, J. E., Schaid, D. J. and Rocca, W. A. (2001). "Hysterectomy, menopause, and estrogen use preceding Parkinson's disease: an exploratory case-control study." *Mov Disord* 16(5): 830-837.

Benson, B. C., Shaw, P. J., Azzouz, M., Highley, J. R. and Hautbergue, G. M. (2021). "Proteinopathies as Hallmarks of Impaired Gene Expression, Proteostasis and Mitochondrial Function in Amyotrophic Lateral Sclerosis." *Front Neurosci* 15: 783624.

Borlongan, C. V., Koutouzis, T. K., Randall, T. S., Freeman, T. B., Cahill, D. W. and Sanberg, P. R. (1995). "Systemic 3-nitropropionic acid: behavioral deficits and striatal damage in adult rats." *Brain Res Bull* 36(6): 549-556.

Borthwick, G. M., Johnson, M. A., Ince, P. G., Shaw, P. J. and Turnbull, D. M. (1999). "Mitochondrial enzyme activity in amyotrophic lateral sclerosis: implications for the role of mitochondria in neuronal cell death." *Ann Neurol* 46(5): 787-790.

Bosco, D., Plastino, M., Cristiano, D., Colica, C., Ermio, C., De Bartolo, M., Mungari, P., Fonte, G., Consoli, D., Consoli, A. and Fava, A. (2012). "Dementia is associated with insulin resistance in patients with Parkinson's disease." *J Neurol Sci* 315(1-2): 39-43.

Boulet, L., Karpati, G. and Shoubridge, E. A. (1992). "Distribution and threshold expression of the tRNA(Lys) mutation in skeletal muscle of patients with myoclonic epilepsy and ragged-red fibers (MERRF)." *Am J Hum Genet* 51(6): 1187-1200.

Brischigliaro, M. and Zeviani, M. (2021). "Cytochrome c oxidase deficiency." *Biochim Biophys Acta Bioenerg* 1862(1): 148335.

Brockmann, K., Srulijes, K., Pflederer, S., Hauser, A. K., Schulte, C., Maetzler, W., Gasser, T. and Berg, D. (2015). "GBA-associated Parkinson's disease: reduced survival and more rapid progression in a prospective longitudinal study." *Mov Disord* 30(3): 407-411.

Brown, M. D., Voljavec, A. S., Lott, M. T., MacDonald, I. and Wallace, D. C. (1992). "Leber's hereditary optic neuropathy: a model for mitochondrial neurodegenerative diseases." *FASEB J* 6(10): 2791-2799.

Browne, S. E., Bowling, A. C., Baik, M. J., Gurney, M., Brown, R. H., Jr. and Beal, M. F. (1998). "Metabolic dysfunction in familial, but not sporadic, amyotrophic lateral sclerosis." *J Neurochem* 71(1): 281-287.

Bruno, C., Martinuzzi, A., Tang, Y., Andreu, A. L., Pallotti, F., Bonilla, E., Shanske, S., Fu, J., Sue, C. M., Angelini, C., DiMauro, S. and Manfredi, G. (1999). "A stop-codon mutation in the human mtDNA cytochrome c oxidase I gene disrupts the functional structure of complex IV." *Am J Hum Genet* 65(3): 611-620.

Burke, P. V., Raitt, D. C., Allen, L. A., Kellogg, E. A. and Poyton, R. O. (1997). "Effects of oxygen concentration on the expression of cytochrome c and cytochrome c oxidase genes in yeast." *J Biol Chem* 272(23): 14705-14712.

Cadonic, C., Sabbir, M. G. and Albensi, B. C. (2016). "Mechanisms of Mitochondrial Dysfunction in Alzheimer's Disease." *Mol Neurobiol* 53(9): 6078-6090.

Campos, Y., Garcia-Redondo, A., Fernandez-Moreno, M. A., Martinez-Pardo, M., Goda, G., Rubio, J. C., Martin, M. A., del Hoyo, P., Cabello, A., Bornstein, B., Garesse, R.

and Arenas, J. (2001). "Early-onset multisystem mitochondrial disorder caused by a nonsense mutation in the mitochondrial DNA cytochrome C oxidase II gene." *Ann Neurol* 50(3): 409-413.

Cardoso, S. M., Santana, I., Swerdlow, R. H. and Oliveira, C. R. (2004). "Mitochondria dysfunction of Alzheimer's disease cybrids enhances Abeta toxicity." *J Neurochem* 89(6): 1417-1426.

Carter, R. J., Lione, L. A., Humby, T., Mangiarini, L., Mahal, A., Bates, G. P., Dunnett, S. B. and Morton, A. J. (1999). "Characterization of progressive motor deficits in mice transgenic for the human Huntington's disease mutation." *J Neurosci* 19(8): 3248-3257.

Chandrasena, R. E., Edirisinghe, P. D., Bolton, J. L. and Thatcher, G. R. (2008). "Problematic detoxification of estrogen quinones by NAD(P)H-dependent quinone oxidoreductase and glutathione-S-transferase." *Chem Res Toxicol* 21(7): 1324-1329.

Chen, D. Z., Y.; Lyons, K. E.; Pahwa, R.; Reddy, M.B (2015). "Green Tea Consumption Reduces Oxidative Stress in Parkinson's Disease Patients." *Journal of Behavioral and Brain Science* 5(6): 194-202.

Chen, X. C., Chen, Y., Zhu, Y. G., Fang, F. and Chen, L. M. (2002). "Protective effect of ginsenoside Rg1 against MPTP-induced apoptosis in mouse substantia nigra neurons." *Acta Pharmacol Sin* 23(9): 829-834.

Cheng, S., Hou, J., Zhang, C., Xu, C., Wang, L., Zou, X., Yu, H., Shi, Y., Yin, Z. and Chen, G. (2015). "Minocycline reduces neuroinflammation but does not ameliorate neuron loss in a mouse model of neurodegeneration." *Sci Rep* 5: 10535.

Cherny, R. A., Atwood, C. S., Xilinas, M. E., Gray, D. N., Jones, W. D., McLean, C. A., Barnham, K. J., Volitakis, I., Fraser, F. W., Kim, Y., Huang, X., Goldstein, L. E., Moir, R. D., Lim, J. T., Beyreuther, K., Zheng, H., Tanzi, R. E., Masters, C. L. and Bush, A. I. (2001). "Treatment with a copper-zinc chelator markedly and rapidly inhibits beta-amyloid accumulation in Alzheimer's disease transgenic mice." *Neuron* 30(3): 665-676.

Clark, K. M., Taylor, R. W., Johnson, M. A., Chinnery, P. F., Chrzanowska-Lightowlers, Z. M., Andrews, R. M., Nelson, I. P., Wood, N. W., Lamont, P. J., Hanna, M. G., Lightowlers, R. N. and Turnbull, D. M. (1999). "An mtDNA mutation in the initiation codon of the cytochrome C oxidase subunit II gene results in lower levels of the protein and a mitochondrial encephalomyopathy." *Am J Hum Genet* 64(5): 1330-1339.

Cottrell, D. A., Borthwick, G. M., Johnson, M. A., Ince, P. G. and Turnbull, D. M. (2002). "The role of cytochrome c oxidase deficient hippocampal neurones in Alzheimer's disease." *Neuropathol Appl Neurobiol* 28(5): 390-396.

Curtis, A. F., Masellis, M., Hsiung, G. R., Moineddin, R., Zhang, K., Au, B., Millett, G., Mackenzie, I., Rogaeva, E. and Tierney, M. C. (2017). "Sex differences in the prevalence of genetic mutations in FTD and ALS: A meta-analysis." *Neurology* 89(15): 1633-1642.

Das, J., Miller, S. T. and Stern, D. L. (2004). "Comparison of diverse protein sequences of the nuclear-encoded subunits of cytochrome C oxidase suggests conservation of structure underlies evolving functional sites." *Mol Biol Evol* 21(8): 1572-1582.

Dauer, W. and Przedborski, S. (2003). "Parkinson's disease: mechanisms and models." *Neuron* 39(6): 889-909.

Davey, D. A. (2013). "Alzheimer's disease, dementia, mild cognitive impairment and the menopause: a 'window of opportunity'?" *Womens Health (Lond)* 9(3): 279-290.

de Freitas, L. F. and Hamblin, M. R. (2016). "Proposed Mechanisms of Photobiomodulation or Low-Level Light Therapy." *IEEE J Sel Top Quantum Electron* 22(3).

de la Monte, S. M. (2014). "Type 3 diabetes is sporadic Alzheimers disease: mini-review." *Eur Neuropsychopharmacol* 24(12): 1954-1960.

de la Torre, J. C. (2012). "Cardiovascular risk factors promote brain hypoperfusion leading to cognitive decline and dementia." *Cardiovasc Psychiatry Neurol* 2012: 367516.

De Taboada, L., Yu, J., El-Amouri, S., Gattoni-Celli, S., Richieri, S., McCarthy, T., Streeter, J. and Kindy, M. S. (2011). "Transcranial laser therapy attenuates amyloid-beta peptide neuropathology in amyloid-beta protein precursor transgenic mice." *J Alzheimers Dis* 23(3): 521-535.

Delic, V., Kurien, C., Cruz, J., Zivkovic, S., Barretta, J., Thomson, A., Hennessey, D., Joseph, J., Ehrhart, J., Willing, A. E., Bradshaw, P. and Garbuzova-Davis, S. (2018). "Discrete mitochondrial aberrations in the spinal cord of sporadic ALS patients." *J Neurosci Res* 96(8): 1353-1366.

Deschauer, M., Swalwell, H., Strauss, M., Zierz, S. and Taylor, R. W. (2006). "Novel mitochondrial transfer RNA(Phe) gene mutation associated with late-onset neuromuscular disease." *Arch Neurol* 63(6): 902-905.

Dhar, P., Das, S. K., Barhwal, K., Hota, S. K., Mishra, K. P. and Singh, S. B. (2019). "Trans-Himalayan Phytococktail Confers Protection Against Hypobaric Hypoxia-Induced Hippocampal Neurodegeneration and Memory Impairment in Male Sprague Dawley Rats." *High Alt Med Biol* 20(3): 279-292.

Digicaylioglu, M., Garden, G., Timberlake, S., Fletcher, L. and Lipton, S. A. (2004). "Acute neuroprotective synergy of erythropoietin and insulin-like growth factor I." *Proc Natl Acad Sci U S A* 101(26): 9855-9860.

DiMauro, S., Nicholson, J. F., Hays, A. P., Eastwood, A. B., Papadimitriou, A., Koenigsberger, R. and DeVivo, D. C. (1983). "Benign infantile mitochondrial myopathy due to reversible cytochrome c oxidase deficiency." *Ann Neurol* 14(2): 226-234.

DiMauro, S., Servidei, S., Zeviani, M., DiRocco, M., DeVivo, D. C., DiDonato, S., Uziel, G., Berry, K., Hoganson, G., Johnsen, S. D. and et al. (1987). "Cytochrome c oxidase deficiency in Leigh syndrome." *Ann Neurol* 22(4): 498-506.

Dineley, K. E., Votyakova, T. V. and Reynolds, I. J. (2003). "Zinc inhibition of cellular energy production: implications for mitochondria and neurodegeneration." *J Neurochem* 85(3): 563-570.

Dixon, S. J. and Stockwell, B. R. (2014). "The role of iron and reactive oxygen species in cell death." *Nat Chem Biol* 10(1): 9-17.

Dykens, J. A., Moos, W. H. and Howell, N. (2005). "Development of 17alpha-estradiol as a neuroprotective therapeutic agent: rationale and results from a phase I clinical study." *Ann N Y Acad Sci* 1052: 116-135.

Ebanks, B. and Chakrabarti, L. (2022). "Mitochondrial ATP Synthase is a Target of Oxidative Stress in Neurodegenerative Diseases." *Front Mol Biosci* 9: 854321.

Farahvash, A. and Stuchebrukhov, A. (2018). "Investigating the Many Roles of Internal Water in Cytochrome c Oxidase." *J Phys Chem B* 122(31): 7625-7635.

Farrer, L. A. (1985). "Diabetes mellitus in Huntington disease." *Clin Genet* 27(1): 62-67.

Flannery, P. J. and Trushina, E. (2019). "Mitochondrial dynamics and transport in Alzheimer's disease." *Mol Cell Neurosci* 98: 109-120.

Fornuskova, D., Stiburek, L., Wenchich, L., Vinsova, K., Hansikova, H. and Zeman, J. (2010). "Novel insights into the assembly and function of human nuclear-encoded cytochrome c oxidase subunits 4, 5a, 6a, 7a and 7b." *Biochem J* 428(3): 363-374.

Fox, J. H., Kama, J. A., Lieberman, G., Chopra, R., Dorsey, K., Chopra, V., Volitakis, I., Cherny, R. A., Bush, A. I. and Hersch, S. (2007). "Mechanisms of copper ion mediated Huntington's disease progression." *PLoS One* 2(3): e334.

Fox, T. D. (1979). "Genetic and physical analysis of the mitochondrial gene for subunit II of yeast cytochrome c oxidase." *Journal of Molecular Biology* 130(1): 63-82.

Francisci, S., Bohn, C., Frontali, L. and Bolotin-Fukuhara, M. (1998). "Ts mutations in mitochondrial tRNA genes: characterization and effects of two point mutations in the mitochondrial gene for tRNAphe in Saccharomyces cerevisiae." *Curr Genet* 33(2): 110-116.

Gan, M., Moussaud, S., Jiang, P. and McLean, P. J. (2015). "Extracellular ATP induces intracellular alpha-synuclein accumulation via P2X1 receptor-mediated lysosomal dysfunction." *Neurobiol Aging* 36(2): 1209-1220.

Gao, X. and Xing, D. (2009). "Molecular mechanisms of cell proliferation induced by low power laser irradiation." *J Biomed Sci* 16: 4.

Gao, Y. L., Wang, N., Sun, F. R., Cao, X. P., Zhang, W. and Yu, J. T. (2018). "Tau in neurodegenerative disease." *Ann Transl Med* 6(10): 175.

Gattermann, N., Retzlaff, S., Wang, Y. L., Hofhaus, G., Heinisch, J., Aul, C. and Schneider, W. (1997). "Heteroplasmic Point Mutations of Mitochondrial DNA Affecting Subunit I of Cytochrome c Oxidase in Two Patients With Acquired Idiopathic Sideroblastic Anemia." *Blood* 90(12): 4961-4972.

Gawryluk, R. M. R., Kamikawa, R., Stairs, C. W., Silberman, J. D., Brown, M. W. and Roger, A. J. (2016). "The Earliest Stages of Mitochondrial Adaptation to Low Oxygen Revealed in a Novel Rhizarian." *Curr Biol* 26(20): 2729-2738.

Glerum, D. M. and Tzagoloff, A. (1997). "Submitochondrial distributions and stabilities of subunits 4, 5, and 6 of yeast cytochrome oxidase in assembly defective mutants." *FEBS Letters* 412(3): 410-414.

Glerum, D. M. and Tzagoloff, A. (1998). "Affinity purification of yeast cytochrome oxidase with biotinylated subunits 4, 5, or 6." *Anal Biochem* 260(1): 38-43.

Goto, Y. I., Nonaka, I. and Horai, S. (1990). "A mutation in the tRNALeu(UUR) gene associated with the MELAS subgroup of mitochondrial encephalomyopathies." *Nature* 348(6302): 651-653.

Greene, J. G., Porter, R. H., Eller, R. V. and Greenamyre, J. T. (1993). "Inhibition of succinate dehydrogenase by malonic acid produces an "excitotoxic" lesion in rat striatum." *J Neurochem* 61(3): 1151-1154.

Grieb, P. (2016). "Intracerebroventricular Streptozotocin Injections as a Model of Alzheimer's Disease: in Search of a Relevant Mechanism." *Mol Neurobiol* 53(3): 1741-1752.

Gu, X. S., Wang, F., Zhang, C. Y., Mao, C. J., Yang, J., Yang, Y. P., Liu, S., Hu, L. F. and Liu, C. F. (2016). "Neuroprotective Effects of Paeoniflorin on 6-OHDA-Lesioned Rat Model of Parkinson's Disease." *Neurochem Res* 41(11): 2923-2936.

Guedes-Dias, P., Pinho, B. R., Soares, T. R., de Proenca, J., Duchen, M. R. and Oliveira, J. M. (2016). "Mitochondrial dynamics and quality control in Huntington's disease." *Neurobiol Dis* 90: 51-57.

Hallmann, K., Kudin, A. P., Zsurka, G., Kornblum, C., Reimann, J., Stuve, B., Waltz, S., Hattingen, E., Thiele, H., Nurnberg, P., Rub, C., Voos, W., Kopatz, J., Neumann, H. and Kunz, W. S. (2016). "Loss of the smallest subunit of cytochrome c oxidase, COX8A, causes Leigh-like syndrome and epilepsy." *Brain* 139(Pt 2): 338-345.

Hamblin, M. (2008). The role of nitric oxide in low level light therapy, *SPIE*.

Hamblin, M. and Demidova, T. (2006). Mechanisms of low level light therapy, *SPIE*.

Hanna, M. G., Nelson, I. P., Rahman, S., Lane, R. J., Land, J., Heales, S., Cooper, M. J., Schapira, A. H., Morgan-Hughes, J. A. and Wood, N. W. (1998). "Cytochrome c oxidase deficiency associated with the first stop-codon point mutation in human mtDNA." *Am J Hum Genet* 63(1): 29-36.

Hao, H. and Moraes, C. T. (1996). "Functional and molecular mitochondrial abnormalities associated with a C --> T transition at position 3256 of the human mitochondrial genome. The effects of a pathogenic mitochondrial tRNA point mutation in organelle translation and RNA processing." *J Biol Chem* 271(4): 2347-2352.

Hara, Y., Yuk, F., Puri, R., Janssen, W. G., Rapp, P. R. and Morrison, J. H. (2014). "Presynaptic mitochondrial morphology in monkey prefrontal cortex correlates with working memory and is improved with estrogen treatment." *Proc Natl Acad Sci U S A* 111(1): 486-491.

Heetveld, S., Rizzu, P. and Heutink, P. (2015). Genetics of Frontotemporal Dementia. Movement Disorder Genetics. S. A. Schneider and J. M. T. Brás. Cham, *Springer International Publishing*: 75-105.

Henderson, V. W. and Lobo, R. A. (2012). "Hormone therapy and the risk of stroke: perspectives 10 years after the Women's Health Initiative trials." *Climacteric* 15(3): 229-234.

Hentosh, S., Zhu, L., Patino, J., Furr, J. W., Rocha, N. P. and Furr Stimming, E. (2021). "Sex Differences in Huntington's Disease: Evaluating the Enroll-HD Database." *Mov Disord Clin Pract* 8(3): 420-426.

Hill, B. C. (1994). "Modeling the sequence of electron transfer reactions in the single turnover of reduced, mammalian cytochrome c oxidase with oxygen." *J Biol Chem* 269(4): 2419-2425.

Hirai, K., Hayashi, T., Chan, P. H., Zeng, J., Yang, G. Y., Basus, V. J., James, T. L. and Litt, L. (2004). "PI3K inhibition in neonatal rat brain slices during and after hypoxia reduces phospho-Akt and increases cytosolic cytochrome c and apoptosis." *Brain Res Mol Brain Res* 124(1): 51-61.

Hodge, M. R., Kim, G., Singh, K. and Cumsky, M. G. (1989). "Inverse regulation of the yeast COX5 genes by oxygen and heme." *Mol Cell Biol* 9(5): 1958-1964.

Hogg, E., Athreya, K., Basile, C., Tan, E. E., Kaminski, J. and Tagliati, M. (2018). "High Prevalence of Undiagnosed Insulin Resistance in Non-Diabetic Subjects with Parkinson's Disease." *J Parkinsons Dis* 8(2): 259-265.

Horvath, R., Lochmuller, H., Scharfe, C., Do, B. H., Oefner, P. J., Muller-Hocker, J., Schoser, B. G., Pongratz, D., Auer, D. P. and Jaksch, M. (2003). "A tRNA(Ala) mutation causing mitochondrial myopathy clinically resembling myotonic dystrophy." *J Med Genet* 40(10): 752-757.

Horvath, R., Scharfe, C., Hoeltzenbein, M., Do, B. H., Schroder, C., Warzok, R., Vogelgesang, S., Lochmuller, H., Muller-Hocker, J., Gerbitz, K. D., Oefner, P. J. and Jaksch, M. (2002). "Childhood onset mitochondrial myopathy and lactic acidosis caused by a stop mutation in the mitochondrial cytochrome c oxidase III gene." *J Med Genet* 39(11): 812-816.

Huttemann, M., Frank, V. and Kadenbach, B. (1999). "The possible role of isoforms of cytochrome c oxidase subunit VIa in mammalian thermogenesis." *Cell Mol Life Sci* 55(11): 1482-1490.

Huttemann, M., Helling, S., Sanderson, T. H., Sinkler, C., Samavati, L., Mahapatra, G., Varughese, A., Lu, G., Liu, J., Ramzan, R., Vogt, S., Grossman, L. I., Doan, J. W., Marcus, K. and Lee, I. (2012). "Regulation of mitochondrial respiration and apoptosis through cell signaling: cytochrome c oxidase and cytochrome c in ischemia/reperfusion injury and inflammation." *Biochim Biophys Acta* 1817(4): 598-609.

Huttemann, M., Kadenbach, B. and Grossman, L. I. (2001). "Mammalian subunit IV isoforms of cytochrome c oxidase." *Gene* 267(1): 111-123.

Huttemann, M., Klewer, S., Lee, I., Pecinova, A., Pecina, P., Liu, J., Lee, M., Doan, J. W., Larson, D., Slack, E., Maghsoodi, B., Erickson, R. P. and Grossman, L. I. (2012). "Mice deleted for heart-type cytochrome c oxidase subunit 7a1 develop dilated cardiomyopathy." *Mitochondrion* 12(2): 294-304.

Huttemann, M., Lee, I., Gao, X., Pecina, P., Pecinova, A., Liu, J., Aras, S., Sommer, N., Sanderson, T. H., Tost, M., Neff, F., Aguilar-Pimentel, J. A., Becker, L., Naton, B., Rathkolb, B., Rozman, J., Favor, J., Hans, W., Prehn, C., Puk, O., Schrewe, A., Sun, M., Hofler, H., Adamski, J., Bekeredjian, R., Graw, J., Adler, T., Busch, D. H., Klingenspor, M., Klopstock, T., Ollert, M., Wolf, E., Fuchs, H., Gailus-Durner, V., Hrabe de Angelis, M., Weissmann, N., Doan, J. W., Bassett, D. J. and Grossman, L. I. (2012). "Cytochrome c oxidase subunit 4 isoform 2-knockout mice show reduced enzyme activity, airway hyporeactivity, and lung pathology." *FASEB J* 26(9): 3916-3930.

Indrieri, A., van Rahden, V. A., Tiranti, V., Morleo, M., Iaconis, D., Tammaro, R., D'Amato, I., Conte, I., Maystadt, I., Demuth, S., Zvulunov, A., Kutsche, K., Zeviani, M. and Franco, B. (2012). "Mutations in COX7B cause microphthalmia with linear skin lesions, an unconventional mitochondrial disease." *Am J Hum Genet* 91(5): 942-949.

Inoue, K., Nakada, K., Ogura, A., Isobe, K., Goto, Y., Nonaka, I. and Hayashi, J. I. (2000). "Generation of mice with mitochondrial dysfunction by introducing mouse mtDNA carrying a deletion into zygotes." *Nat Genet* 26(2): 176-181.

Inoue, M., Uchino, S., Iida, A., Noguchi, S., Hayashi, S., Takahashi, T., Fujii, K., Komaki, H., Takeshita, E., Nonaka, I., Okada, Y., Yoshizawa, T., Van Lommel, L., Schuit, F., Goto, Y. I., Mimaki, M. and Nishino, I. (2019). "COX6A2 variants cause a muscle-specific cytochrome c oxidase deficiency." *Ann Neurol* 86(2): 193-202.

Irwin, R. W., Yao, J., Hamilton, R. T., Cadenas, E., Brinton, R. D. and Nilsen, J. (2008). "Progesterone and estrogen regulate oxidative metabolism in brain mitochondria." *Endocrinology* 149(6): 3167-3175.

Ishigami, I., Lewis-Ballester, A., Echelmeier, A., Brehm, G., Zatsepin, N. A., Grant, T. D., Coe, J. D., Lisova, S., Nelson, G., Zhang, S., Dobson, Z. F., Boutet, S., Sierra, R. G., Batyuk, A., Fromme, P., Fromme, R., Spence, J. C. H., Ros, A., Yeh, S. R. and Rousseau, D. L. (2019). "Snapshot of an oxygen intermediate in the catalytic reaction of cytochrome c oxidase." *Proc Natl Acad Sci U S A* 116(9): 3572-3577.

Itoh, K., Weis, S., Mehraein, P. and Muller-Hocker, J. (1996). "Cytochrome c oxidase defects of the human substantia nigra in normal aging." *Neurobiol Aging* 17(6): 843-848.

Iwata, S., Ostermeier, C., Ludwig, B. and Michel, H. (1995). "Structure at 2.8 A resolution of cytochrome c oxidase from Paracoccus denitrificans." *Nature* 376(6542): 660-669.

Jaksch, M., Hofmann, S., Kleinle, S., Liechti-Gallati, S., Pongratz, D. E., Muller-Hocker, J., Jedele, K. B., Meitinger, T. and Gerbitz, K. D. (1998). "A systematic mutation screen of 10 nuclear and 25 mitochondrial candidate genes in 21 patients with cytochrome c oxidase (COX) deficiency shows tRNA(Ser) (UCN) mutations in a subgroup with syndromal encephalopathy." *J Med Genet* 35(11): 895-900.

Jellinger, K. A. (2009). "Recent advances in our understanding of neurodegeneration." *J Neural Transm (Vienna)* 116(9): 1111-1162.

Jellinger, K. A. (2010). "Basic mechanisms of neurodegeneration: a critical update." *J Cell Mol Med* 14(3): 457-487.

Jimenez Del Rio, M. and Velez-Pardo, C. (2000). "17 beta-estradiol protects lymphocytes against dopamine and iron-induced apoptosis by a genomic-independent mechanism. Implication in Parkinson's disease." *Gen Pharmacol* 35(1): 1-9.

Johann, S., Dahm, M., Kipp, M., Beyer, C. and Arnold, S. (2010). "Oestrogen regulates mitochondrial respiratory chain enzyme transcription in the mouse spinal cord." *J Neuroendocrinol* 22(8): 926-935.

Jung, J. Y., Han, C. R., Jeong, Y. J., Kim, H. J., Lim, H. S., Lee, K. H., Park, H. O., Oh, W. M., Kim, S. H. and Kim, W. J. (2007). "Epigallocatechin gallate inhibits nitric oxide-induced apoptosis in rat PC12 cells." *Neurosci Lett* 411(3): 222-227.

Jurcau, A. (2021). "Insights into the Pathogenesis of Neurodegenerative Diseases: Focus on Mitochondrial Dysfunction and Oxidative Stress." *Int J Mol Sci* 22(21).

Kadenbach, B., Huttemann, M., Arnold, S., Lee, I. and Bender, E. (2000). "Mitochondrial energy metabolism is regulated via nuclear-coded subunits of cytochrome c oxidase." *Free Radic Biol Med* 29(3-4): 211-221.

Kagi, J. H. and Ulmer, D. D. (1968). "Hydrogen-deuterium exchange of cytochrome c. II. Effect of pH." *Biochemistry* 7(8): 2718-2723.

Kaliszewska, A., Allison, J., Martini, M. and Arias, N. (2021). "Improving Age-Related Cognitive Decline through Dietary Interventions Targeting Mitochondrial Dysfunction." *Int J Mol Sci* 22(7).

Kang, J., Lim, L. and Song, J. (2019). "ATP binds and inhibits the neurodegeneration-associated fibrillization of the FUS RRM domain." *Commun Biol* 2: 223.

Karu, T. (2000). Mechanisms of low-power laser light action on cellular level, *SPIE*.

Karu, T. I. (1986). "(Molecular mechanism of the therapeutic effect of low-intensity laser irradiation)." *Dokl Akad Nauk SSSR* 291(5): 1245-1249.

Karu, T. I. and Kolyakov, S. F. (2005). "Exact action spectra for cellular responses relevant to phototherapy." *Photomed Laser Surg* 23(4): 355-361.

Kasahara, A., Ishikawa, K., Yamaoka, M., Ito, M., Watanabe, N., Akimoto, M., Sato, A., Nakada, K., Endo, H., Suda, Y., Aizawa, S. and Hayashi, J. (2006). "Generation of trans-mitochondrial mice carrying homoplasmic mtDNAs with a missense mutation in a structural gene using ES cells." *Hum Mol Genet* 15(6): 871-881.

Kauppila, J. H. K., Baines, H. L., Bratic, A., Simard, M. L., Freyer, C., Mourier, A., Stamp, C., Filograna, R., Larsson, N. G., Greaves, L. C. and Stewart, J. B. (2016). "A Phenotype-Driven Approach to Generate Mouse Models with Pathogenic mtDNA Mutations Causing Mitochondrial Disease." *Cell Rep* 16(11): 2980-2990.

Keightley, J. A., Hoffbuhr, K. C., Burton, M. D., Salas, V. M., Johnston, W. S., Penn, A. M., Buist, N. R. and Kennaway, N. G. (1996). "A microdeletion in cytochrome c oxidase (COX) subunit III associated with COX deficiency and recurrent myoglobinuria." *Nat Genet* 12(4): 410-416.

Kemppainen, K. K., Rinne, J., Sriram, A., Lakanmaa, M., Zeb, A., Tuomela, T., Popplestone, A., Singh, S., Sanz, A., Rustin, P. and Jacobs, H. T. (2014). "Expression of alternative oxidase in Drosophila ameliorates diverse phenotypes due to cytochrome oxidase deficiency." *Hum Mol Genet* 23(8): 2078-2093.

Keshtkar, S., Kaviani, M., Jabbarpour, Z., Al-Abdullah, I. H., Aghdaei, M. H., Nikeghbalian, S., Shamsaeefar, A., Geramizadeh, B., Azarpira, N. and Ghahremani, M. H. (2020). "Significant reduction of apoptosis induced via hypoxia and oxidative stress in isolated human islet by resveratrol." *Nutr Metab Cardiovasc Dis* 30(7): 1216-1226.

Khalimonchuk, O., Bird, A. and Winge, D. R. (2007). "Evidence for a pro-oxidant intermediate in the assembly of cytochrome oxidase." *J Biol Chem* 282(24): 17442-17449.

Kim, J., Moody, J. P., Edgerly, C. K., Bordiuk, O. L., Cormier, K., Smith, K., Beal, M. F. and Ferrante, R. J. (2010). "Mitochondrial loss, dysfunction and altered dynamics in Huntington's disease." *Hum Mol Genet* 19(20): 3919-3935.

Kim, J. H., Choi, W., Lee, J. H., Jeon, S. J., Choi, Y. H., Kim, B. W., Chang, H. I. and Nam, S. W. (2009). "Astaxanthin inhibits H2O2-mediated apoptotic cell death in mouse neural progenitor cells via modulation of P38 and MEK signaling pathways." *J Microbiol Biotechnol* 19(11): 1355-1363.

Klichko, V., Sohal, B. H., Radyuk, S. N., Orr, W. C. and Sohal, R. S. (2014). "Decrease in cytochrome c oxidase reserve capacity diminishes robustness of Drosophila melanogaster and shortens lifespan." *Biochem J* 459(1): 127-135.

Kobayashi, Y., Momoi, M. Y., Tominaga, K., Momoi, T., Nihei, K., Yanagisawa, M., Kagawa, Y. and Ohta, S. (1990). "A point mutation in the mitochondrial tRNA(Leu)(UUR) gene in MELAS (mitochondrial myopathy, encephalopathy, lactic acidosis and stroke-like episodes)." *Biochem Biophys Res Commun* 173(3): 816-822.

Koh, S. H., Kim, S. H., Kwon, H., Park, Y., Kim, K. S., Song, C. W., Kim, J., Kim, M. H., Yu, H. J., Henkel, J. S. and Jung, H. K. (2003). "Epigallocatechin gallate protects nerve growth factor differentiated PC12 cells from oxidative-radical-stress-induced apoptosis through its effect on phosphoinositide 3-kinase/Akt and glycogen synthase kinase-3." *Brain Res Mol Brain Res* 118(1-2): 72-81.

Koh, S. H., Lee, S. M., Kim, H. Y., Lee, K. Y., Lee, Y. J., Kim, H. T., Kim, J., Kim, M. H., Hwang, M. S., Song, C., Yang, K. W., Lee, K. W., Kim, S. H. and Kim, O. H. (2006). "The effect of epigallocatechin gallate on suppressing disease progression of ALS model mice." *Neurosci Lett* 395(2): 103-107.

Korte, N., Nortley, R. and Attwell, D. (2020). "Cerebral blood flow decrease as an early pathological mechanism in Alzheimer's disease." *Acta Neuropathol* 140(6): 793-810.

Kraytsberg, Y., Kudryavtseva, E., McKee, A. C., Geula, C., Kowall, N. W. and Khrapko, K. (2006). "Mitochondrial DNA deletions are abundant and cause functional impairment in aged human substantia nigra neurons." *Nat Genet* 38(5): 518-520.

Kumar, A., Ahmad, I., Shukla, S., Singh, B. K., Patel, D. K., Pandey, H. P. and Singh, C. (2010). "Effect of zinc and paraquat co-exposure on neurodegeneration: Modulation of oxidative stress and expression of metallothioneins, toxicant responsive and transporter genes in rats." *Free Radic Res* 44(8): 950-965.

Kumar, A., Chaudhary, T. and Mishra, J. (2013). "Minocycline modulates neuroprotective effect of hesperidin against quinolinic acid induced Huntington's disease like symptoms in rats: behavioral, biochemical, cellular and histological evidences." *Eur J Pharmacol* 720(1-3): 16-28.

Kumar, P., Kale, R. K. and Baquer, N. Z. (2011). "Estradiol modulates membrane-linked ATPases, antioxidant enzymes, membrane fluidity, lipid peroxidation, and lipofuscin in aged rat liver." *J Aging Res* 2011: 580245.

Kumar, V., Singh, B. K., Chauhan, A. K., Singh, D., Patel, D. K. and Singh, C. (2016). "Minocycline Rescues from Zinc-Induced Nigrostriatal Dopaminergic Neurodegeneration: Biochemical and Molecular Interventions." *Mol Neurobiol* 53(5): 2761-2777.

Kushnareva, Y., Murphy, A. N. and Andreyev, A. (2002). "Complex I-mediated reactive oxygen species generation: modulation by cytochrome c and NAD(P)+ oxidation-reduction state." *Biochem J* 368(Pt 2): 545-553.

LaMarche, A. E., Abate, M. I., Chan, S. H. and Trumpower, B. L. (1992). "Isolation and characterization of COX12, the nuclear gene for a previously unrecognized subunit of Saccharomyces cerevisiae cytochrome c oxidase." *Journal of Biological Chemistry* 267(31): 22473-22480.

Lane, N. (2006). "Cell biology: power games." *Nature* 443(7114): 901-903.

Lee, I., Huttemann, M., Liu, J., Grossman, L. I. and Malek, M. H. (2012). "Deletion of heart-type cytochrome c oxidase subunit 7a1 impairs skeletal muscle angiogenesis and oxidative phosphorylation." *J Physiol* 590(20): 5231-5243.

Lee, J. K., Ding, Y., Conrad, A. L., Cattaneo, E., Epping, E., Mathews, K., Gonzalez-Alegre, P., Cahill, L., Magnotta, V., Schlaggar, B. L., Perlmutter, J. S., Kim, R. E., Dawson, J. D. and Nopoulos, P. (2017). "Sex-specific effects of the Huntington gene on normal neurodevelopment." *J Neurosci Res* 95(1-2): 398-408.

Lehmann, D., Schubert, K., Joshi, P. R., Hardy, S. A., Tuppen, H. A., Baty, K., Blakely, E. L., Bamberg, C., Zierz, S., Deschauer, M. and Taylor, R. W. (2015). "Pathogenic mitochondrial mt-tRNA(Ala) variants are uniquely associated with isolated myopathy." *Eur J Hum Genet* 23(12): 1735-1738.

Lemaire, C., Robineau, S. and Netter, P. (1998). "Molecular and biochemical analysis of Saccharomyces cerevisiae cox1 mutants." *Curr Genet* 34(2): 138-145.

Lepp, H. and Brzezinski, P. (2009). "Internal charge transfer in cytochrome c oxidase at a limited proton supply: Proton pumping ceases at high pH." *Biochimica et Biophysica Acta (BBA) - General Subjects* 1790(6): 552-557.

Leung, K. W., Yung, K. K., Mak, N. K., Chan, Y. S., Fan, T. P. and Wong, R. N. (2007). "Neuroprotective effects of ginsenoside-Rg1 in primary nigral neurons against rotenone toxicity." *Neuropharmacology* 52(3): 827-835.

Li, Y., Trush, M. A. and Yager, J. D. (1994). "DNA damage caused by reactive oxygen species originating from a copper-dependent oxidation of the 2-hydroxy catechol of estradiol." *Carcinogenesis* 15(7): 1421-1427.

Li, Y., Yang, C., Wang, S., Yang, D., Zhang, Y., Xu, L., Ma, L., Zheng, J., Petersen, R. B., Zheng, L., Chen, H. and Huang, K. (2020). "Copper and iron ions accelerate the prion-like propagation of alpha-synuclein: A vicious cycle in Parkinson's disease." *Int J Biol Macromol* 163: 562-573.

Lin, A. M., Fan, S. F., Yang, D. M., Hsu, L. L. and Yang, C. H. (2003). "Zinc-induced apoptosis in substantia nigra of rat brain: neuroprotection by vitamin D3." *Free Radic Biol Med* 34(11): 1416-1425.

Lin, H. F., Liao, K. F., Chang, C. M., Lin, C. L. and Lai, S. W. (2018). "Tamoxifen usage correlates with increased risk of Parkinson's disease in older women with breast cancer: a case-control study in Taiwan." *Eur J Clin Pharmacol* 74(1): 99-107.

Linder, D., Freund, R. and Kadenbach, B. (1995). "Species-specific expression of cytochrome c oxidase isozymes." *Comparative Biochemistry and Physiology Part B: Biochemistry and Molecular Biology* 112(3): 461-469.

Little, A. G., Kocha, K. M., Lougheed, S. C. and Moyes, C. D. (2010). "Evolution of the nuclear-encoded cytochrome oxidase subunits in vertebrates." *Physiol Genomics* 42(1): 76-84.

Little, A. G., Lau, G., Mathers, K. E., Leary, S. C. and Moyes, C. D. (2018). "Comparative biochemistry of cytochrome c oxidase in animals." *Comp Biochem Physiol B Biochem Mol Biol* 224: 170-184.

Little, A. G., Lougheed, S. C. and Moyes, C. D. (2010). "Evolutionary affinity of billfishes (Xiphiidae and Istiophoridae) and flatfishes (Plueronectiformes): Independent and trans-subordinal origins of endothermy in teleost fishes." *Mol Phylogenet Evol* 56(3): 897-904.

Liu, H., Zhang, J., Yang, Y., Zhang, L. and Zeng, X. (2012). "Decreased cerebral perfusion and oxidative stress result in acute and delayed cognitive impairment." *Curr Neurovasc Res* 9(3): 152-158.

Liu, J. F., Yan, X. D., Qi, L. S., Li, L., Hu, G. Y., Li, P. and Zhao, G. (2015). "Ginsenoside Rd attenuates Abeta25-35-induced oxidative stress and apoptosis in primary cultured hippocampal neurons." *Chem Biol Interact* 239: 12-18.

Liu, L. X., Chen, W. F., Xie, J. X. and Wong, M. S. (2008). "Neuroprotective effects of genistein on dopaminergic neurons in the mice model of Parkinson's disease." *Neurosci Res* 60(2): 156-161.

Liu, W., Gnanasambandam, R., Benjamin, J., Kaur, G., Getman, P. B., Siegel, A. J., Shortridge, R. D. and Singh, S. (2007). "Mutations in cytochrome c oxidase subunit

VIa cause neurodegeneration and motor dysfunction in Drosophila." *Genetics* 176(2): 937-946.

Lobo-Jarne, T. and Ugalde, C. (2018). "Respiratory chain supercomplexes: Structures, function and biogenesis." *Semin Cell Dev Biol* 76: 179-190.

Lu, Y., Wang, R., Dong, Y., Tucker, D., Zhao, N., Ahmed, M. E., Zhu, L., Liu, T. C., Cohen, R. M. and Zhang, Q. (2017). "Low-level laser therapy for beta amyloid toxicity in rat hippocampus." *Neurobiol Aging* 49: 165-182.

Lucioli, S., Hoffmeier, K., Carrozzo, R., Tessa, A., Ludwig, B. and Santorelli, F. M. (2006). "Introducing a novel human mtDNA mutation into the Paracoccus denitrificans COX I gene explains functional deficits in a patient." *Neurogenetics* 7(1): 51-57.

Macheda, T., Roberts, K., Lyons, D. N., Higgins, E., Ritter, K. J., Lin, A. L., Alilain, W. J. and Bachstetter, A. D. (2019). "Chronic Intermittent Hypoxia Induces Robust Astrogliosis in an Alzheimer's Disease-Relevant Mouse Model." *Neuroscience* 398: 55-63.

Mahul-Mellier, A. L., Burtscher, J., Maharjan, N., Weerens, L., Croisier, M., Kuttler, F., Leleu, M., Knott, G. W. and Lashuel, H. A. (2020). "The process of Lewy body formation, rather than simply alpha-synuclein fibrillization, is one of the major drivers of neurodegeneration." *Proc Natl Acad Sci U S A* 117(9): 4971-4982.

Malpartida, A. B., Williamson, M., Narendra, D. P., Wade-Martins, R. and Ryan, B. J. (2021). "Mitochondrial Dysfunction and Mitophagy in Parkinson's Disease: From Mechanism to Therapy." *Trends Biochem Sci* 46(4): 329-343.

Mancuso, M., Filosto, M., Mootha, V. K., Rocchi, A., Pistolesi, S., Murri, L., DiMauro, S. and Siciliano, G. (2004). "A novel mitochondrial tRNAPhe mutation causes MERRF syndrome." *Neurology* 62(11): 2119-2121.

Mandal, S., Guptan, P., Owusu-Ansah, E. and Banerjee, U. (2005). "Mitochondrial regulation of cell cycle progression during development as revealed by the tenured mutation in Drosophila." *Dev Cell* 9(6): 843-854.

Maniura-Weber, K., Taylor, R. W., Johnson, M. A., Chrzanowska-Lightowlers, Z., Morris, A. A., Charlton, C. P., Turnbull, D. M. and Bindoff, L. A. (2004). "A novel point mutation in the mitochondrial tRNA(Trp) gene produces a neurogastrointestinal syndrome." *Eur J Hum Genet* 12(6): 509-512.

Martini, M. and Arias, N. (2021). "Near-infrared light spectroscopy and stimulation in cognitive neuroscience: the need for an integrative view?" *J Integr Neurosci* 20(4): 1105-1109.

Massa, V., Fernandez-Vizarra, E., Alshahwan, S., Bakhsh, E., Goffrini, P., Ferrero, I., Mereghetti, P., D'Adamo, P., Gasparini, P. and Zeviani, M. (2008). "Severe infantile encephalomyopathy caused by a mutation in COX6B1, a nucleus-encoded subunit of cytochrome c oxidase." *Am J Hum Genet* 82(6): 1281-1289.

Mendez, N. V., Wharton, J. A., Leclerc, J. L., Blackburn, S. L., Douglas-Escobar, M. V., Weiss, M. D., Seubert, C. N. and Dore, S. (2013). "Clinical Implications of Bilirubin-Associated Neuroprotection and Neurotoxicity." *Int J Clin Anesthesiol* 1(2).

Menze, E. T., Esmat, A., Tadros, M. G., Khalifa, A. E. and Abdel-Naim, A. B. (2016). "Genistein improves sensorimotor gating: Mechanisms related to its neuroprotective effects on the striatum." *Neuropharmacology* 105: 35-46.

Millar, A. H., Eubel, H., Jansch, L., Kruft, V., Heazlewood, J. L. and Braun, H. P. (2004). "Mitochondrial cytochrome c oxidase and succinate dehydrogenase complexes contain plant specific subunits." *Plant Mol Biol* 56(1): 77-90.

Mkaouar-Rebai, E., Chamkha, I., Kammoun, T., Alila-Fersi, O., Aloulou, H., Hachicha, M. and Fakhfakh, F. (2013). "A novel MT-CO1 m.6498C>A variation associated with the m.7444G>A mutation in the mitochondrial COI/tRNA(Ser(UCN)) genes in a patient with hearing impairment, diabetes and congenital visual loss." *Biochem Biophys Res Commun* 430(2): 585-591.

Moraes, C. T., Ciacci, F., Bonilla, E., Ionasescu, V., Schon, E. A. and DiMauro, S. (1993). "A mitochondrial tRNA anticodon swap associated with a muscle disease." *Nat Genet* 4(3): 284-288.

Moraes, C. T., Ricci, E., Bonilla, E., DiMauro, S. and Schon, E. A. (1992). "The mitochondrial tRNA(Leu(UUR)) mutation in mitochondrial encephalomyopathy, lactic acidosis, and strokelike episodes (MELAS): genetic, biochemical, and morphological correlations in skeletal muscle." *Am J Hum Genet* 50(5): 934-949.

Morleo, M. and Franco, B. (1993). Microphthalmia with Linear Skin Defects Syndrome. GeneReviews((R)). M. P. Adam, H. H. Ardinger, R. A. Pagon et al. Seattle (WA).

Mosconi, L., Rahman, A., Diaz, I., Wu, X., Scheyer, O., Hristov, H. W., Vallabhajosula, S., Isaacson, R. S., de Leon, M. J. and Brinton, R. D. (2018). "Increased Alzheimer's risk during the menopause transition: A 3-year longitudinal brain imaging study." *PLoS One* 13(12): e0207885.

Muftuoglu, M., Elibol, B., Dalmizrak, O., Ercan, A., Kulaksiz, G., Ogus, H., Dalkara, T. and Ozer, N. (2004). "Mitochondrial complex I and IV activities in leukocytes from patients with parkin mutations." *Mov Disord* 19(5): 544-548.

Nelson, I., Hanna, M. G., Alsanjari, N., Scaravilli, F., Morgan-Hughes, J. A. and Harding, A. E. (1995). "A new mitochondrial DNA mutation associated with progressive dementia and chorea: a clinical, pathological, and molecular genetic study." *Ann Neurol* 37(3): 400-403.

Nilsen, J. (2008). "Estradiol and neurodegenerative oxidative stress." *Front Neuroendocrinol* 29(4): 463-475.

Nilsen, J., Irwin, R. W., Gallaher, T. K. and Brinton, R. D. (2007). "Estradiol in vivo regulation of brain mitochondrial proteome." *J Neurosci* 27(51): 14069-14077.

Noble, W., Garwood, C., Stephenson, J., Kinsey, A. M., Hanger, D. P. and Anderton, B. H. (2009). "Minocycline reduces the development of abnormal tau species in models of Alzheimer's disease." *FASEB J* 23(3): 739-750.

Northington, F. J., Ferriero, D. M., Graham, E. M., Traystman, R. J. and Martin, L. J. (2001). "Early Neurodegeneration after Hypoxia-Ischemia in Neonatal Rat Is Necrosis while Delayed Neuronal Death Is Apoptosis." *Neurobiol Dis* 8(2): 207-219.

Novack, G. V., Galeano, P., Castano, E. M. and Morelli, L. (2020). "Mitochondrial Supercomplexes: Physiological Organization and Dysregulation in Age-Related Neurodegenerative Disorders." *Front Endocrinol (Lausanne)* 11: 600.

Nunnari, J. and Suomalainen, A. (2012). "Mitochondria: in sickness and in health." *Cell* 148(6): 1145-1159.

Olsson, P.-E., Kling, P. and Hogstrand, C. (1998). Mechanisms of heavy metal accumulation and toxicity in fish. Metal Metabolism in Aquatic Environments. W. J. Langston and M. J. Bebianno. Boston, MA, *Springer US*: 321-350.

Pacelli, C., Latorre, D., Cocco, T., Capuano, F., Kukat, C., Seibel, P. and Villani, G. (2011). "Tight control of mitochondrial membrane potential by cytochrome c oxidase." *Mitochondrion* 11(2): 334-341.

Pahnke, J., Frohlich, C., Krohn, M., Schumacher, T. and Paarmann, K. (2013). "Impaired mitochondrial energy production and ABC transporter function-A crucial interconnection in dementing proteopathies of the brain." *Mech Ageing Dev* 134(10): 506-515.

Pan, H., Chen, J., Shen, K., Wang, X., Wang, P., Fu, G., Meng, H., Wang, Y. and Jin, B. (2015). "Mitochondrial modulation by Epigallocatechin 3-Gallate ameliorates cisplatin induced renal injury through decreasing oxidative/nitrative stress, inflammation and NF-kB in mice." *PLoS One* 10(4): e0124775.

Pandit, S. K., Westendorp, B. and de Bruin, A. (2013). "Physiological significance of polyploidization in mammalian cells." *Trends Cell Biol* 23(11): 556-566.

Pandya, A., Xia, X. J., Erdenetungalag, R., Amendola, M., Landa, B., Radnaabazar, J., Dangaasuren, B., Van Tuyle, G. and Nance, W. E. (1999). "Heterogenous point mutations in the mitochondrial tRNA Ser(UCN) precursor coexisting with the A1555G mutation in deaf students from Mongolia." *Am J Hum Genet* 65(6): 1803-1806.

Pandya, V. A. and Patani, R. (2021). "Region-specific vulnerability in neurodegeneration: lessons from normal ageing." *Ageing Res Rev* 67: 101311.

Pastore, D., Greco, M. and Passarella, S. (2000). "Specific helium-neon laser sensitivity of the purified cytochrome c oxidase." *Int J Radiat Biol* 76(6): 863-870.

Patel, M. R., Miriyala, G. K., Littleton, A. J., Yang, H., Trinh, K., Young, J. M., Kennedy, S. R., Yamashita, Y. M., Pallanck, L. J. and Malik, H. S. (2016). "A mitochondrial DNA hypomorph of cytochrome oxidase specifically impairs male fertility in Drosophila melanogaster." *Elife* 5.

Petrosillo, G., Portincasa, P., Grattagliano, I., Casanova, G., Matera, M., Ruggiero, F. M., Ferri, D. and Paradies, G. (2007). "Mitochondrial dysfunction in rat with nonalcoholic fatty liver Involvement of complex I, reactive oxygen species and cardiolipin." *Biochim Biophys Acta* 1767(10): 1260-1267.

Pierron, D., Wildman, D. E., Huttemann, M., Markondapatnaikuni, G. C., Aras, S. and Grossman, L. I. (2012). "Cytochrome c oxidase: evolution of control via nuclear subunit addition." *Biochim Biophys Acta* 1817(4): 590-597.

Pierron, F., Bourret, V., St-Cyr, J., Campbell, P. G., Bernatchez, L. and Couture, P. (2009). "Transcriptional responses to environmental metal exposure in wild yellow perch (Perca flavescens) collected in lakes with differing environmental metal concentrations (Cd, Cu, Ni)." *Ecotoxicology* 18(5): 620-631.

Pietrangelo, A. (2017). "Ferroportin disease: pathogenesis, diagnosis and treatment." *Haematologica* 102(12): 1972-1984.

Pillai, N. R., AlDhaheri, N. S., Ghosh, R., Lim, J., Streff, H., Nayak, A., Graham, B. H., Hanchard, N. A., Elsea, S. H. and Scaglia, F. (2019). "Biallelic variants in COX4I1 associated with a novel phenotype resembling Leigh syndrome with developmental regression, intellectual disability, and seizures." *Am J Med Genet A* 179(10): 2138-2143.

Pitceathly, R. D., Rahman, S., Wedatilake, Y., Polke, J. M., Cirak, S., Foley, A. R., Sailer, A., Hurles, M. E., Stalker, J., Hargreaves, I., Woodward, C. E., Sweeney, M. G., Muntoni, F., Houlden, H., Taanman, J. W., Hanna, M. G. and Consortium, U. K. (2013). "NDUFA4 mutations underlie dysfunction of a cytochrome c oxidase subunit linked to human neurological disease." *Cell Rep* 3(6): 1795-1805.

Pitceathly, R. D. S. and Taanman, J. W. (2018). "NDUFA4 (Renamed COXFA4) Is a Cytochrome-c Oxidase Subunit." *Trends Endocrinol Metab* 29(7): 452-454.

Prabu, S. K., Anandatheerthavarada, H. K., Raza, H., Srinivasan, S., Spear, J. F. and Avadhani, N. G. (2006). "Protein kinase A-mediated phosphorylation modulates cytochrome c oxidase function and augments hypoxia and myocardial ischemia-related injury." *J Biol Chem* 281(4): 2061-2070.

Puchkov, M. N., Vassarais, R. A., Korepanova, E. A. and Osipov, A. N. (2013). "Cytochrome c produces pores in cardiolipin-containing planar bilayer lipid membranes in the presence of hydrogen peroxide." *Biochim Biophys Acta* 1828(2): 208-212.

Purushothuman, S., Johnstone, D. M., Nandasena, C., Mitrofanis, J. and Stone, J. (2014). "Photobiomodulation with near infrared light mitigates Alzheimer's disease-related pathology in cerebral cortex - evidence from two transgenic mouse models." *Alzheimers Res Ther* 6(1): 2.

Quinn, N. P. and Marsden, C. D. (1986). "Menstrual-related fluctuations in Parkinson's disease." *Mov Disord* 1(1): 85-87.

Radford, N. B., Wan, B., Richman, A., Szczepaniak, L. S., Li, J. L., Li, K., Pfeiffer, K., Schagger, H., Garry, D. J. and Moreadith, R. W. (2002). "Cardiac dysfunction in mice lacking cytochrome-c oxidase subunit VIaH." *Am J Physiol Heart Circ Physiol* 282(2): H726-733.

Rahman, S., Taanman, J. W., Cooper, J. M., Nelson, I., Hargreaves, I., Meunier, B., Hanna, M. G., Garcia, J. J., Capaldi, R. A., Lake, B. D., Leonard, J. V. and Schapira, A. H. (1999). "A missense mutation of cytochrome oxidase subunit II causes defective assembly and myopathy." *Am J Hum Genet* 65(4): 1030-1039.

Raitio, M., Jalli, T. and Saraste, M. (1987). "Isolation and analysis of the genes for cytochrome c oxidase in Paracoccus denitrificans." *EMBO J* 6(9): 2825-2833.

Ramalingam, M. and Kim, S. J. (2015). "Insulin exerts neuroprotective effects via Akt/Bcl-2 signaling pathways in differentiated SH-SY5Y cells." *J Recept Signal Transduct Res* 35(1): 1-7.

Ramalingam, M. and Kim, S. J. (2016). "The Neuroprotective Role of Insulin Against MPP(+) -Induced Parkinson's Disease in Differentiated SH-SY5Y Cells." *J Cell Biochem* 117(4): 917-926.

Reddy, P. H. (2008). "Mitochondrial medicine for aging and neurodegenerative diseases." *Neuromolecular Med* 10(4): 291-315.

Reddy, P. H., Tripathi, R., Troung, Q., Tirumala, K., Reddy, T. P., Anekonda, V., Shirendeb, U. P., Calkins, M. J., Reddy, A. P., Mao, P. and Manczak, M. (2012). "Abnormal mitochondrial dynamics and synaptic degeneration as early events in

Alzheimer's disease: implications to mitochondria-targeted antioxidant therapeutics." *Biochim Biophys Acta* 1822(5): 639-649.
Regenboog, M., van Kuilenburg, A. B., Verheij, J., Swinkels, D. W. and Hollak, C. E. (2016). "Hyperferritinemia and iron metabolism in Gaucher disease: Potential pathophysiological implications." *Blood Rev* 30(6): 431-437.
Richards, T. A. and Archibald, J. M. (2011). "Cell evolution: gene transfer agents and the origin of mitochondria." *Curr Biol* 21(3): R112-114.
Rikani, A. A., Choudhry, Z., Choudhry, A. M., Rizvi, N., Ikram, H., Mobassarah, N. J. and Tulli, S. (2014). "The mechanism of degeneration of striatal neuronal subtypes in Huntington disease." *Ann Neurosci* 21(3): 112-114.
Rivera-Mancia, S., Perez-Neri, I., Rios, C., Tristan-Lopez, L., Rivera-Espinosa, L. and Montes, S. (2010). "The transition metals copper and iron in neurodegenerative diseases." *Chem Biol Interact* 186(2): 184-199.
Rohou, H., Francisci, S., Rinaldi, T., Frontali, L. and Bolotin-Fukuhara, M. (2001). "Reintroduction of a characterized Mit tRNA glycine mutation into yeast mitochondria provides a new tool for the study of human neurodegenerative diseases." *Yeast* 18(3): 219-227.
Rojas, J. C., Bruchey, A. K. and Gonzalez-Lima, F. (2012). "Low-level light therapy improves cortical metabolic capacity and memory retention." *J Alzheimers Dis* 32(3): 741-752.
Rojas, J. C., Lee, J., John, J. M. and Gonzalez-Lima, F. (2008). "Neuroprotective effects of near-infrared light in an in vivo model of mitochondrial optic neuropathy." *J Neurosci* 28(50): 13511-13521.
Roodhooft, A. M., Van Acker, K. J., Martin, J. J., Ceuterick, C., Scholte, H. R. and Luyt-Houwen, I. E. (1986). "Benign mitochondrial myopathy with deficiency of NADH-CoQ reductase and cytochrome c oxidase." *Neuropediatrics* 17(4): 221-226.
Rossmanith, W., Raffelsberger, T., Roka, J., Kornek, B., Feucht, M. and Bittner, R. E. (2003). "The expanding mutational spectrum of MERRF substitution G8361A in the mitochondrial tRNALys gene." *Ann Neurol* 54(6): 820-823.
Russo, J., Hasan Lareef, M., Balogh, G., Guo, S. and Russo, I. H. (2003). "Estrogen and its metabolites are carcinogenic agents in human breast epithelial cells." *J Steroid Biochem Mol Biol* 87(1): 1-25.
Sabolic, I. (2006). "Common mechanisms in nephropathy induced by toxic metals." *Nephron Physiol* 104(3): p107-114.
Sacconi, S., Salviati, L., Nishigaki, Y., Walker, W. F., Hernandez-Rosa, E., Trevisson, E., Delplace, S., Desnuelle, C., Shanske, S., Hirano, M., Schon, E. A., Bonilla, E., De Vivo, D. C., DiMauro, S. and Davidson, M. M. (2008). "A functionally dominant mitochondrial DNA mutation." *Hum Mol Genet* 17(12): 1814-1820.
Sagan, L. (1967). "On the origin of mitosing cells." *J Theor Biol* 14(3): 255-274.
Salehpour, F., Mahmoudi, J., Kamari, F., Sadigh-Eteghad, S., Rasta, S. H. and Hamblin, M. R. (2018). "Brain Photobiomodulation Therapy: a Narrative Review." *Mol Neurobiol* 55(8): 6601-6636.
Salkovic-Petrisic, M., Osmanovic-Barilar, J., Bruckner, M. K., Hoyer, S., Arendt, T. and Riederer, P. (2011). "Cerebral amyloid angiopathy in streptozotocin rat model of

sporadic Alzheimer's disease: a long-term follow up study." *J Neural Transm (Vienna)* 118(5): 765-772.

Salo, M. K., Rapola, J., Somer, H., Pihko, H., Koivikko, M., Tritschler, H. J. and DiMauro, S. (1992). "Reversible mitochondrial myopathy with cytochrome c oxidase deficiency." *Arch Dis Child* 67(8): 1033-1035.

Sanderson, T. H., Mahapatra, G., Pecina, P., Ji, Q., Yu, K., Sinkler, C., Varughese, A., Kumar, R., Bukowski, M. J., Tousignant, R. N., Salomon, A. R., Lee, I. and Huttemann, M. (2013). "Cytochrome C is tyrosine 97 phosphorylated by neuroprotective insulin treatment." *PLoS One* 8(11): e78627.

Santana-Blank, L., Rodriguez-Santana, E. and Santana-Rodriguez, K. (2010). "Theoretic, experimental, clinical bases of the water oscillator hypothesis in near-infrared photobiomodulation." *Photomed Laser Surg* 28 Suppl 1: S41-52.

Santillo, A. F., Skoglund, L., Lindau, M., Eeg-Olofsson, K. E., Tovi, M., Engler, H., Brundin, R. M., Ingvast, S., Lannfelt, L., Glaser, A. and Kilander, L. (2009). "Frontotemporal dementia-amyotrophic lateral sclerosis complex is simulated by neurodegeneration with brain iron accumulation." *Alzheimer Dis Assoc Disord* 23(3): 298-300.

Santorelli, F. M., Mak, S. C., El-Schahawi, M., Casali, C., Shanske, S., Baram, T. Z., Madrid, R. E. and DiMauro, S. (1996). "Maternally inherited cardiomyopathy and hearing loss associated with a novel mutation in the mitochondrial tRNA(Lys) gene (G8363A)." *Am J Hum Genet* 58(5): 933-939.

Santorelli, F. M., Tanji, K., Sano, M., Shanske, S., El-Shahawi, M., Kranz-Eble, P., DiMauro, S. and De Vivo, D. C. (1997). "Maternally inherited encephalopathy associated with a single-base insertion in the mitochondrial tRNATrp gene." *Ann Neurol* 42(2): 256-260.

Sapin, E., Peyron, C., Roche, F., Gay, N., Carcenac, C., Savasta, M., Levy, P. and Dematteis, M. (2015). "Chronic Intermittent Hypoxia Induces Chronic Low-Grade Neuroinflammation in the Dorsal Hippocampus of Mice." *Sleep* 38(10): 1537-1546.

Sarti, P., Forte, E., Mastronicola, D., Giuffre, A. and Arese, M. (2012). "Cytochrome c oxidase and nitric oxide in action: molecular mechanisms and pathophysiological implications." *Biochim Biophys Acta* 1817(4): 610-619.

Selkoe, D. J. and Hardy, J. (2016). "The amyloid hypothesis of Alzheimer's disease at 25 years." *EMBO Mol Med* 8(6): 595-608.

Sen, T., Sen, N., Tripathi, G., Chatterjee, U. and Chakrabarti, S. (2006). "Lipid peroxidation associated cardiolipin loss and membrane depolarization in rat brain mitochondria." *Neurochem Int* 49(1): 20-27.

Seneca, S., Goemans, N., Van Coster, R., Givron, P., Reybrouck, T., Sciot, R., Meulemans, A., Smet, J. and Van Hove, J. L. (2005). "A mitochondrial tRNA aspartate mutation causing isolated mitochondrial myopathy." *Am J Med Genet A* 137(2): 170-175.

Sharma, D., Maurya, A. K. and Singh, R. (1993). "Age-related decline in multiple unit action potentials of CA3 region of rat hippocampus: correlation with lipid peroxidation and lipofuscin concentration and the effect of centrophenoxine." *Neurobiol Aging* 14(4): 319-330.

Sharma, S. K., Kharkwal, G. B., Sajo, M., Huang, Y. Y., De Taboada, L., McCarthy, T. and Hamblin, M. R. (2011). "Dose response effects of 810 nm laser light on mouse primary cortical neurons." *Lasers Surg Med* 43(8): 851-859.

Shimada, A., Hara, F., Shinzawa-Itoh, K., Kanehisa, N., Yamashita, E., Muramoto, K., Tsukihara, T. and Yoshikawa, S. (2021). "Critical roles of the CuB site in efficient proton pumping as revealed by crystal structures of mammalian cytochrome c oxidase catalytic intermediates." *J Biol Chem* 297(3): 100967.

Shimizu, A., Mito, T., Hayashi, C., Ogasawara, E., Koba, R., Negishi, I., Takenaga, K., Nakada, K. and Hayashi, J. (2014). "Transmitochondrial mice as models for primary prevention of diseases caused by mutation in the tRNA(Lys) gene." *Proc Natl Acad Sci U S A* 111(8): 3104-3109.

Shoffner, J. M. (1997). "Oxidative phosphorylation defects and Alzheimer's disease." *Neurogenetics* 1(1): 13-19.

Shoffner, J. M., Lott, M. T., Lezza, A. M. S., Seibel, P., Ballinger, S. W. and Wallace, D. C. (1990). "Myoclonic epilepsy and ragged-red fiber disease (MERRF) is associated with a mitochondrial DNA tRNALys mutation." *Cell* 61(6): 931-937.

Shteyer, E., Saada, A., Shaag, A., Al-Hijawi, F. A., Kidess, R., Revel-Vilk, S. and Elpeleg, O. (2009). "Exocrine pancreatic insufficiency, dyserythropoeitic anemia, and calvarial hyperostosis are caused by a mutation in the COX4I2 gene." *Am J Hum Genet* 84(3): 412-417.

Silvestri, G., Moraes, C. T., Shanske, S., Oh, S. J. and DiMauro, S. (1992). "A new mtDNA mutation in the tRNA(Lys) gene associated with myoclonic epilepsy and ragged-red fibers (MERRF)." *Am J Hum Genet* 51(6): 1213-1217.

Silvestri, G., Rana, M., DiMuzio, A., Uncini, A., Tonali, P. and Servidei, S. (1998). "A late-onset mitochondrial myopathy is associated with a novel mitochondrial DNA (mtDNA) point mutation in the tRNA(Trp) gene." *Neuromuscul Disord* 8(5): 291-295.

Silvestri, G., Santorelli, F. M., Shanske, S., Whitley, C. B., Schimmenti, L. A., Smith, S. A. and DiMauro, S. (1994). "A new mtDNA mutation in the tRNA(Leu(UUR)) gene associated with maternally inherited cardiomyopathy." *Hum Mutat* 3(1): 37-43.

Singh, S., Misiak, M., Beyer, C. and Arnold, S. (2009). "Cytochrome c oxidase isoform IV-2 is involved in 3-nitropropionic acid-induced toxicity in striatal astrocytes." *Glia* 57(14): 1480-1491.

Sinkler, C. A., Kalpage, H., Shay, J., Lee, I., Malek, M. H., Grossman, L. I. and Huttemann, M. (2017). "Tissue- and Condition-Specific Isoforms of Mammalian Cytochrome c Oxidase Subunits: From Function to Human Disease." *Oxid Med Cell Longev* 2017: 1534056.

Smits, P., Mattijssen, S., Morava, E., van den Brand, M., van den Brandt, F., Wijburg, F., Pruijn, G., Smeitink, J., Nijtmans, L., Rodenburg, R. and van den Heuvel, L. (2010). "Functional consequences of mitochondrial tRNA Trp and tRNA Arg mutations causing combined OXPHOS defects." *Eur J Hum Genet* 18(3): 324-329.

Sommer, A. P. (2019). "Mitochondrial cytochrome c oxidase is not the primary acceptor for near infrared light-it is mitochondrial bound water: the principles of low-level light therapy." *Ann Transl Med* 7(Suppl 1): S13.

Sommer, A. P., Bieschke, J., Friedrich, R. P., Zhu, D., Wanker, E. E., Fecht, H. J., Mereles, D. and Hunstein, W. (2012). "670 nm laser light and EGCG complementarily reduce amyloid-beta aggregates in human neuroblastoma cells: basis for treatment of Alzheimer's disease?" *Photomed Laser Surg* 30(1): 54-60.

Spagnolo, M., Tomelleri, G., Vattemi, G., Filosto, M., Rizzuto, N. and Tonin, P. (2001). "A new mutation in the mitochondrial tRNA(Ala) gene in a patient with ophthalmoplegia and dysphagia." *Neuromuscul Disord* 11(5): 481-484.

Sparks, D. L., Friedland, R., Petanceska, S., Schreurs, B. G., Shi, J., Perry, G., Smith, M. A., Sharma, A., Derosa, S., Ziolkowski, C. and Stankovic, G. (2006). "Trace copper levels in the drinking water, but not zinc or aluminum influence CNS Alzheimer-like pathology." *J Nutr Health Aging* 10(4): 247-254.

Squitti, R., Lupoi, D., Pasqualetti, P., Dal Forno, G., Vernieri, F., Chiovenda, P., Rossi, L., Cortesi, M., Cassetta, E. and Rossini, P. M. (2002). "Elevation of serum copper levels in Alzheimer's disease." *Neurology* 59(8): 1153-1161.

Srinivasan, S. and Avadhani, N. G. (2012). "Cytochrome c oxidase dysfunction in oxidative stress." *Free Radic Biol Med* 53(6): 1252-1263.

Steinrucke, P., Steffens, G. C., Panskus, G., Buse, G. and Ludwig, B. (1987). "Subunit II of cytochrome c oxidase from Paracoccus denitrificans. DNA sequence, gene expression and the protein." *Eur J Biochem* 167(3): 431-439.

Sterky, F. H., Hoffman, A. F., Milenkovic, D., Bao, B., Paganelli, A., Edgar, D., Wibom, R., Lupica, C. R., Olson, L. and Larsson, N. G. (2012). "Altered dopamine metabolism and increased vulnerability to MPTP in mice with partial deficiency of mitochondrial complex I in dopamine neurons." *Hum Mol Genet* 21(5): 1078-1089.

Sundar Boyalla, S., Barbara Victor, M., Roemgens, A., Beyer, C. and Arnold, S. (2011). "Sex- and brain region-specific role of cytochrome c oxidase in 1-methyl-4-phenylpyridinium-mediated astrocyte vulnerability." *J Neurosci Res* 89(12): 2068-2082.

Swalwell, H., Deschauer, M., Hartl, H., Strauss, M., Turnbull, D. M., Zierz, S. and Taylor, R. W. (2006). "Pure myopathy associated with a novel mitochondrial tRNA gene mutation." *Neurology* 66(3): 447-449.

Swerdlow, R. H. (2018). "Mitochondria and Mitochondrial Cascades in Alzheimer's Disease." *J Alzheimers Dis* 62(3): 1403-1416.

Szklarczyk, R. and Huynen, M. A. (2010). "Mosaic origin of the mitochondrial proteome." *Proteomics* 10(22): 4012-4024.

Taanman, J. W. and Capaldi, R. A. (1993). "Subunit VIa of yeast cytochrome c oxidase is not necessary for assembly of the enzyme complex but modulates the enzyme activity. Isolation and characterization of the nuclear-coded gene." *Journal of Biological Chemistry* 268(25): 18754-18761.

Tamilselvam, K., Braidy, N., Manivasagam, T., Essa, M. M., Prasad, N. R., Karthikeyan, S., Thenmozhi, A. J., Selvaraju, S. and Guillemin, G. J. (2013). "Neuroprotective effects of hesperidin, a plant flavanone, on rotenone-induced oxidative stress and apoptosis in a cellular model for Parkinson's disease." *Oxid Med Cell Longev* 2013: 102741.

Tamiya, G., Makino, S., Hayashi, M., Abe, A., Numakura, C., Ueki, M., Tanaka, A., Ito, C., Toshimori, K., Ogawa, N., Terashima, T., Maegawa, H., Yanagisawa, D., Tooyama, I., Tada, M., Onodera, O. and Hayasaka, K. (2014). "A mutation of

COX6A1 causes a recessive axonal or mixed form of Charcot-Marie-Tooth disease." *Am J Hum Genet* 95(3): 294-300.

Tian, J., Zhang, S., Li, G., Liu, Z. and Xu, B. (2009). "20(S)-ginsenoside Rg3, a neuroprotective agent, inhibits mitochondrial permeability transition pores in rat brain." *Phytother Res* 23(4): 486-491.

Tiranti, V., Carrara, F., Confalonieri, P., Mora, M., Maffei, R. M., Lamantea, E. and Zeviani, M. (1999). "A novel mutation (8342G-->A) in the mitochondrial tRNA(Lys) gene associated with progressive external ophthalmoplegia and myoclonus." *Neuromuscul Disord* 9(2): 66-71.

Tiranti, V., Corona, P., Greco, M., Taanman, J. W., Carrara, F., Lamantea, E., Nijtmans, L., Uziel, G. and Zeviani, M. (2000). "A novel frameshift mutation of the mtDNA COIII gene leads to impaired assembly of cytochrome c oxidase in a patient affected by Leigh-like syndrome." *Hum Mol Genet* 9(18): 2733-2742.

Toompuu, M., Yasukawa, T., Suzuki, T., Hakkinen, T., Spelbrink, J. N., Watanabe, K. and Jacobs, H. T. (2002). "The 7472insC mitochondrial DNA mutation impairs the synthesis and extent of aminoacylation of tRNASer(UCN) but not its structure or rate of turnover." *J Biol Chem* 277(25): 22240-22250.

Trigo, D., Avelar, C., Fernandes, M., Sa, J. and da Cruz, E. S. O. (2022). "Mitochondria, energy, and metabolism in neuronal health and disease." *FEBS Lett*.

Trimmer, P. A., Schwartz, K. M., Borland, M. K., De Taboada, L., Streeter, J. and Oron, U. (2009). "Reduced axonal transport in Parkinson's disease cybrid neurites is restored by light therapy." *Mol Neurodegener* 4: 26.

Trueblood, C. E. and Poyton, R. O. (1987). "Differential effectiveness of yeast cytochrome c oxidase subunit genes results from differences in expression not function." *Mol Cell Biol* 7(10): 3520-3526.

Tsukihara, T., Aoyama, H., Yamashita, E., Tomizaki, T., Yamaguchi, H., Shinzawa-Itoh, K., Nakashima, R., Yaono, R. and Yoshikawa, S. (1996). "The whole structure of the 13-subunit oxidized cytochrome c oxidase at 2.8 A." *Science* 272(5265): 1136-1144.

Tulinius, M., Moslemi, A. R., Darin, N., Westerberg, B., Wiklund, L. M., Holme, E. and Oldfors, A. (2003). "Leigh syndrome with cytochrome-c oxidase deficiency and a single T insertion nt 5537 in the mitochondrial tRNATrp gene." *Neuropediatrics* 34(2): 87-91.

Van Laar, V. S. and Berman, S. B. (2013). "The interplay of neuronal mitochondrial dynamics and bioenergetics: implications for Parkinson's disease." *Neurobiol Dis* 51: 43-55.

Varlamov, D. A., Kudin, A. P., Vielhaber, S., Schroder, R., Sassen, R., Becker, A., Kunz, D., Haug, K., Rebstock, J., Heils, A., Elger, C. E. and Kunz, W. S. (2002). "Metabolic consequences of a novel missense mutation of the mtDNA CO I gene." *Hum Mol Genet* 11(16): 1797-1805.

Vaz, A. R., Delgado-Esteban, M., Brito, M. A., Bolanos, J. P., Brites, D. and Almeida, A. (2010). "Bilirubin selectively inhibits cytochrome c oxidase activity and induces apoptosis in immature cortical neurons: assessment of the protective effects of glycoursodeoxycholic acid." *J Neurochem* 112(1): 56-65.

Vielhaber, S., Kunz, D., Winkler, K., Wiedemann, F. R., Kirches, E., Feistner, H., Heinze, H. J., Elger, C. E., Schubert, W. and Kunz, W. S. (2000). "Mitochondrial DNA abnormalities in skeletal muscle of patients with sporadic amyotrophic lateral sclerosis." *Brain* 123 (Pt 7): 1339-1348.

Villani, G. and Attardi, G. (1997). "In vivo control of respiration by cytochrome c oxidase in wild-type and mitochondrial DNA mutation-carrying human cells." *Proc Natl Acad Sci U S A* 94(4): 1166-1171.

Violi, F., Solovyev, N., Vinceti, M., Mandrioli, J., Lucio, M. and Michalke, B. (2020). "The study of levels from redox-active elements in cerebrospinal fluid of amyotrophic lateral sclerosis patients carrying disease-related gene mutations shows potential copper dyshomeostasis." *Metallomics* 12(5): 668-681.

Vondrackova, A., Vesela, K., Hansikova, H., Docekalova, D. Z., Rozsypalova, E., Zeman, J. and Tesarova, M. (2012). "High-resolution melting analysis of 15 genes in 60 patients with cytochrome-c oxidase deficiency." *J Hum Genet* 57(7): 442-448.

Wang, D., Liu, L., Li, S. and Wang, C. (2018). "Effects of paeoniflorin on neurobehavior, oxidative stress, brain insulin signaling, and synaptic alterations in intracerebroventricular streptozotocin-induced cognitive impairment in mice." *Physiol Behav* 191: 12-20.

Wang, X., Zhu, S., Drozda, M., Zhang, W., Stavrovskaya, I. G., Cattaneo, E., Ferrante, R. J., Kristal, B. S. and Friedlander, R. M. (2003). "Minocycline inhibits caspase-independent and -dependent mitochondrial cell death pathways in models of Huntington's disease." *Proc Natl Acad Sci U S A* 100(18): 10483-10487.

Wang, X., Zhu, S., Pei, Z., Drozda, M., Stavrovskaya, I. G., Del Signore, S. J., Cormier, K., Shimony, E. M., Wang, H., Ferrante, R. J., Kristal, B. S. and Friedlander, R. M. (2008). "Inhibitors of cytochrome c release with therapeutic potential for Huntington's disease." *J Neurosci* 28(38): 9473-9485.

Wang, Y., Reis, C., Applegate, R., 2nd, Stier, G., Martin, R. and Zhang, J. H. (2015). "Ischemic conditioning-induced endogenous brain protection: Applications pre-, per- or post-stroke." *Exp Neurol* 272: 26-40.

Wassertheil-Smoller, S., Hendrix, S. L., Limacher, M., Heiss, G., Kooperberg, C., Baird, A., Kotchen, T., Curb, J. D., Black, H., Rossouw, J. E., Aragaki, A., Safford, M., Stein, E., Laowattana, S., Mysiw, W. J. and Investigators, W. H. I. (2003). "Effect of estrogen plus progestin on stroke in postmenopausal women: the Women's Health Initiative: a randomized trial." *JAMA* 289(20): 2673-2684.

Watanabe, T., Inoue, S., Hiroi, H., Orimo, A., Kawashima, H. and Muramatsu, M. (1998). "Isolation of estrogen-responsive genes with a CpG island library." *Mol Cell Biol* 18(1): 442-449.

Weinreb, O., Amit, T., Mandel, S. and Youdim, M. B. (2009). "Neuroprotective molecular mechanisms of (-)-epigallocatechin-3-gallate: a reflective outcome of its antioxidant, iron chelating and neuritogenic properties." *Genes Nutr* 4(4): 283-296.

Weiss-Brummer, B., Guba, R., Haid, A. and Schweyen, R. J. (1979). "Fine structure of OXI1, the mitochondrial gene coding for subunit II of yeast cytochrome c oxidase." *Curr Genet* 1(1): 75-83.

Wikström, M. and Sharma, V. (2018). "Proton pumping by cytochrome c oxidase – A 40 year anniversary." *Biochimica et Biophysica Acta (BBA) - Bioenergetics* 1859(9): 692-698.

Wilhelm, J. and Herget, J. (1999). "Hypoxia induces free radical damage to rat erythrocytes and spleen: analysis of the fluorescent end-products of lipid peroxidation." *Int J Biochem Cell Biol* 31(6): 671-681.

Wong-Riley, M. T., Bai, X., Buchmann, E. and Whelan, H. T. (2001). "Light-emitting diode treatment reverses the effect of TTX on cytochrome oxidase in neurons." *Neuroreport* 12(14): 3033-3037.

Wong-Riley, M. T., Liang, H. L., Eells, J. T., Chance, B., Henry, M. M., Buchmann, E., Kane, M. and Whelan, H. T. (2005). "Photobiomodulation directly benefits primary neurons functionally inactivated by toxins: role of cytochrome c oxidase." *J Biol Chem* 280(6): 4761-4771.

Wong, L. J., Dai, P., Tan, D., Lipson, M., Grix, A., Sifry-Platt, M., Gropman, A. and Chen, T. J. (2001). "Severe lactic acidosis caused by a novel frame-shift mutation in mitochondrial-encoded cytochrome c oxidase subunit II." *Am J Med Genet* 102(1): 95-99.

Wood-Kaczmar, A., Gandhi, S., Yao, Z., Abramov, A. Y., Miljan, E. A., Keen, G., Stanyer, L., Hargreaves, I., Klupsch, K., Deas, E., Downward, J., Mansfield, L., Jat, P., Taylor, J., Heales, S., Duchen, M. R., Latchman, D., Tabrizi, S. J. and Wood, N. W. (2008). "PINK1 is necessary for long term survival and mitochondrial function in human dopaminergic neurons." *PLoS One* 3(6): e2455.

Wooten, G. F., Currie, L. J., Bovbjerg, V. E., Lee, J. K. and Patrie, J. (2004). "Are men at greater risk for Parkinson's disease than women?" *J Neurol Neurosurg Psychiatry* 75(4): 637-639.

Wu, J. J., Quijano, C., Chen, E., Liu, H., Cao, L., Fergusson, M. M., Rovira, II, Gutkind, S., Daniels, M. P., Komatsu, M. and Finkel, T. (2009). "Mitochondrial dysfunction and oxidative stress mediate the physiological impairment induced by the disruption of autophagy." *Aging (Albany NY)* 1(4): 425-437.

Xiao, G., Fan, Q., Wang, X. and Zhou, B. (2013). "Huntington disease arises from a combinatory toxicity of polyglutamine and copper binding." *Proc Natl Acad Sci U S A* 110(37): 14995-15000.

Xu, H., DeLuca, S. Z. and O'Farrell, P. H. (2008). "Manipulating the metazoan mitochondrial genome with targeted restriction enzymes." *Science* 321(5888): 575-577.

Xu, L., Chen, W. F. and Wong, M. S. (2009). "Ginsenoside Rg1 protects dopaminergic neurons in a rat model of Parkinson's disease through the IGF-I receptor signalling pathway." *Br J Pharmacol* 158(3): 738-748.

Xu, Q., Langley, M., Kanthasamy, A. G. and Reddy, M. B. (2017). "Epigallocatechin Gallate Has a Neurorescue Effect in a Mouse Model of Parkinson Disease." *J Nutr* 147(10): 1926-1931.

Xu, R., Chen, J., Cong, X., Hu, S. and Chen, X. (2008). "Lovastatin protects mesenchymal stem cells against hypoxia- and serum deprivation-induced apoptosis by activation of PI3K/Akt and ERK1/2." *J Cell Biochem* 103(1): 256-269.

Xu, Z., Guo, X., Yang, Y., Tucker, D., Lu, Y., Xin, N., Zhang, G., Yang, L., Li, J., Du, X., Zhang, Q. and Xu, X. (2017). "Low-Level Laser Irradiation Improves Depression-Like Behaviors in Mice." *Mol Neurobiol* 54(6): 4551-4559.

Yang, J., Liu, X., Bhalla, K., Kim, C. N., Ibrado, A. M., Cai, J., Peng, T. I., Jones, D. P. and Wang, X. (1997). "Prevention of apoptosis by Bcl-2: release of cytochrome c from mitochondria blocked." *Science* 275(5303): 1129-1132.

Yang, L., Dong, Y., Wu, C., Li, Y., Guo, Y., Yang, B., Zong, X., Hamblin, M. R., Liu, T. C. and Zhang, Q. (2019). "Photobiomodulation preconditioning prevents cognitive impairment in a neonatal rat model of hypoxia-ischemia." *J Biophotonics* 12(6): e201800359.

Yang, S. H., Liu, R., Perez, E. J., Wen, Y., Stevens, S. M., Jr., Valencia, T., Brun-Zinkernagel, A. M., Prokai, L., Will, Y., Dykens, J., Koulen, P. and Simpkins, J. W. (2004). "Mitochondrial localization of estrogen receptor beta." *Proc Natl Acad Sci U S A* 101(12): 4130-4135.

Ying, R., Liang, H. L., Whelan, H. T., Eells, J. T. and Wong-Riley, M. T. (2008). "Pretreatment with near-infrared light via light-emitting diode provides added benefit against rotenone- and MPP+-induced neurotoxicity." *Brain Res* 1243: 167-173.

Yoneda, M., Tanno, Y., Horai, S., Ozawa, T., Miyatake, T. and Tsuji, S. (1990). "A common mitochondrial DNA mutation in the t-RNA(Lys) of patients with myoclonus epilepsy associated with ragged-red fibers." *Biochem Int* 21(5): 789-796.

Yousuf, S., Atif, F., Ahmad, M., Hoda, N., Ishrat, T., Khan, B. and Islam, F. (2009). "Resveratrol exerts its neuroprotective effect by modulating mitochondrial dysfunctions and associated cell death during cerebral ischemia." *Brain Res* 1250: 242-253.

Yu, L., Lin, Z., Cheng, X., Chu, J., Li, X., Chen, C., Zhu, T., Li, W., Lin, W. and Tang, W. (2022). "Thorium inhibits human respiratory chain complex IV (cytochrome c oxidase)." *J Hazard Mater* 424(Pt B): 127546.

Yuan, H., Qian, Y., Xu, Y., Cao, J., Bai, L., Shen, W., Ji, F., Zhang, X., Kang, D., Mo, J. Q., Greinwald, J. H., Han, D., Zhai, S., Young, W. Y. and Guan, M. X. (2005). "Cosegregation of the G7444A mutation in the mitochondrial COI/tRNA(Ser(UCN)) genes with the 12S rRNA A1555G mutation in a Chinese family with aminoglycoside-induced and nonsyndromic hearing loss." *Am J Med Genet A* 138A(2): 133-140.

Yue, L. and Humayun, M. S. (2015). "Monte Carlo analysis of the enhanced transcranial penetration using distributed near-infrared emitter array." *J Biomed Opt* 20(8): 88001.

Zeng, H., Chen, Q. and Zhao, B. (2004). "Genistein ameliorates beta-amyloid peptide (25-35)-induced hippocampal neuronal apoptosis." *Free Radic Biol Med* 36(2): 180-188.

Zeviani, M., Gellera, C., Antozzi, C., Rimoldi, M., Morandi, L., Villani, F., Tiranti, V. and DiDonato, S. (1991). "Maternally inherited myopathy and cardiomyopathy: association with mutation in mitochondrial DNA tRNA(Leu) (UUR)." *Lancet* 338(8760): 143-147.

Zeviani, M., Muntoni, F., Savarese, N., Serra, G., Tiranti, V., Carrara, F., Mariotti, C. and DiDonato, S. (1993). "A MERRF/MELAS overlap syndrome associated with a new point mutation in the mitochondrial DNA tRNA(Lys) gene." *Eur J Hum Genet* 1(1): 80-87.

Zeviani, M., Peterson, P., Servidei, S., Bonilla, E. and DiMauro, S. (1987). "Benign reversible muscle cytochrome c oxidase deficiency: a second case." *Neurology* 37(1): 64-67.

Zhao, L. and Brinton, R. D. (2007). "WHI and WHIMS follow-up and human studies of soy isoflavones on cognition." *Expert Rev Neurother* 7(11): 1549-1564.

Zhao, L., Mao, Z. and Brinton, R. D. (2009). "A select combination of clinically relevant phytoestrogens enhances estrogen receptor beta-binding selectivity and neuroprotective activities in vitro and in vivo." *Endocrinology* 150(2): 770-783.

Zhao, L., Wu, T. W. and Brinton, R. D. (2004). "Estrogen receptor subtypes alpha and beta contribute to neuroprotection and increased Bcl-2 expression in primary hippocampal neurons." *Brain Res* 1010(1-2): 22-34.

Zhao, Y. and Gong, C. X. (2015). "From chronic cerebral hypoperfusion to Alzheimer-like brain pathology and neurodegeneration." *Cell Mol Neurobiol* 35(1): 101-110.

Zheng, M., Liu, C., Fan, Y., Yan, P., Shi, D. and Zhang, Y. (2017). "Neuroprotection by Paeoniflorin in the MPTP mouse model of Parkinson's disease." *Neuropharmacology* 116: 412-420.

Zhu, S., Stavrovskaya, I. G., Drozda, M., Kim, B. Y., Ona, V., Li, M., Sarang, S., Liu, A. S., Hartley, D. M., Wu, D. C., Gullans, S., Ferrante, R. J., Przedborski, S., Kristal, B. S. and Friedlander, R. M. (2002). "Minocycline inhibits cytochrome c release and delays progression of amyotrophic lateral sclerosis in mice." *Nature* 417(6884): 74-78.

Zimmer, C. (2009). "Origins. On the origin of eukaryotes." *Science* 325(5941): 666-668.

Zomorrodi, R., Loheswaran, G., Pushparaj, A. and Lim, L. (2019). "Pulsed Near Infrared Transcranial and Intranasal Photobiomodulation Significantly Modulates Neural Oscillations: a pilot exploratory study." *Scientific Reports* 9(1): 6309.

Zong, S., Wu, M., Gu, J., Liu, T., Guo, R. and Yang, M. (2018). "Structure of the intact 14-subunit human cytochrome c oxidase." *Cell Res* 28(10): 1026-1034.

Zsurka, G., Hampel, K. G., Nelson, I., Jardel, C., Mirandola, S. R., Sassen, R., Kornblum, C., Marcorelles, P., Lavoue, S., Lombes, A. and Kunz, W. S. (2010). "Severe epilepsy as the major symptom of new mutations in the mitochondrial tRNA(Phe) gene." *Neurology* 74(6): 507-512.

Zwolak, I. (2021). "Epigallocatechin Gallate for Management of Heavy Metal-Induced Oxidative Stress: Mechanisms of Action, Efficacy, and Concerns." *Int J Mol Sci* 22(8).

Chapter 2

Factors Determining the Orientation of Ligated Histidines in Hemoproteins, Cytochromes, and Model Systems

Dragan M. Popović* and Ivana S. Đorđević
University of Belgrade, Institute of Chemistry, Technology and Metallurgy, National Institute of the Republic of Serbia, Belgrade, Serbia

Abstract

Native heme proteins cover a broad range of biological functions, including those which show redox-active ligand complexes. Therefore, it is of great interest to understand the role of the protein environment in modulating the heme cofactor's properties. Different orientations of the imidazole rings relative to heme can shift the heme redox potential and control the coordination of substrates to heme proteins and cytochromes. The molecular model systems built with the minimalistic structural requirements can help further to sort out the different factors and control mechanisms. The entire RCSB PDB database was searched to reveal these factors, and the heme proteins with at least one histidine ligand were analyzed. We analyzed a total of 693 hemes in 432 different crystal structures of heme proteins. The exhaustive analysis of available crystal structures of heme proteins has shown that two major factors determine the orientation of axial histidine ligands: (1) the interaction of the imidazole ring with the heme propionic acids and (2) the interaction with its histidine backbone. The results of the PDB study have been interpreted by a molecular force field and quantum-chemical computations, confirming that the emphasized interactions are mainly of electrostatic origin. The protein environment offers a sufficient number

* Corresponding Author's Email: dragan.popovic@ihtm.bg.ac.rs.

In: Structure and Function of Cytochromes
Editor: Elaine J. Mathis
ISBN: 979-8-88697-102-6

and different oriented hydrogen-bond acceptor groups, as was found in several detailed case studies. Surprisingly, based on the same analysis, the orientation of the NδH group of imidazole axially coordinated to heme does not often depend on a specific hydrogen-bond pattern with acceptor groups of the protein. Thus, the hydrogen-bond pattern is not often decisive, and it is probably the natural pathway to fine-tune the orientation of the ligated imidazoles. DFT computations revealed two main factors determining the imidazole orientation: (1) the direct intramolecular electrostatic interactions of propionic groups with the polar NδH groups of imidazole, and (2) the electrostatic interaction of the total dipole moment of the imidazole-heme complex with the reaction field. The theoretical and computational outstanding challenges in this area were also discussed. One of the purposes of the presented study is to shed some light on the histidine orientation in proteins and find a set of rules that control or govern histidine orientation.

Keywords: cheminformatics, PDB data mining, molecular mechanics, force-field, DFT, heme, coordinated His, cytochromes, propionic groups, backbone, electrostatics, hydrogen bonds

Abbreviations

Cyt*b*, C*b*	cytochrome *b*
Cytbc_1	cytochrome bc_1
Cytb_6f	cytochrome b_6f
bis-C*c*	bis-histidine cytochrome *c*
mono-C*c*	mono-histidine cytochrome *c*
C*c*Po	cytochrome *c* peroxidase
mono-C*c*'	cytochrome *c*'
mono-C*c*124	cytochrome *c* isozyme 1 and 2
*c*550	cytochrome *c*2
PRA	propionic group
PRD	propionic group
DFT	Density Functional Theory
RESP	Restrained Electrostatic Potential
vdW	van der Waals
PDB	Protein Data Base
FF	Force Field
Hb	Hemoglobin
Mb	Myoglobin

Introduction

The heme (iron protoporphyrin IX) is ubiquitous and one of the principal pigments of life. Heme cofactors belong to the highly conserved superfamily of tetrapyrroles. The heme group is widely distributed throughout biological systems and performs many roles as active sites in biomolecules (Mense and Zhang, 2006). Since its discovery, the structural and functional aspects of heme proteins have continuously fascinated the scientific community (Reedy and Gibney, 2004). The heme-binding sites evolved within a variety of protein scaffolds (Martin et al., 1998) to accomplish such diverse tasks as electrons transfer by cytochromes (Gray and Winkler, 1996), transport, storage, and detection of small molecules by myoglobin or hemoglobin (Sowole, Vuong and Konermann, 2015; Padmanaban, Venkateswar and Rangarajan, 1989; Hon et al., 2003; Mense and Zhang, 2006), metal ion storage (Reedy and Gibney, 2004), catalysis by catalases and peroxidases (Karakus and Yonca Y., 2020; Adam, Waldemar et al., 1999), control of gene expression (Mense and Zhang, 2006; Faller et al., 2007), and even programmed cell death (Narula et al., 1999).

The diversity in the function of the heme group arises in particular from the various possibilities of axial ligation and interactions with the protein surrounding. The heme is typically a hexacoordinated iron complex with porphyrin as a tetradentate ligand and two axial ligands, but it can also be a pentacoordinate system. The two axial positions are occupied mainly by amino acids histidine, methionine, cysteine, and tyrosine or small molecules such as water, O_2, CO, CO_2, NO, CN^- or other substrate molecules. A different axially ligated atom or group affects heme affinity, conformational stability, and electrochemical properties (Nesset et al., 1996; Zhuang et al., 2006).

Since heme proteins belong to an extensively spread group of proteins with different biological functions, it is crucial to understand the influence of apoproteins in modulating and tuning the properties of the hemes. For that purpose, various synthetic polypeptides and proteins, with heme as a cofactor, have been synthesized (Mutter et al., 1988; Bryson et al., 1995; Choma et al., 1994; Rau and Haehnel, 1998; Gibney et al., 2000). In the beginning, the focus of the synthetic analog approach was small models of natural or synthetic metalloporphyrins with various axial ligands and their chemical properties (Yu et al., 2014; Liu et al., 2014). Hereafter, the application of genetically engineered proteins provides new insights into the structural chemistry of heme proteins, but at the same time, knowledge about the orientation of axially coordinated histidines could be helpful in the design of artificial proteins.

Extensive theoretical (Vangberg and Ghosh, 1999; Jewsbury et al., 1994; Harris, Loew and Waskell, 1998; Rovira and Parinello, 1999) and experimental studies (Walker, 1999; Safo et al., 1994; Nakamura et al., 1996; Walker et al., 1996; Shokhirev and Walker, 1995) on heme model systems provided an insight into the influence of axial ligands on spectral properties and redox potentials of hemes (Wallace and Clark-Lewis, 1992; Lloyd et al., 1995; Pond et al., 1999). It has been found that not only the type but also the conformation and orientation of the axial ligands can influence the heme properties.

Still, the open question is defining the factors that determine the conformation and orientation of the imidazole rings relative to each other and relative to the heme. It is well-known that different conformations can cause a shift in the heme redox potential (Walker et al., 1986). Furthermore, the orientation of the coordinated histidine is considered to have a strong influence on the function and spectroscopic properties of hemes (Walker, 1999) and can control the coordination of substrates to heme proteins (Menyhard and Keseru, 1998).

Experimental data on heme model systems and molecular mechanics studies revealed that the porphyrins macrocycle displays a range of distorted non-planar shapes. At least four groups of factors have been identified to cause the ruffling of the porphyrin ring: the central metal, axial ligation, peripheral substitution, and the protein environment (Song et al., 1998; Shelnutt et al., 1998; Anđelković et al., 2010). The orientations of axial ligands with bulky substituents can result in significant conformational changes in the porphyrin ring. The porphyrin ring remains planar in complexes with the parallel orientation of two planar axial ligands. For porphyrin complexes, which have two planar axial ligands in perpendicular orientation, the porphyrin ring is almost always distorted from planarity (Walker, 1999; Safo et al., 1994; Shelnutt et al., 2000).

Experimental data for $[Fe(TMP)(5\text{-}MeHIm)_2]ClO_4$ (Munro, 1999) and theoretical, quantum-chemical calculations under Density Functional Theory (DFT) on $[Fe(por)(py)_2]$ and $[Fe(por)(py)_2]^+$ systems (Ghosh, Gonzales and Vangberg, 1999) have shown that there is no preference for parallel or perpendicular mutual orientation of axially coordinated planar ligands. The conclusion is that the two conformational isomers are almost isoenergetic. The energy balance between the two forms results from the crystal field stabilization effects favoring the parallel form, and steric effects originate from substituents of axial ligands and the porphyrin that favor the perpendicular form. The estimates of two opposite energetic effects are both less than 3 kcal mol^{-1} (Munro et al., 1995 and 1999; Marques et al., 1995). Moreover, experimental and theoretical investigations on model systems (Momot and Walker, 1997a; Shokhirev et al., 1997; Polam et al., 1997) demonstrate that there is no rotation barrier for imidazole ligands axially coordinated to porphyrin without bulky substituents. It means that the mutual interactions between two axial histidines or their interactions with the porphyrin atom skeleton are negligible.

Similarly, the orientation of axial ligands in the heme model systems can also depend on crystal field stabilization effects or steric effects caused by bulky substituents on axial ligands and the porphyrin (Walker, 1999; Munro et al., 1999; Safo et al., 1997). In heme proteins, a couple of different factors can influence the orientation of coordinated imidazoles. Among them are hydrogen bonds between the imidazole NδH group and H-bond acceptors of the protein (Hu et al., 2008 and 2018), nonbonded interactions of the imidazole ring with the protein backbone and side chains (Berry and Walker, 2008; Negron, Fufezan and Koder, 2009), nonbonded interactions of the imidazole with the porphyrin atom skeleton (Olafson and Goddard, 1977) and the side chains of cysteines covalently bound to heme (in cytochrome *c* heme proteins). Also, electrostatic interactions and the influence of charged amino acid side chains should be considered (Jenkins et al., 2010).

This chapter discusses the general and the most significant factors determining the axial imidazole orientation. The different types of PDB data mining analysis have been performed in the past, for example, from studying the cation-π interactions in proteins (Zarić, Popović, and Knapp, 2000) to assisting the drug design and discovery of new medicaments (Willett, 2011), from searching whether histidines prefer a particular conformation (Berman et al., 2000) to examine the factors determining the orientation of coordinated histidines in heme proteins (Zarić, Popović, and Knapp, 2000) or to facilitating the modeling of the synthetic protein structures with desired properties (Popović et al., 2001). It has already been shown that analyzing the crystal structures can provide information about interactions in proteins (Gallivani and Dougherty, 1999). The theoretical methods were applied to evaluate several possible types of interactions between the imidazole ring of histidine ligated to heme and the porphyrin atom skeleton, the propionic acids, and if available with the cysteines covalently bound to heme. Also, the impacts of histidine backbone on the orientation of the imidazole concerning the heme plane have been studied. Finally, the hydrogen-bonding pattern of imidazoles, axially coordinated to heme, was analyzed. In a few cases, it was found that a specific hydrogen bond conditioned a particular orientation of axially ligated imidazole rings.

Method

Data Mining in the PDB

The Protein Data Bank (PDB) search has offered 697 hemes belonging to 432 different crystal structures of proteins, with at least one histidine ligated to heme (Zarić, Popović and Knapp, 2001). The primary focus of the search was the determination of the general rules governing the relative orientation of imidazole axially ligated to heme, so it was not necessary to limit the search to proteins with the same amino-acid sequence. The selected heme proteins were classified into six groups:

1. 138 mono-histidine-ligated hemes in 133 proteins of the myoglobin (Mb) group
2. 323 mono-histidine-ligated hemes in 121 proteins of the hemoglobin (Hb) group
3. 72 mono-histidine-ligated hemes in 72 proteins of the cytochrome *c* peroxidase (C*c*Po) group
4. 99 mono-histidine-ligated hemes in 68 proteins of the cytochrome *c* (mono-C*c*) type
5. 39 bis-histidine-ligated hemes in 17 proteins of the cytochrome *c* (bis-C*c*) type and
6. 26 bis-histidine-ligated hemes in 21cytochrome *b* (Cb) type proteins.

The list of the PDB codes for different groups of heme protein can be found elsewhere (Zarić, Popović, and Knapp, 2001). In heme proteins of the cytochrome *c* group, the pyrrole rings of heme are located opposite the propionic acid groups and bound to cysteines covalently (see Figure 1).

For a more detailed insight into the factor determining the imidazole ring orientation, the following characterizations have been made:

- The orientation of ligated histidines relative to heme and relative to the second histidine (in bis-histidine ligated hemes).

- The orientation of axially ligated imidazoles concerning their histidine backbone conformation and the position and orientation of the histidine backbone relative to the heme plane.
- The orientation of hydrogen bonds involving imidazole that is axially ligated to heme.

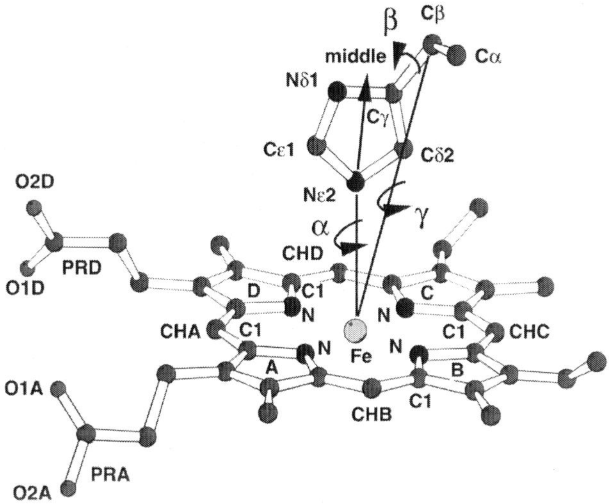

Figure 1. Definition of the three torsion angles α, β, and γ used to characterize the conformation of imidazole ligated to heme. The pseudo-torsion angle α (Nδ1–Hδ1–Cϵ1–Hϵ1) with rotation axis Nϵ2–Fe denotes the orientation of the Nδ1H-group of the imidazole relative to the center of the two propionic acid groups PRA and PRD defined by the atom CHA. The torsion angle β (Cα–Cβ–Cγ–Nδ1) characterizes the orientation of the imidazole ring relative to the Cα–Cβ bond of the histidine backbone. The orientation of the histidine backbone with respect to the center of the propionic acids of the heme is measured by the pseudo-torsion angle γ (Cα–Cβ–FE–CHA). For the structure displayed, the values of the torsion angles are $\alpha \approx 0°$, $\beta \approx 180°$ and $\gamma \approx 180°$.

Characterizing the Orientation of Ligated Imidazole Relative to Heme

Analyzed crystal structures of heme proteins retrieved from PDB revealed that the N(His)–Fe distances are close to 2.0 Å. The bond angles N(His)–Fe–N(heme) do not significantly deviate from the ideal values of 90°, so the coordinated imidazole rings are nearly orthogonal to the corresponding heme planes. To evaluate possible differences in the histidine orientation, axially ligated to heme, the following parameters were defined and schematized in Figure 1:

1. The pseudo-torsion angle α (Nδ1–Hδ1–Cϵ1–Hϵ1) was used to define the orientation of the imidazole ring concerning the propionic acids of the heme and monitor the imidazole ring rotation around the axis of the Nϵ2–FE bond of the imidazole with the heme iron.
2. The torsion angle β (Cα–Cβ–Cγ–Nδ1) was selected to identify the orientation of the imidazole ring of the histidine relative to the polypeptide backbone.
3. The pseudo-torsion angle γ (Cα–Cβ–FE–CHA) was used to describe the histidine backbone orientation concerning the heme propionic acids.
4. The mutual orientation of the imidazole planes of two ligated histidines (1 and 2) is characterized by the difference in the corresponding torsion angles $\Delta\alpha = |\alpha_2 - \alpha_1|$.

The *cis*-conformation defines the zero point value of the corresponding torsion angle. The positive rotation sense of the torsion angles α, β, and γ is opposite to clockwise.

Geometrically, it can be shown that these three torsion angles are not independent and fulfill the equation 1:

$$\gamma = \alpha + \beta \tag{1}$$

The obtained data were grouped according to the different heme protein families, and the corresponding torsion angle distributions were represented graphically by generating histograms with 36 bins each of 10° width. For a clearer representation of the data, the values of the angular distribution functions were connected by continuous lines.

Characterizing Hydrogen Bonds Involving Imidazole Axially Ligated to Heme

Two criteria were used to determine whether a hydrogen bond is possible between the Nδ1–Hδ1 group of the imidazole ring and a suitable hydrogen bond acceptor.

It is most likely that The NδH group of coordinated imidazole forms a hydrogen bond with a CO group of the protein backbone. However, the other protein polar groups can be involved in these hydrogen bonds. The one search of hydrogen bonds was carried out by the criterion that a distance between the Nδ1 atom of the imidazole ring and the hydrogen bond acceptor atom must be less than 3.5 Å. The spatial surrounding was analyzed to find all possible hydrogen bond motifs by rotating the imidazole ring around its Nε2–Fe bond with heme without changing other parts of the crystal structure. Another search included a hydrogen bond for a distance less than 4.0 Å between the Nδ1–Hδ group and different hydrogen bond acceptor atoms of surrounding amino-acid residues of proteins. To account for all possible hydrogen bonds, a slightly larger distance was chosen without considering a change in protein conformation due to the formation of potentially new hydrogen bonds. The obtained data were presented graphically with the inverse of the atom pair distances of the possible hydrogen bonding partners, given as a function of the pseudo-torsion angle α, characterizing the imidazole orientation. The inverse of this atom pair distance was chosen as a parameter since it provides a rough measure of the electrostatic interactions, which are relevant for the strength of the hydrogen bond.

Molecular Force Field Computations

To rationalize the experimentally observed preference for histidine ligated to heme orientation, the force field calculations were performed with the CHARMM22 program (MacKerell et al., 1998). The multi-orientation methodology was applied for the imidazole ring ligated to heme by varying the torsion angle α. The energies of different imidazole-heme conformations were calculated by setting the dielectric constant to $\epsilon = 1$, typically used for the force field. Different groups of atoms and interaction types were considered to find the most relevant ones. In the case of cytochrome *c*, the interactions of the imidazole with the covalently bound cysteines were also considered. Using the atomic coordinates from appropriate crystal structures of heme proteins provided a more realistic estimate of the interaction energy between the imidazole ring and the heme and covalently bound cysteine atoms. In these computations, two propionic groups of heme were considered negatively charged, if not otherwise stated.

The influence of the protein backbone conformation was estimated by the calculated interaction energy between the histidine's imidazole ring and its protein backbone. For this purpose, all computations were done on model systems with the C-terminus being amidated and N-terminus being methylated, so artifacts from the bare charges are avoided. Different histidine conformations, relative to its backbone, were simulated by variation of three torsion angles between the imidazole ring and the protein backbone of histidine, as shown in Figure 1:

β_{-1} (N–Cα–Cβ–Cγ), β (Cα–Cβ–Cγ–Nδ1) and β_{+1} (Cβ–Cγ–Nδ1–Cε1)

Among these angles, torsion angle β is the most relevant for characterizing the different orientations of the imidazole ring toward the protein backbone. Therefore, during the minimization of the energy, the angle β was assigned different fixed values, while all other degrees of freedom, including torsion angles β_{-1} and β_{+1}, were allowed to relax.

Quantum-Chemical Computations

In the presented Density Functional Theory (DFT) study, interactions of the heme imidazole and propionic groups were considered in vacuum, protein, and water environments (Galstyan, Zarić, and Knapp, 2005). Quantum-chemical computations were performed on the model system of the imidazole-heme complex extracted from the crystal structure of protein cytochrome *c*3, PDB code 3cyr (Simoes et al., 1998). The geometries of the reduced form of the model complex $[Fe^{II}(por)(Im)2]^{2-}$ were optimized in the vacuum, followed by the single-point electronic energy calculations. All quantum-chemical calculations were performed with JAGUAR program package version 4.2 (JAGUAR 4.2, 1991). Under DFT formalism, the hybrid B3LYP functional was considered, which includes exchange on the Hartree-Fock level, the Slater local exchange functional (Slater John C., 1974), Becke's 1988 nonlocal gradient correction (Becke, 1988), as a correlation term the Vosko-Wilk-Nusair local functional (Vosko, Wilk and Nusair, 1980) and the Lee-Yang-Parr local and nonlocal functional (Lee, Yang, and Paar, 1988) implemented as previously described (Miehlich et al., 1989). For a reliable description of transition metal atom, LACVP (Hay and Wadt, 1985) with effective core potential was applied, while 6-31G basis functions were used for carbon, nitrogen, oxygen, and hydrogen atoms. The initial calculation of the starting conformation (α_1, α_2) = (0°, 0°) of the imidazole-heme complex in a vacuum revealed that the singlet state is by 8 kcal mol^{-1} lower in energy than the quintet state. Appling extended basis sets (LACV3P** + and 6-311+G**) did not lead to significant changes apart from slightly smaller differences between the heme quintet and singlet energies of 7.3 kcal mol^{-1}. Therefore, only the low-spin heme model complex was considered for further computations.

Computation of the Solution-Phase Energy by One-Step Procedure

One way to evaluate the solution-phase energy G_s is a one-step procedure, where the molecular system is embedded in a dielectric continuum. After a geometry optimization of the imidazole-heme complex in the presence of the reaction field, the solution-phase energy is evaluated by simultaneously calculating electronic and electrostatic solvation energies. The electrostatic energy is calculated within a dielectric continuum model, while the reaction field contribution is accounted for using the JAGUAR implemented Poisson equation solver (Tannor et al., 1994; Marten et al., 1996). The boundaries between solute and solvent are defined using the same van der Waals (vdW) radii in both one-step and two-step procedures.

Computation of Atomic Partial Charges

In a two-step procedure, molecular solvation energies were calculated by atomic charges derived using the restrained electrostatic potential (RESP) method (Bayly et al., 1993; Cornell et al., 1993). The RESP method has become the standard approach for fitting the partial charges of metal-protein complexes, as well as developing force fields for biomolecules (Li and Merz, 2017; Đorđević and Niketić, 2012; Đorđević et al., 2015; Dupradeau et al., 2010; Popović et al., 2001). In this method, partial atomic charges were obtained by iterative optimization between DFT-derived electrostatic potential in a vacuum (V_{QM}) and the Coulomb monopole approximation model, by applying a hyperbolic restraint function to moderate atomic charges. In the first stage of the RESP fitting procedure, the charges of the oxygen atoms belonging to the carboxyl groups were set up to be equivalent using a restraining weight of 0.0005 au, while the charges of all other atoms were left unconstrained. In the second stage, the charges of carbon atoms and their hydrogen atoms were left free for fitting, but charges of hydrogens atoms of the CH_2 group belonging to the propionic acids were constrained to be equivalent. Also, charges of all other atoms were set up at their values obtained from the first stage of the RESP procedure. A restraining weight in the second stage was set at 0.001 au.

Computation of Solvation Energy

The initial conformation of the imidazole-heme complex with pseudo-torsion angles (α_1, α_2) = (0, 0) was used to evaluate the energetics of conformational changes in a dielectric medium. Solution-phase energy difference ΔG_s is evaluated following the thermodynamic cycle (Figure 2) and according to equation 2:

$$\Delta G_s = \Delta E_{vac} + \Delta G^2_{solv} - \Delta G^1_{solv} \quad (2)$$

where ΔE_{vac} is the conformational energy in vacuum calculated as the quantum-chemical electronic energy, ΔG^1_{solv} and ΔG^2_{solv} are the solvation free energies of conformations 1 and 2, respectively, where conformation 2 is considered as the reference conformation.

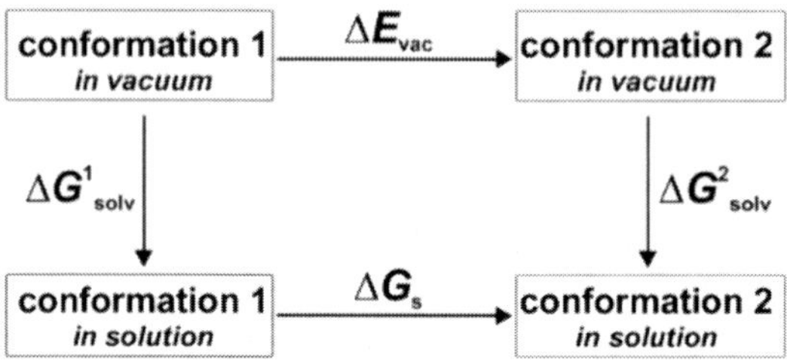

Figure 2. Thermodynamic cycle to calculate the solution-phase energy difference, ΔG_s, in Eq. 2.

In calculations of solvation energy within the protein and the water environment, a dielectric continuum model was used with a dielectric constant of $\epsilon_p = 10$ and $\epsilon_w = 80$, respectively. The general dependence of imidazole-heme conformational energies on the value of the dielectric constant was additionally estimated. Applying different dielectric constants of 40, 20, 4, and 2 did not affect qualitative conclusions.

The program module SOLVATE of the MEAD program suite (Bashford and Gerwert, 1992; Bashford Donald, 1997) was applied for the solvation energies evaluation. In this approach, the solute is simulated as a set of atomic point charges embedded in the solute volume of dielectric constant $\epsilon = 1$. The environment outside of the solute was represented by dielectric constants of $\epsilon_w = 80$ and $\epsilon_p = [2\text{–}40]$ for water and protein, respectively. For these simulations, the ionic strength was set to zero. The solute volume was defined from the atomic vdW radii as incorporated in the program JAGUAR. The only exception was made for the negatively charged oxygen atoms, where 1.8 Å instead of 1.6 Å was used to account for the more extended wave function at the negatively charged oxygen (H 1.15 Å, C 1.9 Å, N 1.6 Å, O 1.8 Å, Fe 1.456 Å). A similar procedure has been successfully used to calculate the pKa values of protonable groups, redox potentials of cofactors, and bioenergetics of electron transfer and proton transfer reactions in different enzymes and proteins (Popović et al., 2001; Quenneville, Popović and Stuchebrukhov, 2004; Popović and Stuchebrukhov, 2005; Quenneville, Popović, and

Stuchebrukhov, 2006; Popović and Stuchebrukhov, 2012; Popović and Đorđević, 2020).

Solvent Probe Radius

For water as a solvent, the mainly accepted value of a probe radius is 1.4 Å. However, for the protein environment, there is not such a generally accepted value. A suitable probe radius was estimated on dipeptides and tripeptides of the 20 different natural amino acids. The probe radius calculator, implemented in the program JAGUAR, calculates the probe radius according to the molecular weight and the density of the solvent. The densities were determined by the ACD/ChemSketch program (ACD/ChemSketch, 1994). This procedure shows that the calculated solvent probe radii r_{solv} lie within 3–4 Å. To exclude the dependence of the imidazole-heme conformational energies on r_{solv}, test computations were performed for two conformations [(α_1, α_2) at (0°, 0°) and (180°, 180°)] for r_{solv} in the interval [1.4 Å, 8.0 Å] (Figure 13). As the calculations have shown, uncertainties in r_{solv} of 3–4 Å have no significant impact on the solvation energy of the imidazole-heme complex. Contrary, the value of the protein dielectric constant has a significant influence on calculating the solvation energy of the same complex. Based on the results, the value of r_{solv} = 3 Å was selected for further calculations.

Modeling of Propionic Group Interaction

The two negatively charged heme propionates in the protein environment are the most common solvent-exposed or form a salt bridge. Rarely the acidic oxygens of heme propionates are buried in the protein and/or neutralized by protonation. How the interactions of such forms of propionate (salt-bridge formation or protonated) cannot be modeled by a continuum dielectric alone, additional electrostatic energy calculations on the imidazole-heme complexes were necessary. For that purpose, a positive unit charge with the vdW radius of hydrogen was placed close to the acidic oxygens of the heme propionates.

Computation of Dipole Moment

There are several suitable points to calculate the molecule's dipole moment, such as the center of mass, the center of symmetry, or the center of charges. By default, the JAGUAR program calculates molecular dipole moments concerning the center of mass. As the center of mass and the center of charges vary with the imidazole-heme conformation, the heme iron was selected as a reference point in computations of dipole moment.

Results and Discussion

General Overview of PDB Data

A systematic analysis of structural data of heme proteins retrieved from the Protein Data Bank (PDB) was reported by Zarić et al. (Zarić, Popović and Knapp, 2001). The distribution of the pseudo-torsion angle α for all analyzed heme proteins is graphically illustrated in Figure 3. The solid line shows data of all heme proteins, except the data from the Hb and Mb families. Collected data of the Hb and Mb groups, which are very large and consist of proteins with a high degree of sequence homology, are illustrated separately by dashed-dotted lines, scaled down by the factor 4. Also, for better visibility, the two groups of cytochrome *c* heme proteins, mono-C*c* (dashed line) and bis-C*c* (dotted line), have the angle *α* around 0°, are shown individually. In many crystal structures of heme proteins, the value of the angle *α* is close to 0°, where the projection of the imidazole N*δ*–H bond vector on the heme plane is placed between the two propionic groups PRA (+45°) and PRD (–45°). It is obvious that the N*δ*H group of imidazole is oriented toward the center of both propionic acids ($\alpha = 0°$) or at least toward one propionic acid group ($\alpha = 45°$) in most crystal structures. Within the cytochrome *c* peroxidase (C*c*Po) group, many structures have the angles *α* around 120°.

Further, in the Hb and Mb proteins, the angle *α* adopts values between 30° and 70° (Figure 3, dashed-dotted line). In a small number of crystal structures, the angle *α* can also take other values, but there are very few structures with an *α* value close to ±180°. Therefore, it is reasonable to expect an attractive interaction between the imidazole ring and the propionic acid groups, which favors such orientations of imidazole relative to the heme. Since the orientation of the imidazole was not in all cases toward the propionic groups, other factors should be additionally considered.

Another study, based on a PDB search of heme proteins, found that of all 1503 heme-binding proteins, 83% belong to proteins with His ligated to heme (Fufezan, Zhang and Gunner, 2008). The ligands were classified into three groups, and within these groups, the distributions of the angles *α* were analyzed. The first group of heme proteins with ligating *b*-type hemes (51 hemes) revealed a relatively even distribution of *α* angle values with a maximal at –170° for nearly 9% of proteins. The second group containing ligated His to *c*-type heme, where the histidine ligand is not part of the Cys–X–X–Cys–His (i.e., CXXCH) motif (42 hemes), has shown the distribution of

α over the entire range of angles ($\alpha \approx -10°$, 8.5%; 80°, 33.3%; and 110°, 9.5%). Opposed to the previous groups, the third group with ligating *c*-type hemes where the histidine ligand is part of the CXXCH binding motif (58 hemes) has shown the asymmetric alpha distribution. Most of these proteins (86%) have the angles α around 0°, of which 51.7% have values at –15°. The orientation with $\alpha \approx 90°$ has only 7% of all proteins in the third group. Independent of the heme type, the orientations with values of $\alpha \approx \pm 45°$ or $\pm 135°$ are not significantly abounded.

Based on the results of this PDB search, the authors concluded that *b*- or *c*-type heme-binding motifs (not in CXXCH) did not show a significant preference for a particular α value. However, histidine ligands in the *c*-type CXXCH heme-binding motif demonstrated a high preference toward orientation with α near –10° and less commonly near 90°.

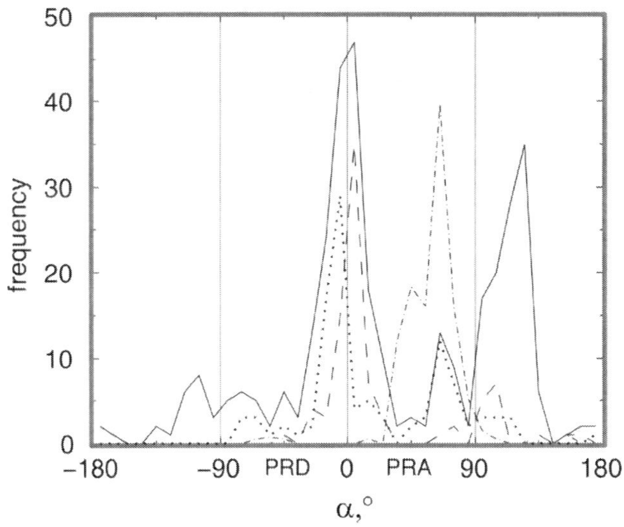

Figure 3. Distributions of the torsion angle α of the imidazole heme conformation for different groups of heme proteins from the PDB. Solid line: imidazoles of all heme proteins except for the Mb and Hb groups; dashed line: heme proteins of the mono-C*c* group; dotted line: heme proteins of the bis-C*c* group; dashed-dotted line: proteins of the Mb and Hb groups scaled down by the factor four. The approximate orientation of the propionic acid groups PRA and PRD is marked on the x-axis.

In a systematic survey of bis-His in the PDB, 1618 heme protein models were found, which include 3623 heme moieties (Merlino et al., 2009). This study also showed that for the great majority of hemoproteins, a histidine residue dominates in the primary coordination sphere of the iron.

Role of Propionic Acids

Analyzing Data from the PDB

Statistic analysis of heme protein structural data (Zarić, Popović, and Knapp, 2001) showed that the orientation of histidine axial ligands is primarily conditioned by the interaction of the imidazole ring with the heme propionic acids. Furthermore, a possible correlation between the conformations of the two heme propionic acid groups and the distribution of the values for torsion angle α was analyzed for the following geometries:

1. both propionic acid groups pointing toward the ligated histidine,
2. both propionic acid groups in the heme plane,
3. both turned away from the ligated histidine,
4. only PRA pointing toward the ligated histidine,
5. only PRD pointing toward the ligated histidine.

The corresponding distributions of the torsion angle α for all bis-histidine heme proteins are displayed in Figure 4a (conformations 1 – 3) and 4b (conformations 4 and 5). A closer inspection of the crystallographic data of bis-C*c* and C*b* proteins revealed that the N*δ*H group of the imidazole ligated to heme is preferentially oriented toward the propionic acid of heme when these groups are facing each other (geometries 1) (Figure 4a). Likewise, this tendency has been observed for orientations in which propionic acid groups are positioned in the heme plane or the opposite side of the plane concerning the imidazole ring (geometries 2 and 3). For geometries 2 and 3, the N*δ*H group of the imidazole ring is also in contact with the propionic acid group, located on the same side of the heme plane as the imidazole ring (Figure 4b). However, a certain number of crystallographic data reveal that *α* values are larger than 90° or smaller than –90°. Such values are observed mainly in geometries with the propionic acid group PRA points toward the imidazole ring and cytochrome *b*5 subgroup proteins.

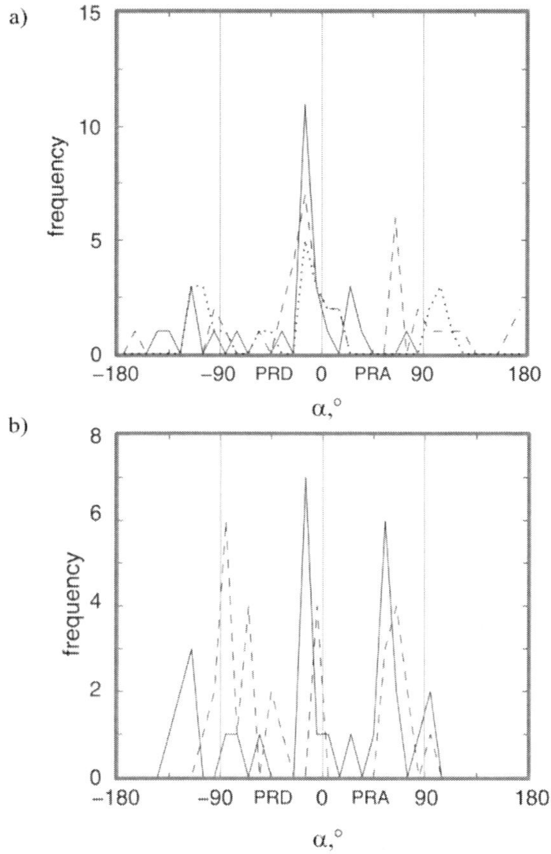

Figure 4. Distribution of the torsion angle α of the imidazole heme conformation for all bis-C*c* and C*b* heme proteins. a) solid line: both propionic acid groups point toward imidazole; dashed line: both propionic acid groups are in the heme plane; dotted line: both propionic acid groups turn away from the imidazole. b) solid line: PRA points toward, PRD away from the imidazole; dashed line: vice versa. The location of the propionic acid groups is marked on the x-axis.

Force Field Computations

To get a deeper insight into the influence of the propionic acid groups on the orientation of a ligated imidazole and, therefore, on the values of angle α, further investigations have been shifted towards theoretical calculations on suitable model systems. Some early attempts to elucidate the interactions involving the propionic acid group and the imidazole ring of heme were achieved by force field calculations (Zarić, Popović, and Knapp, 2001).

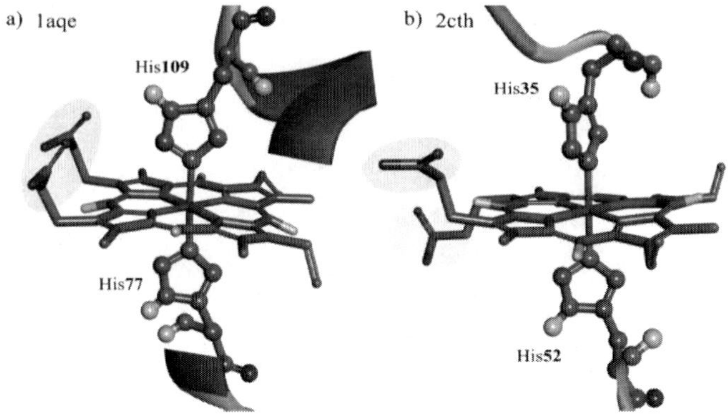

Figure 5. Part of protein crystal structure with PDB code a) 1aqe, and b) 2cth.

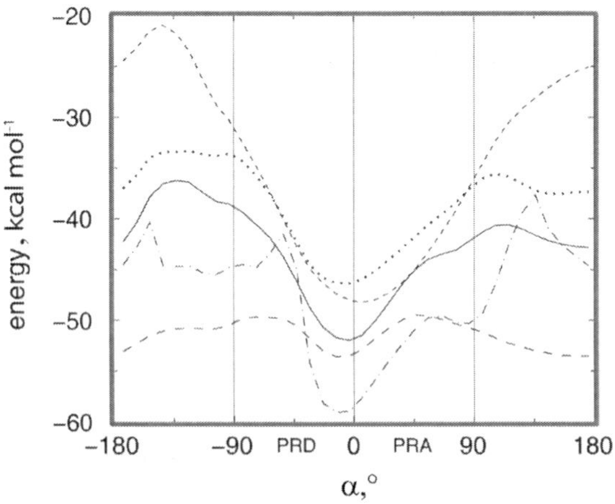

Figure 6. Calculated interaction energies of imidazole with heme for the torsion angle α. Both propionic groups are unprotonated (if not otherwise stated) and point toward the imidazole. The heme coordinates were taken from the crystal structure of cytochrome *c* (PDB code 1aqe) of the group bis-C*c*. The imidazole of His109 was considered. Solid line: total interaction energy between imidazole and heme; dotted line: electrostatic interaction only; short-dashed line: interaction of the polar groups Nδ1–Hδ1 and Cϵ1–Hϵ1 of imidazole with the carboxyl groups of the two propionic acid groups of heme; dashed-dotted line: interaction between imidazole and heme with hydrogen-bonded propionic acid groups; long-dashed line: interaction between imidazole and heme with both propionic groups protonated.

Two crystal structures of the cytochrome *c*3 subgroup, PDB code 1aqe (Aubert et al., 1998), and PDB code 2cth (Simoes et al., 1998; Matias et al., 1996), were selected as an appropriate model to study different orientations between the propionic acid group and the imidazole ring. Namely, these crystal structures have the two imidazole rings coordinated to heme, so the arrangement of propionic groups is directed towards or away from the imidazole ring (Figure 5).

In the structure 1aqe, the histidines His109 and His77 are axially ligated to heme, with both propionic groups located on the same side of the heme plane as His109 (Figure 5a). The calculated interaction energies of the imidazole ring of His109 with heme are displayed in Figure 6, as a function of the pseudo-torsion angle α, describing the imidazole orientation relative to the heme. In agreement with most cytochrome c crystal structures, the total interaction energy (solid line) favors conformations with the angle α close to zero, where the NδH group of the imidazole points toward the center of the two propionic acid groups. In the considered crystal structure, the corresponding value of the angle is $\alpha = 7°$. The calculations also show the dominant electrostatic contribution (dotted line in Figure 6) to the total interaction energy, originating from interactions between the polar groups of the imidazole (Nδ1–Hδ1 and Cϵ1–Hϵ1) and the carboxyl groups (COO^-) of the two propionic acid groups (short dashed line in Figure 6).

The hydrogen-bonding pattern is modeled by considering the structure with PDB code 1aqe (Aubert et al., 1998), where one crystal of water (HOH209) bridges the two oxygen atoms of the different propionic acids. The other water molecule is placed for modeling the hydrogen bond, which the propionic acid PRA forms with Tyr19(Oη).

The conformation of the two water molecules was subsequently energy minimized. The influence of the propionic acids on the orientation of a ligated imidazole becomes small if they are neutralized by protonation, which is, however, unlikely for these acidic groups (long dashed line in Figure 6). The electrostatic potential of the propionic groups may also be shielded by salt bridges, which can be formed with arginine or lysine. In that case, a reduction in the electrostatic interactions between the imidazole ring and propionic groups can be expected, but less than for the protonated propionic groups.

Propionic groups in cytochrome *c*3 with PDB code 1aqe are on the opposite side of the heme plane for His77. Orientation with the imidazole ring and propionic acid groups located on the opposite side of the heme plane has significantly weaker electrostatic interactions. The minimum energy profile can be described as broad within the angular interval of –90° to 90°.

Nevertheless, in the crystal structure, the corresponding value of the angle is $\alpha = 13°$. The total interaction energy is displayed as a solid line in Figure 7.

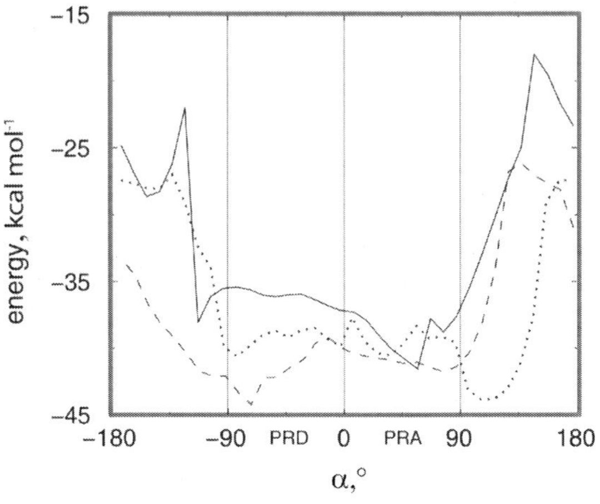

Figure 7. Calculated interaction of the imidazole with the heme as a function of torsion angle α. Both propionic groups are unprotonated and point away from the imidazole. The solid line shows results from His77 ligated to the heme of cytochrome *c*3 structure 1aqe, where the imidazole of His77 is on the opposite side of the heme plane as the two propionic acid groups. The dashed line shows results from His35 ligated to the heme of the cytochrome *c*3 structure 2cth, where PRD points toward the imidazole of His35 and PRA points away. The dotted line also shows results for the cytochrome *c*3 structure 2cth, but here the imidazole of His52 on the opposite side is considered, where PRA points toward the imidazole ligated to heme and PRD points away.

Finally, the third model system included PRD directed towards His35 (dashed line in Figure 7) and PRA directed towards His52 on the opposite sides of the heme (dotted line in Figure 7). Here, the NH group of the imidazole ring adopted the best possible orientation for interaction with both propionic groups. Due to the attractive interactions of the imidazole with the propionic groups, the energy minimum at negative (positive) values of the angle α is deeper for His35 (His52). That partially agrees with the corresponding values of the angle α for the two histidines in the crystal structure, which are $\alpha = -63°$ for His35 and $\alpha = -14°$ for His52.

A similar force field study was reported seven years later, in 2008 (Fufezan, Zhang and Gunner, 2008).

In this study, the influence of the propionic groups on histidine orientation was investigated. The simplified model complex of *c*-type heme proteins with binding motifs CXXCH was used, and the protein environment was simulated by applying a uniform dielectric constant of 6 (for more details, see Fufezan, Zhang and Gunner, 2008). The total interaction energy and its electrostatic contribution were monitored as a function of angle α (i.e., ligand orientation). Force field calculations include three forms of the propionic groups (deprotonated, one propionate deprotonated, and both protonated) and three different ligand orientations:

1. The two ligands are parallel to each other with $\alpha = 0°$, pointing between the propionates;
2. The ligands are parallel to each other and perpendicular to the propionates, both with an $\alpha = 90°$;
3. The two ligands are perpendicular to each other at $\alpha = 0°$ and $\alpha = 90°$.

In a case when both propionates are deprotonated, the energy landscape has a deep and symmetric shape. Independent of initial ligand orientation and after energy minimization, both ligands prefer to be pointed between the two propionic acids.

When one propionate is deprotonated, the energy landscape has slightly asymmetric, a cross-like shape with a local minimum at (0°, 0°). Conformations where at least one ligand points towards the deprotonated acid group ($\alpha \approx 20°$) are preferable.

Finally, if both propionates are protonated, the system exhibits a smooth energy landscape with minima at $\alpha \approx 0°, \pm 90°$, and $\pm 180°$.

DFT Computations

Several experimental studies reported that imidazole-heme model systems without propionic acids or other bulky substituted imidazole ligands have shown minimal and perhaps vanishing rotational barriers (Momot and Walker, 1997b; Shokhirev et al., 1997; Nakamura et al., 1994; Polam et al., 1997). These observations have been confirmed by quantum-chemical calculations on unsubstituted imidazole-heme model systems (Medaković and Zarić, 2003).

Presumably, the first systematic quantum-chemical study of interactions involving heme with ionized propionic acid groups, axially coordinated with two imidazoles, was reported in a paper by Galstyan et al. (Galstyan, Zarić, and Knapp, 2005). This DFT study has considered how the intramolecular interactions of propionic groups and intermolecular interactions of reaction fields impact the imidazole orientation.

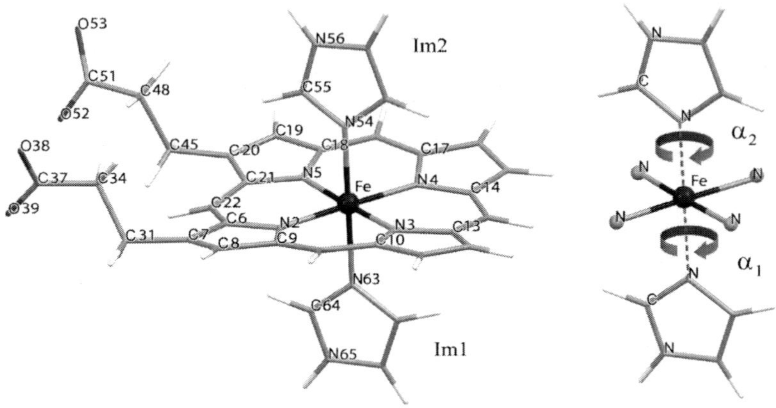

Figure 8. A simplified model system of the heme used to generate all other imidazole-heme conformations. It constitutes the regularized conformation of two propionates and two imidazoles axially coordinated to the iron porphyrin. The propionates belonging to heme are oriented such that both acidic groups are on the same side of the heme-porphyrin plane. This conformation resembles the heme structure in cytochrome *c*3 (PDB entry 3cyr) with a root mean square deviation of 0.32 Å. The orientation of imidazoles in the imidazole-heme complex can be characterized by the pseudotorsion angles α_1 and α_2 defined by the following quartets of atoms C22–Fe–N63–C64 and C22–Fe–N54–C55, respectively. For the imidazole-heme structure displayed, these torsion angles have values (α_1, α_2) = (0°, 0°). The following torsion angles were constrained during geometry optimization of the imidazole-heme model complex as described in the text: C6–C7–C31–C34, C7–C31–C34–C37, C8–C7–C31–C34, C22–C6–C7–C31, C31–C34–C37–O39, C31–C34–C37–O38, C19–C20–C45–C48, C20–C45–C48–C51, C21–C20–C45–C48, C22–C21–C20–C45, C45–C48–C51–O52, C45–C48–C51–O53.

As the initial model, the imidazole-heme complex was extracted from the crystal structure of protein cytochrome *c*3 with PDB code 3cyr. The model complex $[Fe^{II}(por)(Im)2]^{2-}$has the iron(II) ion axially ligated by two imidazoles and the negatively charged propionates located on the same side of the heme plane (Figure 8).

DFT calculations were performed for various orientations of the imidazole planes (defined by torsion angles α_1 and α_2) in a vacuum, protein, and water environment. Also, the one-step and the two-step procedure were applied for the solution-phase energies calculations (for more details, see METHOD).

Imidazole-Heme Conformation in Vacuum

According to DFT results, the minimum of the conformational energy corresponding to the imidazole NH groups points toward the heme propionic acids, with values of torsion angles near zero, $\alpha = 0°$ (Figure 9). As pointed out in the previous section, this is the most abundant imidazole-heme conformations in protein crystal structures. The largest energy barrier corresponds to the pseudo-torsion angles of about $\alpha = \pm 135°$. Such conformations are characterized by the imidazole NH groups pointing away from the propionic acids. The conformational energy of the system is strongly influenced by electrostatic interactions between the partially positive hydrogen of the imidazole NH group and the negatively charged propionic groups of the heme. According to Figure 9, the increment of calculated energy for the imidazole-heme model complex is directly proportional to the increase of the absolute value of the pseudo-torsion angle α. It increases simultaneously with the distance between the propionic groups and the imidazole NH group. Interestingly, a weak electrostatic attraction was noted between the propionic acid groups and the hydrogen atoms of the imidazole carbons opposite the NH group. These interactions lead to stabilization of model system conformations with $\alpha \approx 180°$.

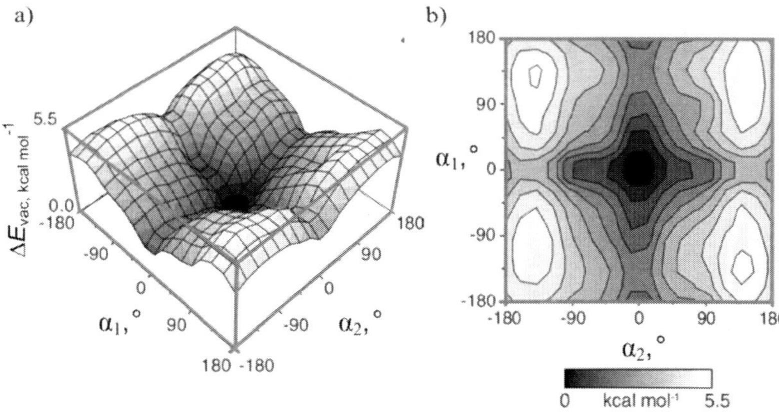

Figure 9. Imidazole-heme conformational energy landscape ΔE_{vac} in a vacuum as a function of mutual orientations of the axially ligated imidazoles defined by the pseudo-torsion angles α_1 and α_2. The quantum-chemical calculations at different imidazole-heme orientations were performed as described in the METHOD section. Surfaces were generated on an 18×18 grid with additional interpolation and smoothing. The actual angular resolution based on the computation is 45°.

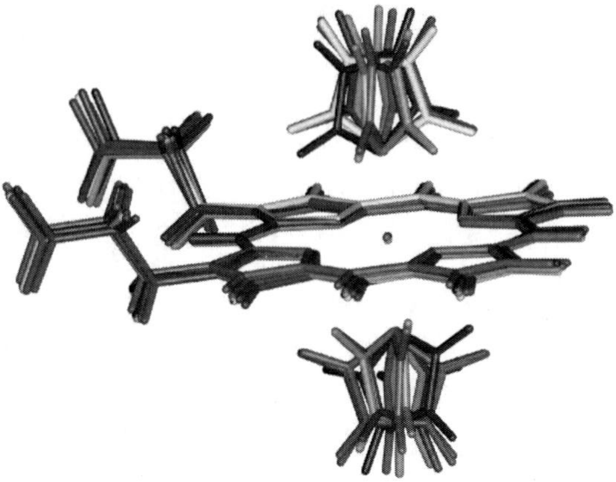

Figure 10. Structural variations as a result of quantum-chemical geometry optimization of the bis-imidazole ligated heme complex depend on the mutual orientations of the imidazoles obtained for constrained-geometry optimization under vacuum conditions.

Depending on the orientation relative to the propionic acids, charges of vinylene group hydrogen atoms of the axially ligated imidazole have values of 0.13 to 0.20 unit charges. The charge of the NH hydrogen is in the range of 0.32–0.34 unit charges, therefore, a similar order of magnitude as charges of vinylene group hydrogen atoms. Only minor variations of the imidazole nitrogen and iron ion (Nε2–Fe) bond length were detected. The maximum deviation of the Nε2–Fe bond has the value of 2.13 Å for (α_1, α_2) = (0°, 0°) and a minimum value of 2.01 Å for (α_1, α_2) = (180°, 90°). The porphyrin ring of the model system exhibits significant variation in ruffling due to the presence of propionic acids. Contrary, all unsubstituted axially ligated imidazole-heme complexes show a small degree of ruffling. Monitoring of the torsion angle C9–N2–N4–C14, C18–N5–N3–C13, C6–N2–N4–C17, and C21–N5–N3–C10 revealed variations between 0 and 16.5°, 0 and 17.2°, 0 and 24.3°, and 0 and 24.4°, respectively. All orientations of the axially ligated imidazoles have shown some degree of ruffling. The more pronounced ruffling was found for the low-energy conformations (α_1, α_2 ϵ [–45°, 45°]) in comparison to the high-energy conformations (α_1, α_2 ϵ [–135°, 135°]). However, a certain fraction of ruffling (about 8°) derives from rigidly connecting the propionic groups with the porphyrin ring atom skeleton. Even for conformation with the energy minimum (α = 0°), additional ruffling is generated by the imidazole NH group dragging the acidic groups of the propionates toward the imidazole. This effect is significantly strong considering that the geometry optimizations of the imidazole-heme model complex were done in a vacuum, where electrostatic screening is absent. The structural changes of heme conditioned by the orientation of the imidazoles are shown in Figure 10. Such conclusions are supported by previously collected crystallographic data of heme proteins (Zarić, Popović, and Knapp, 2001).

Imidazole-Heme Conformation in a Dielectric Continuum: The One-Step Procedure

The influence of protein or solvent environments on the imidazole-heme conformation was included by energetic profile calculations in a dielectric continuum, using $\epsilon_w = 80$ for the water environment and $\epsilon_p = 10$ for the protein environment. Imidazole-heme conformational energy landscape (Figure 11) calculated in the one-step procedure (see METHOD) is significantly different from the one calculated in a vacuum (Figure 9). For calculation in a vacuum, the calculated energy difference between global $(\alpha_1, \alpha_2) = (0°, 0°)$ and local $(\alpha_1, \alpha_2) = (180°, 180°)$ minimum is $\Delta G_s(180°–0°) = 3.84$ kcal mol^{-1}. However, this energy difference is only 0.12 kcal mol^{-1} at $\epsilon_p = 10$ and even changes signs, yielding –0.25 kcal mol^{-1} at $\epsilon_w = 80$ (Table 1), which conflicts with experimental data (the most represented crystal structure conformation $(\alpha_1, \alpha_2) = (0°, 0°)$ is usually the most energetically favorable conformer). A more realistic energy value $\Delta G_s(180°–0°) = 1.96$ kcal mol^{-1} was calculated only for a dielectric constant of $\epsilon_p = 10$. Such a low dielectric constant value is suitable for a medium like the interior of a lipid bilayer filled with aliphatic chains or a rare-gas liquid. For simulating the interior of a protein, the most commonly used dielectric constant value is $\epsilon_p = 4$, which also implies using the atomic point charge model of the protein. As atomic charges were not applied for these calculations, the use of value $\epsilon_p = 10$ is more acceptable.

A comparison of represented calculations approaches leads to the conclusion that two factors have a major impact on imidazole orientation in an imidazole-heme complex, but only the first exists under vacuum conditions:

1. The main stabilization factor for the imidazole-heme complex is attractive electrostatic interactions between the propionic group (negatively charged) and hydrogens of the imidazole (positively charged). The direct consequence of this interaction is the imidazole-hem complex conformer with $\alpha = 0°$ and represents a global energy minimum. The conformer with $\alpha = 180°$

represents a local energy minimum since attractive interactions between the mentioned groups are still present. In a dielectric medium, the impact of this electrostatic interaction is reduced due to polarization effects and decreases with increasing dielectric constant.

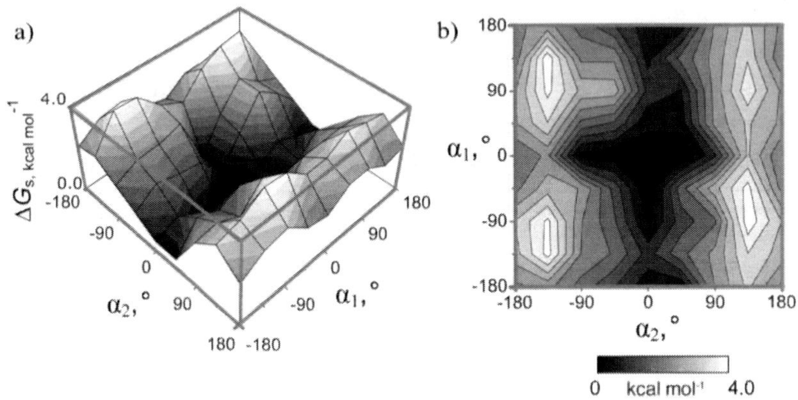

Figure 11. Change of solution-phase energy of the imidazole-heme complex at $\epsilon = 2$ as a function of the mutual orientation of the axially ligated imidazoles characterized by the pseudo-torsion angles α_1 and α_2 as calculated with the one-step procedure.

Table 1. Solution-phase energies Gs obtained with the one-step procedure for the imidazole-heme model complex as a function of the dielectric constant for two different conformations (α_1, α_2) = (0°, 0°) and (α_1, α_2) = (180°, 180°)

ϵ	**Relative Gs (kcal mol^{-1})**		
	(0°, 0°)	(180°,180°)	ΔGs (180°– 0°)
1	–3.84	0	3.84
2	–90.12	–88.15	1.96
4	–130.21	–127.58	0.97
10	–165.28	–165.16	0.12
80	–187.74	–187.99	–0.25

The G_s values are given relative to the (α_1, α_2) = (180°, 180°) conformation at $\epsilon = 1$ using the one-step procedure. In conformation (α_1, α_2) = (0°, 0°), the NH group of the axially ligated imidazole points toward the propionic groups, while at (α_1, α_2) = (180°, 180°), it points away from the propionates.

2. When the dielectric continuum model is applied, the second type of electrostatic interaction appears – the reaction field generated by the dielectric medium reacting with the total dipole moment of the imidazole-heme

complex. Furthermore, the increasing dielectric constant causes a lowering of the conformational energy and the increasing dipole moment of the imidazole-heme complex. For comparison purposes, if the orientation of both imidazoles changes from $\alpha = 0°$ to $\alpha = 180°$ in the one-step procedure, the total dipole moment increases from 71.1 to 78.9 D at $\epsilon_w = 80$ (viz. by 7.8 D) and from 5.4 to 4.1 D at $\epsilon_p = 10$ (viz. by 7.4 D).

It can be concluded that this effect is more pronounced for the imidazole conformer with $\alpha = 180°$ than with $\alpha = 0°$ and lowers the conformational energy for $\alpha = 180°$ in comparison to $\alpha = 0°$. In a simulated water environment $\epsilon_w = 80$, domination of the second factor is a result of the efficient screening of electrostatic interaction between imidazole CH and the NH group and stabilization of the imidazole-heme conformer with $\alpha = 180°$. By changing the dielectric constant from $\epsilon_w = 80$ to $\epsilon_p = 10$, the imidazole-heme orientation is determined by the competition of the two factors but slightly shifted toward the imidazole CH and the NH group. In Table 1 are listed the solution-phase energies for two conformations $(\alpha_1, \alpha_2) = (0°, 0°)$ and $(\alpha_1, \alpha_2) = (180°, 180°)$ at different dielectric constants.

It should be noted that applying the one-step procedure results in an overestimation of the interaction of the imidazole-heme dipole moment and the reaction, and the reason is the appearance of electron leakage effects.

Imidazole-Heme Conformation in A Dielectric Continuum: The Two-Step Procedure

In contrast to the one-step approach, the resulting conformational energy landscape of the imidazole-heme complex in the two-step procedure is similar to that calculated in a vacuum (Figure 12). The global energy minimum corresponds to conformer with the pseudo-torsion angle $\alpha = 0°$. Furthermore, the two-step procedure calculated energy barriers are smaller than those obtained under vacuum conditions.

As expected, the reason is strengthening the interaction of the dipole of the imidazole-heme complex with the reaction field and the simultaneous weakening of the intramolecular electrostatic interaction between imidazole and propionates. In the two-step procedure, the difference in the imidazole-heme dipole moment of the two conformers is only 4.0 D (58.9 D for $(\alpha_1, \alpha_2) = (0°, 0°)$ and 62.9 D for $(\alpha_1, \alpha_2) = (180°, 180°)$.

Also, the difference in solution-phase energies of those two conformers ΔG_s (180°–0°) is 2.35 kcal mol^{-1} in protein and 2.40 kcal mol^{-1} in the water environment. By changing the value of the dielectric constant from a vacuum

($\epsilon = 1$) to protein ($\epsilon_p = 10$), the solution-phase energy ΔG_s (180°–0°) decreases by 1.49 kcal mol^{-1}. By changing the environment from the protein ($\epsilon_p = 10$) to the water ($\epsilon_w = 80$), the solution-phase energy values ΔG_s(180°–0°) remain practically the same.

As Figure 13 shows, the difference in solution-phase energies ΔG_s(180°–0°) decreases monotonically with increasing dielectric constant. The explanation can be found by applying different solvent probe radii r_{solv} of 1.4 and 3.0 Å for water and protein, respectively. Therefore, as a solvent probe radius r_{solv} increases, the value of ΔG_s (180°–0°) decreases. It can be explained by lower solvent accessibility of imidazole and propionic group charges with increasing r_{solv} value (Figure 14).

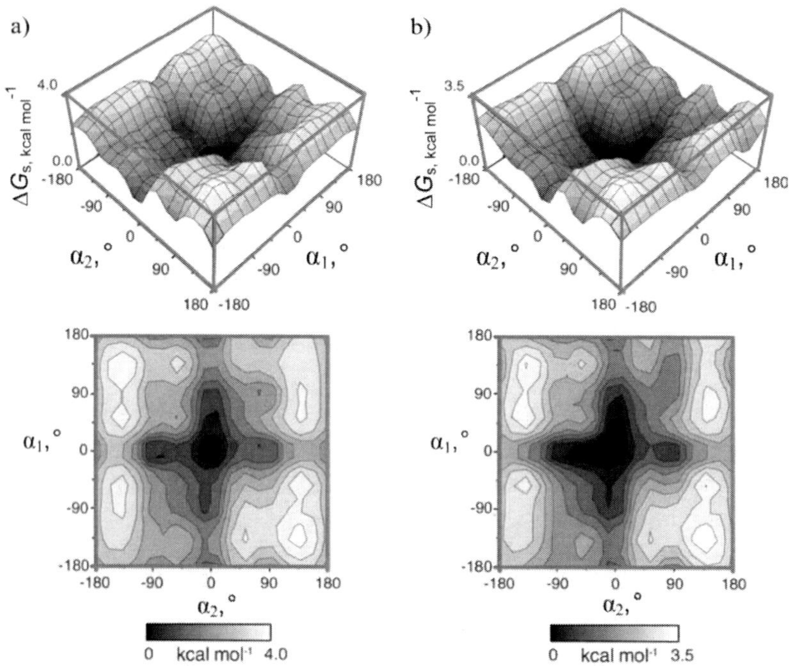

Figure 12. Solution-phase energies G_s of the imidazole-heme complex calculated by the two-step procedure. The G_s energy landscape of the mutual orientation of the axially ligated imidazoles is described by the pseudo-torsion angles α_1 and α_2 with an angular resolution of 45°. a) $\epsilon_p = 10$ and b) $\epsilon_w = 80$. Note the change in the gray shading scale between the left and the right part. The quantum-chemical and electrostatic calculations at different imidazole-heme orientations were performed as described in the METHOD section. Surfaces were generated on the 18×18 grid with additional interpolation and smoothing. The true angular resolution based on the computation is 45°.

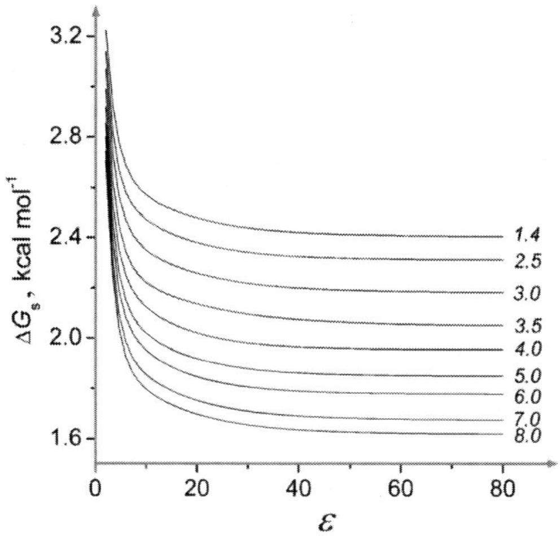

Figure 13. Difference in solution-phase energy $\Delta G_s(180°–0°)$ for the conformations $(\alpha_1, \alpha_2) = (0°, 0°)$ and $(\alpha_1, \alpha_2) = (180°, 180°)$ as a function of the dielectric constant for different solvent probe radii (in angstroms).

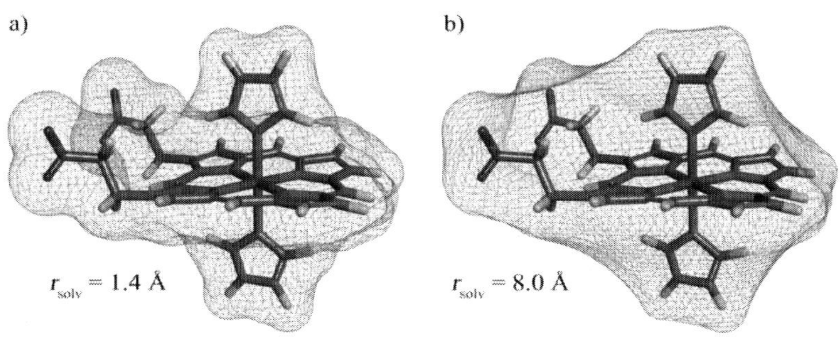

Figure 14. Molecular surface (longitudinal cut) of conformation $(\alpha_1, \alpha_2) = (0°, 0°)$ depending on the value of the solvent probe radius. a) $r_{solv} = 1.4$ Å. b) To demonstrate the effect of solvent probe radius more clearly a very large radius $r_{solv} = 8.0$ Å was used.

Depending on the value and the position of those charges, solvent probe radius r_{solv} has different influences on the solvation energy of both conformers. To better demonstrate this effect, the solvation energy of a single imidazole was monitored in the different dielectric volumes (Figure 15). The decreasing tendency of $\Delta\Delta G_{solv}(180°–0°) = \Delta G_{solv}(180°) - \Delta G_{solv}(0°)$ with an increase in r_{solv} is evident. A positively charged hydrogen atom of the imidazole NH group (with the charge of 0.34 at (0°, 0°) and 0.32 at (180°, 180°)) has a significant contribution to this difference in solvation energy.

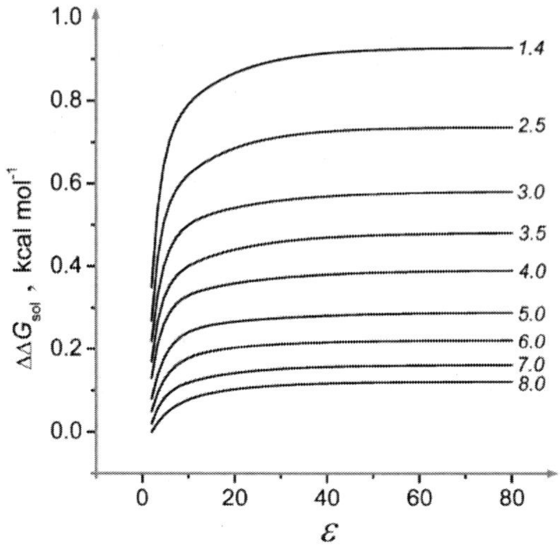

Figure 15. The difference in solvation free energy $\Delta\Delta G_{solv}$(180°–0°) of the imidazole-heme complex for the conformations (α_1, α_2) = (0°, 0°) and (α_1, α_2) = (180°, 180°) as a function of the dielectric constant for different solvent probe radii (in angstroms). In this computation, only the charges of a single imidazole were considered. All other atoms were used to define the dielectric volume of the imidazole-heme complex, while their charges were set to zero.

In conformation (0°, 0°), the hydrogen position between imidazole and the propionic group allows the largest hydrogen influence in changing the solute volume and shape with increasing solvent probe radius. In Figure 14b, the solvent radius was set to r_{solv} = 8.0 Å to show this effect clearly. Hence, there is a direct correlation between the difference in solvation energy $\Delta\Delta G_{solv}$(180°–0°) and the solution phase energy difference ΔG_s(180°–0°), which qualitatively shows the same behavior.

It was frequently observed that the negatively charged acidic groups of propionates form salt bridges with accessible positively charged basic residues in heme proteins. Also, if the acidic groups of the propionates are buried in the protein, they can be in protonated form. Therefore, it is essential to estimate how these changes affect the rotational barrier of imidazole axially ligated to heme. For that purpose, a model system was set up with two counter-charges at distances of 2.0 Å from the acidic oxygens of the respective propionate. For a case when propionates form the salt bridges, two counter-charges at distances of 2.0 Å from the acidic oxygens of the respective propionate were generated.

Table 2. The difference in solution-phase energy of imidazole-heme orientations obtained with the two-step procedure in the presence of two counter-charges (positive unit charge)

ϵ	**ΔG_s (180°–0°) (kcal mol^{-1})[a]**		
	2.0 Å[b]	0.98 Å[c]	None[d]
10	5.94–5.10	5.38	2.35
80	6.42–5.42	5.66	2.40

[a] Difference in solution-phase energy of imidazole-heme orientations located at a distance of 2 or 0.98 Å (first two columns) from the acidic oxygens of each propionate in protein (ϵ_p = 10) and water (ϵ_w = 80) environments. The two orientations are (α_1, α_2) = (0°, 0°) and (α_1, α_2) = (180°, 180°). The third column provides the corresponding energy values without counter-charges. The positive charge at a larger distance mimics the influence of a basic group forming a salt bridge with the propionate; the smaller distance describes a protonated propionic group; [b] Range of energies depending on the exact locations of the unit positive charge; [c] Exact location of the positive charge; [d] Solution-phase energy difference in the absence of a counter charge.

Similarly, two counter-charges at a distance of 0.98 Å from the acidic oxygens of the propionates were generated for the possibility of protonated heme propionates. Interasantly, the presence of these counter-charges causes a significant increase in the energy difference between the two imidazole orientations. This difference is more than 3 kcal mol^{-1} for both geometries corresponding to salt bridges or protonation and is a consequence of the interaction of the total dipole moment of the imidazole-heme complex with the continuum dielectric (Table 2).

In a model without the positive counter-charges, a dipole moment of the negatively charged propionates is directed toward the heme atom skeleton. If the imidazole NH group points away from the propionates, the resulting dipole moment is directed as the dipole moment of the negatively charged propionates –adding those two dipole moments. On the other hand, if the imidazole NH group points towards the propionates, the resulting dipole moment is not directed as the dipole moment of the negatively charged propionates – the subtracting of those two dipole moments. As denoted before, if the imidazole ring orientation leads to the larger dipole moment of the imidazole-heme complex, then the stabilization of this complex is more significant due to strong interaction with the reaction field. This effect reduces the attractive interaction of the imidazole NH group and the acidic oxygens of the propionates. The addition of a positive counter-charge at the propionates inverts the direction of their dipole moments. If the imidazole NH group points toward the propionates, the total dipole moment of the imidazole-heme complex is larger (due to the addition of two partial dipole moments).

Consequently, the interaction between the total dipole moment of the imidazole-heme complex and the reaction field stabilizes this conformer more. Simultaneously the interaction of the imidazole NH group with the propionates diminishes. The change in the interaction of the imidazole NH and the propionates is smaller than the change in the total dipole moment of the imidazole-heme complex and the reaction field. Therefore, the net effect of the adding counter-charges is a stabilization of the imidazole-heme conformer with the imidazole NH group points toward the propionates.

Electron Leakage

Although two types of computational approaches were used to calculate the electrostatic energy of the model systems in dielectric media, only the two-step procedure proved proactive. The one-step procedure delivers conflictive results to data of heme proteins crystal structures. This result can be explained by the so-called electron leakage effect. This effect can be observed for quantum-chemical calculations of a molecular compound in a continuum dielectric medium. More precisely, if an anionic solute is embedded in a high dielectric medium, the electronic wave function may spatially expand beyond the solute cavity. In a real system, explicit solvent molecules surround the solute, and electronic wave functions also try to expand into space. Electrons from explicit solvent molecules push the electrons of the solute back by the Pauli Exclusion Principle.

Consequently, the additional polarized solute magnifies the electrostatic interactions between the solute and the dielectric continuum and artificially lowers the solute energy. This effect is usually small for small, neutral, or cationic molecules, and significant discrepancies have been observed for larger aromatic and thus easy polarizable molecules, especially if they are in the anionic state. Also, this effect enhances with the increase of the dielectric constant.

As the imidazole-heme complex is a polyatomic anion, it can be assumed that the electron leakage effect influences the calculated solution-phase energies of the complex in the one-step procedure. Furthermore, the additionally induced solute polarization causes a stronger interaction of the solute with the reaction field from the dielectric medium. An analogous effect occurs with an increase of the dielectric constant of the surrounding medium.

Thus, calculated energy at the low dielectric constant of $\epsilon_p = 2$ in the one-step procedure is similar to the value obtained from the two-step procedure for a larger dielectric constant $\epsilon_p > 2$. When the dielectric constant exceeds the values of 2, the calculated energies in the one-step procedure deviate significantly from the distribution of imidazole-heme conformers present in protein crystal structures.

Influence of histidine backbone

Analyzing Data from the PDB

As previously discussed, the orientation of the imidazole ring to heme is guided by attractive interactions of the propionic group. Since the imidazole ring is part of the histidine residue (covalently bound to the protein backbone), it can be assumed that the histidine backbone also has a specific orientation relative to the heme plane. To identify a possible orientation pattern of the histidine backbone, available crystallographic data of heme proteins were analyzed by Zarić et al. (Zarić, Popović, and Knapp, 2001).

The distribution of the histidine backbone angle β (Cα–Cβ–Cγ–Nδ1) was selected as a criterion for database search (Figure 16). In the case of non-coordinated histidines, the distribution of angle β exhibited two maxima, around –90° and 90° (dashed line, scaled by a factor 6 in Figure 16). All histidines coordinated to heme, except for the Mb and Hb groups, are shown as a solid line in Figure 16.

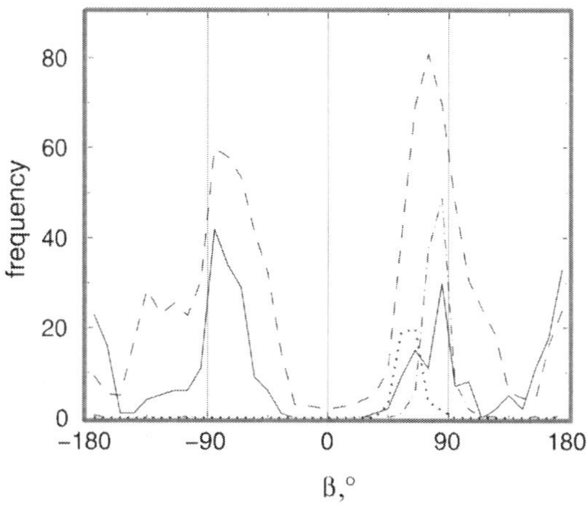

Figure 16. Overview of imidazole histidine backbone angles α derived from the PDB. Dashed line: Data from all non-coordinated histidines of the PDB scaled down by the factor 6. Solid line: Data from all heme proteins possessing hemes with axially coordinated histidines except the Mb and Hb groups. Dotted (dashed-dotted) line: Data from the Mb (Hb) group scaled down by the factor 4.

The distribution of the torsion angle β of all non-coordinated histidines exhibits two maxima, with values between +90° and –90°. In addition, the distribution of the backbone angle of coordinated histidines shows a maximum at about $\beta = \pm 180°$, which primarily arises from the CcPo group. Namely, in the CcPo proteins, the NδH group of coordinated histidine is strongly hydrogen-bonded with a negatively charged Asp.

It determines the orientation of the imidazole ring relative to the heme and enforces the angle β to assume such unusual values (see also section *Imidazole-heme conformations for different groups of hemeproteins*). The heme proteins of the Mb (Hb) group dotted (dash-dotted) line in Figure 16 contribute only to β angles close to +60° (+90°) but not to –90°. It is not surprising since the protein backbone carrying the ligated histidine always comes from the same side of the heme plane in all these structures. If the angle $\beta = -90°$, the imidazole ring will be oriented opposite to the propionates.

Force Field Computations

Force field calculations were performed for the available model system to elaborate on previous experimental findings (Zarić, Popović, and Knapp, 2001). Torsion angle β is the most relevant parameter of the different histidine backbone conformation, so it was assigned fixed values. The energy dependence of backbone conformation was monitored as a function of discrete values beta, while the rest of the molecule was relaxed. Profile of the total energy (solid line) has shown two minima for β values of –90° and 90° (which have a good depth of 5.5 kcal mol^{-1} and 2.5 kcal mol^{-1}, respectively), and two maxima for β values around 0° and ±180° (Figure 17). The calculated extremes of energy function are in good agreement with the experimental angle β distribution for the non-coordinated histidines (Figure 16). The electrostatic contribution in the total energy is dominant, while torsional energy contributes to the difference between the minima and maxima. Shallower minimum around –90° and deeper the minimum at 90° are the consequences of van der Waals's contribution (dotted line) to the total energy. Based on the energy profile, only specific values of angle β (the orientation of histidine backbone to the heme) are acceptable.

As a consequence, only discrete values of angle α (the orientation of the imidazole ring plane to the heme) are possible. Since the values of the angle α and β are interrelated, it is possible then to define angle γ by relations in Eq. 1.

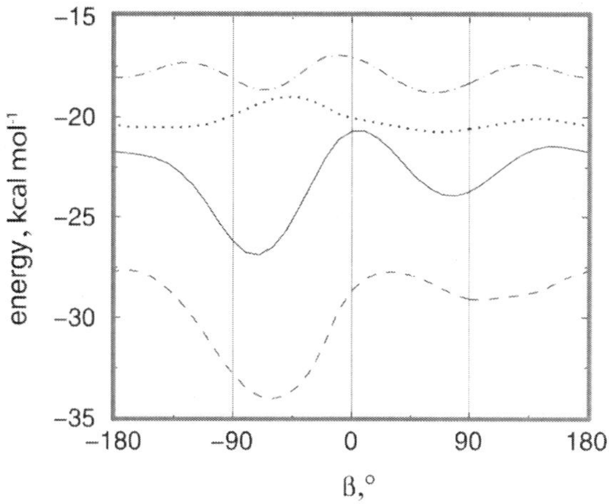

Figure 17. Calculated interaction energies between imidazole and its histidine backbone. The interaction energy is calculated with the CHARMM22 force field and displayed as a function of the torsion angle β. Solid line: total energy, dashed line: electrostatic energy, dotted line: van der Waals (Lennart-Jones) energy, and dashed-dotted line: torsion energy term.

Accordingly, the plane defined by histidine atoms Cα, Cβ, and Cγ, and the imidazole ring plane are nearly parallel if angle β has values of 0° and ±180° (the energy maxima).

Therefore it can be concluded that projections of the Cα–Cβ bond vector of the histidine and the Nδ1–H bond vector of the imidazole on the heme plane are parallel for angle β of ±180° and antiparallel for angle β of 0°.

Another important conclusion emerges from the defined relation of α, β, and γ angels. Namely, the imidazole ring orientations with $\gamma \approx \alpha$ or $\gamma \approx \alpha \pm 180°$ are forbidden.

Imidazole-Heme Conformations for Different Groups of Heme Proteins

Analyzing data from the PDB

Most of the experimentally observed histidine orientations in the heme proteins can be explained by the influence of the propionic groups and/or the histidine backbone on the imidazole ring (Zarić, Popović, and Knapp, 2001). The pseudo-torsion angle γ characterizes the orientation of the histidine backbone relative to the heme. As mentioned before, values of the angle β at about 0° and ±180° are prohibited or at least not very likely. Since the relationship in Eq. 1 conditions the values of angle γ, the imidazole orientations with $\gamma \approx \alpha$ or $\gamma \approx \alpha \pm 180°$ are also prohibited. Consequently, if the protein backbone is directed between the two heme propionates ($\gamma = 0°$), as a result of the unfavorable interactions between the imidazole ring and protein backbone, the typically favorable orientation of the imidazole pointing toward the propionates ($\alpha = 0°$) will be prevented.

On the other hand, the preferred angular regime of the torsion angle β is about +90° and –90° (Figure 16 and Figure 17). Once again, by considering the relation between the angles α, β, and γ (Eq. 1) and the orientation of the histidine protein backbone relative to the heme, it can be concluded that the interaction of the imidazole with the propionic groups determines if the angle β belongs to the allowed regime at +90 or –90. Hence, the angle β adopts the value, which allows the imidazole to orient in such a position that its NδH group points toward the heme propionates. Considering the conditions mentioned above, it was possible to explain the orientations of imidazoles in all groups of heme proteins, except C*c*Po, where a strong hydrogen bond determines the imidazole orientation.

Cytochrome c Groups

In most cytochrome *c* structures, the imidazole NδH group is located between the two propionic groups, PRA and PRD (Figure 3, dashed and dotted lines). Within both cytochrome c groups (mono- and bis-Cc), the angle α is close to 0°, the angle β is typically around 90° or –90° (see Figure 19), while the angle γ adopts the values close to +90° or –90°. For both groups, the orientation of the backbone relative to heme and the interaction of imidazole with its histidine backbone allow the imidazole ring to adopt orientations of angle α around 0° or ±180°. If the angle α is close to 0°, it can be concluded that the interactions of propionic acid groups conditioned the imidazole orientations.

These heme proteins contain the *c*-type heme, which is covalently bound with two cysteines by the pyrrole rings B and C. However, neither structural analysis of heme proteins in the PDB nor force field computations confirmed a significant influence from these cysteines on the orientation of the imidazoles axially coordinated to heme.

Myoglobin and Hemoglobin Groups

The distribution of angle α for the Mb and Hb proteins, characterizing the orientation of the ligated imidazole relative to the heme, is displayed in Figure 18a. For Mb, the α has the value near 45° ($\alpha \approx 45°$, solid line), while for Hb, it is around 65° ($\alpha \approx 65°$, dotted line). These two groups of heme proteins are sequentially and structurally similar.

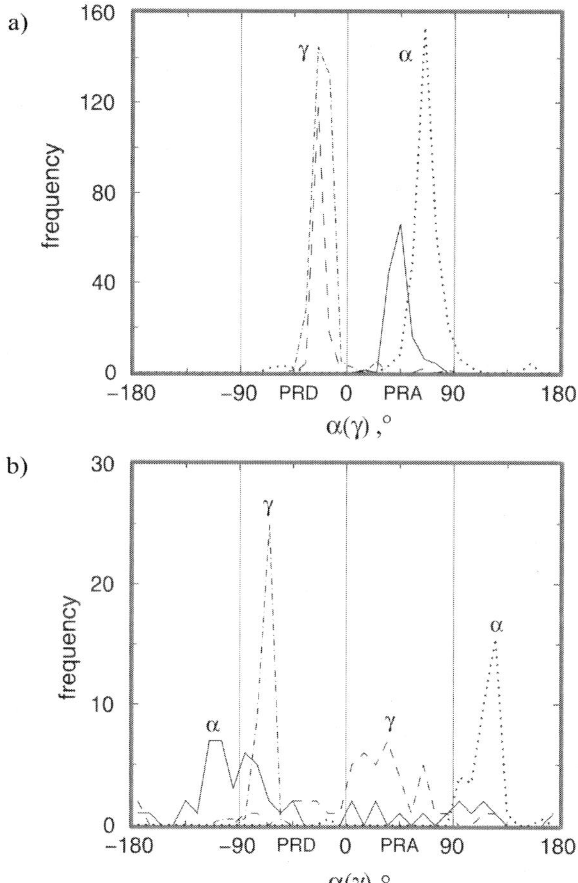

Figure 18. Distribution of the torsion angles α and γ for different heme proteins. The torsion angle α characterizes the orientation of the imidazole relative to the heme, represented by solid and dotted lines. The torsion angle γ characterizes the orientation of the histidine backbone relative to the heme depicted by dashed and dashed-dotted lines. a) Solid and dashed lines refer to the angle distributions of heme proteins from the Mb group. Dotted and dashed-dotted lines refer to the Hb group of heme proteins. b) Solid and dashed line refers to the C*b* group of heme proteins. Dotted and dashed-dotted lines refer to the C*c*Po group of heme proteins.

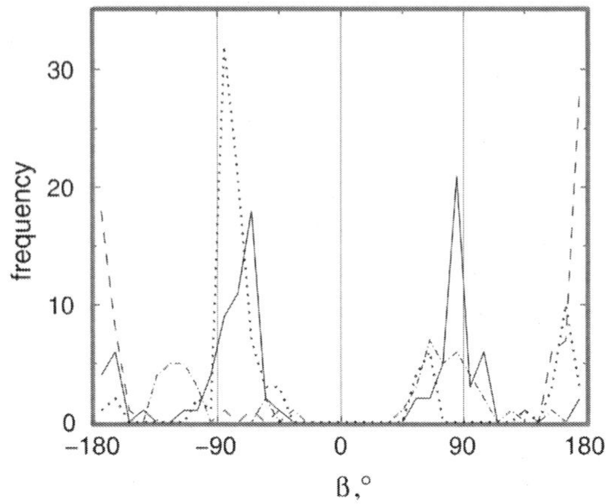

Figure 19. Distribution of the torsion angle β characterizing the orientation of the imidazole ring relative to its histidine backbone for different groups of heme proteins derived from the PDB. Solid line: bis-C*c*, dashed line: C*c*Po, dotted line: mono-C*c*, dashed-dotted line: C*b*.

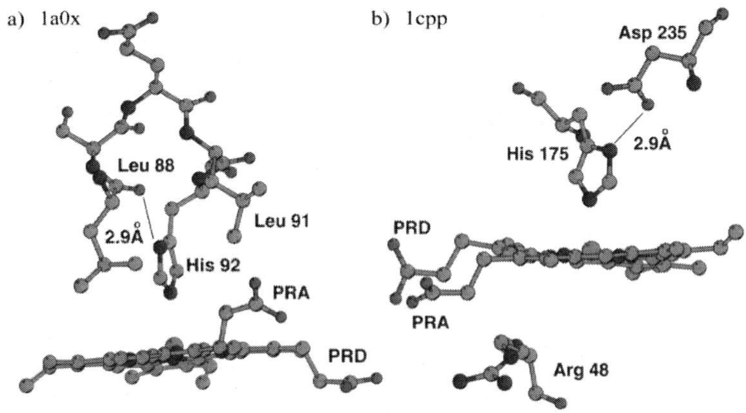

Figure 20. Examples of heme proteins, in which the NδH group of an imidazole ligated to heme is not oriented toward the center of the propionic acid groups. a) The heme neighborhood in the crystal structure of hemoglobin 1a0x, where the orientation of the His92 backbone prevents the generally preferred orientation of the imidazole of His92 toward the propionic acid groups. The hydrogen bond of the imidazole NδH group with the backbone carbonyl of Leu88 is only of secondary importance, b) The heme neighborhood in the crystal structure of cytochrome *c* peroxidase 1cpp (Poulos, Finzel, and Howard, 1987), where the imidazole of His175 is oriented differently because the NδH group is forming a strong hydrogen bond with the negatively charged aspartate Asp235.

Except for a few mutants, in all other structures, the orientation of the imidazole ring and the two propanoic groups are almost uniform. The imidazole and the propionic group PRA are located on the same side of the heme plane and directed towards each other. Since the imidazole and the other propionic group PRD are located on the opposite side of the heme plane, their interaction is weaker.

Additionally, the orientation of the imidazole ring is stabilized by a hydrogen bond between the imidazole NδH group and a backbone CO group (see section *Hydrogen bonding scheme of imidazole ligated to heme*).

A typical crystal structure of Hb protein with PDB code 1a0x (Kavanaugh et al., 1998, and 2005) and the histidine-heme conformation with $\alpha = 67°$ is displayed in Figure 20a. Also, the backbone-heme orientation, represented by the angle γ, for the Mb and Hb proteins, is shown in Figure 18a (the dashed line).

Within the Mb/Hb groups, the angle α adopts a value around –25°, so the preferred imidazole orientation with $\alpha = 0°$ is prevented due to the interaction of the imidazole ring with its histidine backbone. The distribution of the angle α has its maximum slightly shifted toward negative values (Figure 18a), and therefore the preferred orientation of the imidazole-heme angle α has the angle interval around $\alpha = +90°$ rather than at $\alpha = -90°$.

Cytochrome b Group

In the heme proteins of the cytochrome *b* (C*b*) group, the two histidines occupy positions of axial ligands of the heme. The imidazole rings of both histidines are oriented with the preferential angle values α around –90°. In such orientation, the NδH group points away from both propionic acid groups, albeit it is not too far from the propionic acid group PRD (solid line in Figure 18b). Also, this group of heme proteins has the backbone angle γ of the imidazole relative to the heme not far from $\gamma = 0°$ (dashed line in Figure 18b). Hence, the orientation of the imidazole NδH group toward the center of the propionic acid groups is not possible. The distribution of the backbone angle γ has its maximum slightly shifted toward positive values: therefore, the preferential orientation of the imidazole, concerning the heme, has a negative value of the angle α, close to –90°.

Cytochrome c Peroxidase Group

For the C*c*Po group of the heme proteins, a very high degree of sequence identity is observed. Since the backbone Cα–Cβ bond vector of ligated histidine is oriented at angle γ around –90° relative to the heme (Figure 18b, dashed-dotted line), it does not prevent a favorable orientation of the imidazole ring toward the propionic groups. However, the distribution of the torsion angle α adopts its maximum at +120° (Figure 18b, dotted line). In such orientation, the interaction between the imidazole ring and the propionic groups is no longer strong. Since the backbone angle β adopts almost exclusively values close to ±180° (Figure 19, dashed line), the interaction with its histidine backbone is also unfavorable. The distributions of the backbone angle β for different heme protein groups are compared in Figure 19. Hence, the imidazole ring orientation within the *Cc*Po group cannot be explained by interactions with the propionic groups or with the histidine backbone. It can be concluded that the other factor determines the imidazole ring orientation, so a detailed analysis is necessary for the heme surrounding.

An aspartate (Asp235) of the *Cc*Po heme proteins group, located at the heme edge opposite the two propionic groups, forms a hydrogen bond with the NδH group of the imidazole ligated to heme (Figure 20b). As aspartate is negatively charged, this hydrogen bond is strong enough to enforce this orientation of the imidazole. Moreover, in *Cc*Po proteins, both propionic groups are located on the opposite side of the heme plane relative to the ligated histidine, and this propionates conformation weakens their influence on the imidazole orientation. A typical crystal structure of the *Cc*Po group with PDB code 1ccp (Wang et al., 1990) with α = 124° is shown in Figure 20b. A salt bridge formed between the acidic group of PRA and a positively charged arginine (Arg48) partially neutralizes the negative charge of that propionic acid group. As a result, the influence of the propionic acid group on the orientation of the ligated imidazole is decreasing furthermore. Based on these considerations, instead of a propionic group, the specific very strong hydrogen bond controls the imidazole ring orientation of the *Cc*Po protein group.

Mutual Orientation of Two Axially Coordinated Histidines

Analyzing Data from the PDB

Finally, the mutual orientation of the two axial histidines in the groups of the bis-histidine ligated heme proteins were analyzed from collected PDB data (Zarić, Popović and Knapp, 2001). It was accomplished by measuring the difference between the angle α (defined as $\Delta\alpha = |\alpha_2 - \alpha_2|$, see METHOD), which the projection of the NδH groups of two coordinated histidines on the heme plane form. The results are displayed in Figure 21.

Two different groups of heme proteins (C*b* and bis-C*c*) axially coordinated (as ligands) two histidines. As Figure 21 shows, within the bis-C*c* group (solid line), most of the imidazole planes ligated to heme are parallel to each other. That is not unusual since the preferred orientation of the imidazoles for the whole group of cytochrome *c* is to point toward the center of two propionic acid groups. For the group of cytochrome *b* heme proteins, the mutual orientation of the imidazole planes of the axially coordinated histidines does not show such a clear trend, although smaller values of the angle $\Delta\alpha$ are preferred (dashed line in Figure 21). By analyzing PDB data, it was not found that any of these proteins have $\Delta\alpha \approx 180°$, where the imidazole planes are parallel but with oppositely oriented N–H groups. That is highly unlikely since at least one histidine is oriented toward propionates, so the second one should be oriented oppositely.

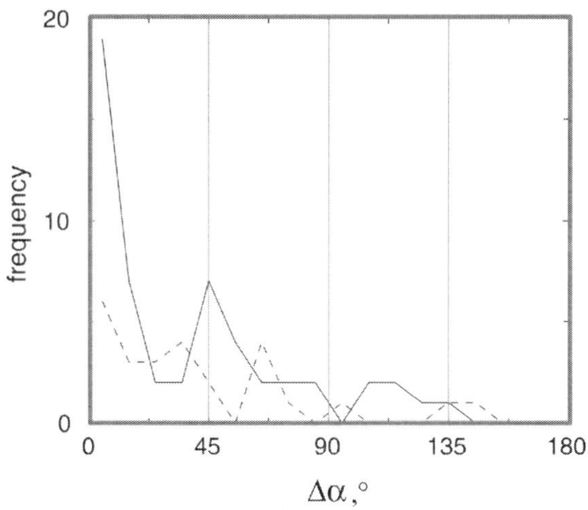

Figure 21. The relative orientation of the imidazole planes of histidines axially coordinated to heme, as measured by the torsion angle difference $\Delta\alpha = |\alpha_2 - \alpha_2|$. The distributions were derived from heme proteins of the PDB, where the two histidines axially coordinate the heme. Solid line: From the cytochrome *c* group (bis-C*c*). Dashed line: From the cytochrome *b* group (C*b*).

A few years later, statistical analyses of the Cambridge Structural Database (CSD) have indicated that parallel or close to parallel the orientations of imidazoles are favorable in bis-His complexes (Rakić, Medaković, and Zarić, 2006). However, according to the analysis of PDB data, ligands in bis-His heme proteins were found more frequently perpendicular rather than parallel to each other (Fufezan, Zhang, and Gunner, 2008). Another analysis of PDB structures in bis-His complexes has revealed that the two histidines occupied almost all the relative orientations, with only a slight preference for the parallel one (Merlino et al., 2009).

Hydrogen Bonding Scheme of Imidazole Ligated to Heme

Analyzing Data from the PDB

Since the nitrogen atom NδH of histidine coordinated to heme is always protonated, it acts as a donor group in a hydrogen bond. For more than 98% of all histidines coordinated to heme, suitable hydrogen bonding partners were found at less than 3.5 Å away from the NδH nitrogen. For the remaining 2%, the distance between a suitable hydrogen bonding partner atom and the NδH nitrogen is more than 3.5 Å. In 7.5% of all cases, an oxygen atom of the water participates as the hydrogen bonding partner. For the imidazole NδH group, the CO group of the protein backbone is the most probable hydrogen bonding partner. Thereby, the residue type and number forming the hydrogen bond vary, depending on the considered protein structure. This section will discuss the results obtained from the PDB data mining.

H-Bonds of Axially Coordinated Imidazoles in Myoglobin and Hemoglobin

Since Mb and Hb proteins show a very high degree of sequential and structural similarity, the hydrogen bond pattern involving coordinated histidines is very similar within the whole group. The fourth residue in the sequence below the coordinated histidine (its backbone CO group) forms a hydrogen bond with the imidazole NδH group. Also, the propionic group PRA points toward the imidazole ring, while the PRD propionic group is located on the opposite side of the heme plane in both heme protein families.

Mb

The hydrogen bond is formed simultaneously between the Hδ atom of coordinated His93, and Leu89 and Ser92, for most proteins of the Mb group. The part of the sequence Leu89-Ala90-X91-Ser92-His93 with X = Gln, Glu, or Asn involved in hydrogen bonding is majorly conserved. The bifurcated hydrogen bond of the imidazole is often asymmetric and has a fine-tuning effect on the imidazole ring orientation over a larger angular regime, from

α = +20° to +78°. However, the value of the pseudo-torsion angle α is close to 45° for most Mb crystal structures. This value of angle α corresponds to the orientation with the imidazole NδH group and the propionic group PRA point toward each other.

Smerdon et al. (1993) studied the effects of the mutation of the conserved residue Ser92 to Ala, Val, and Leu in Mb of a pig. This study has shown that the binding affinity for O_2, CO, and CN ligands increased with the mutations. Although the H-bond network in the heme-binding pocket involving the proximal His97 changes and the Nδ1–H group of the ligated His93 loses one H-bonding partner, the His93 imidazole ring orientation does not change significantly. Also, a mutation in the Mb structure with PDB code 1rse (Lloyd et al., 1996) results in the replacement of Ser92 with Asp. Again, the imidazole ring-maintained orientation with the torsion angle α = 50° remains close to the most probable value for the native Mb. Replacement of Ser92 by Asp92 causes movement of the negatively charged side chain of Asp92 away from the imidazole ring of His93. As a result, electrostatic repulsive interactions appear between the imidazole ring of His93 and the propionic group PRA, pointed toward the imidazole ring. Thus, Asp92 does not form an H-bond with His93 in contrast to Ser92. In a few crystal structures of Mb from sea hare, the ninety-first residue is phenylalanine. Here the only hydrogen bond is formed between the carbonyl backbone oxygen and His95, axially ligated to heme. The imidazole ring orientation corresponds to the value of angle α between 64° and 78°, slightly larger than the most probable value of 45° for Mb. This change in the orientation is probably the result of a repulsive interaction of the negatively charged acidic group of the glutamate Glu94 and the propionic acid group PRA. Consequently, the propionic acid group moves away from the imidazole ring, so its influence on the orientation of the imidazole decreases.

Hb

Similarly, in the Hb family of heme proteins, the NδH group of His87 (or His92) ligated to heme forms a hydrogen bond with the backbone CO group at sequence position 83 (or 88), the fourth residue away from His87 (or His92). Although Leu is predominantly the residue type involved in hydrogen bonding, Phe, Val, and Lys residues are also present. In contrast to the proteins of the Mb family, in most Hb structures, the corresponding serine is missing, so the formatting of the bifurcated hydrogen bond is not possible. The value of pseudo torsion angle α is around 65°.

Some Hb structures, for example, with PDB code 1eca, 1ecd, 1ecn, 1eco (Steigemann and Weber, 1979), have the binding pocket with the heme turned around the CHA–CHC axis (see Figure 1) by 180°. Consequently, the conformation corresponds to the angle α of –50° with the propionic group PRD oriented toward the axially ligated imidazole ring. In these Hb structures, although serine is available at position 86th and phenylalanine at position 83rd, the ligated His87 makes a stronger hydrogen bond with the Oγ oxygen of Ser86 than with the backbone oxygen of Phe83. Among the Hb structures with bifurcated H-bonds, these few structures are the only exceptions. Also, in the Hb structure with PDB code 1ith (Kolatkar et al., 1992), the ligated His94 is a neighbor of Ser93 residue. However, in this structure, only the backbone CO group of Lys90 is involved in a hydrogen bond with His94. In some structures, threonine replaced Ser93 at the same position. Nevertheless, there is no evidence that it forms an H-bond with His.

In a few Hb structures, other residues form a hydrogen bond with the NδH group of the ligated histidine. For example, in the Hb structures with PDB codes 1vhb and 2vhb (Tarricone et al., 1997), instead of the backbone CO group of Ile81, the acidic oxygen of Glu137 forms a strong hydrogen bond with His85. As a consequence, the torsion angle α is unusually large, with a value of 150°. In a mutant Hb structure from sea cucumber with PDB code 1hlb (Mitchell, Kitto, and Hackert, 1995), there are two histidines (His104 and His73) axially coordinated to heme, so the Oη oxygen of Tyr114 forms a hydrogen bond with His104 (α = 125°) and water oxygen forms a hydrogen bond with His73 (α = –163°). In the corresponding wild type structure with PDB code 1hlm (Mauri et al., 1991), His104 is involved in the hydrogen bond with Leu100, and the value of the pseudo-torsion angle is α = 40°.

H-Bonds of Axially Ligated Imidazoles in Cytochrome c Peroxidase

This group also represents a very compact heme protein family with high similarity in their structures and sequences. In the native C*c*Po proteins, the NδH group of the ligated imidazole and the acidic oxygen of an aspartate form a strong hydrogen bond. There are two different subsets of C*c*Po heme proteins. One subset contains hydrogen-bonded Asp235 to the imidazole ring of the coordinated His175, and in one subset, Asp246 forms a hydrogen bond with the coordinated His184. The corresponding values of the pseudo-torsion angle α are in the interval from 115° to 125° and from 100° to 125°, respectively. In the C*c*Po structure with PDB code 1ccc (Goodin and McRee, 1993), Asp235 is mutated to alanine. Since the negatively charged Asp235 is absent, forming the corresponding hydrogen bond is no longer possible. Instead of that, His175 forms the hydrogen bond with a water molecule.

Consequently, the angle *α* becomes smaller (*α* = 100°). The resulting imidazole orientation is toward the propionate PRA, but the effect is insignificant since the formation of a salt bridge with Arg48 causes the reduction of the propionate charges. Also, in the *Cc*Po mutant structures, the presence of Asp235Asn and Asp235Glu (*α* = 122° and *α* = 97°, respectively) cause only minor changes in the orientation of the imidazole ring relative to heme.

H-Bonds of Axially Ligated Imidazoles in Mono-Histidine Ligated Cytochrome c

The main characteristic of *c*-type hemes is two covalently bound cysteines. The histidine and the two cysteines attached to the heme and the part of the sequence Cys-X-X-Cys-His are highly conserved within this group. Concerning the structure, and sequence homology, a group of 68 mono-histidine heme proteins with ligated cytochrome *c* (mono-C*c*), can be divided into a few subgroups. The largest subgroup (mono-C*c*124) contains cytochrome *c* isozyme 1 and 2, cytochrome *c*2 (formerly *c*550), cytochrome *c*4, cytochrome *c*551I, and cytochrome *c*1 from the cytochrome *bc*1 complex. The second subgroup (mono-C*c*6) consists of cytochrome *c*6 (also called *c*553) heme proteins. The members in the first two subgroups are globular proteins with large *α*-helical content. The cytochrome *c*' (mono-C*c*') heme proteins constitute the third subgroup. Their structure is a four-helix bundle with one mono-histidine coordinated heme.

Mono-Cc124

All of the members of this subgroup exhibit significant similarities in the tertiary structure. The carbonyl backbone oxygen of a proline, typically located as the 12th or 18th residue from the coordinated histidine, forms a hydrogen bond with the histidine. However, in some cases, the proline is the 70th (Cyt *c*1 from the Cyt *bc*1 complex) or the 9th residue (cytochrome *c*551I) from the coordinating histidine. A loop structure is formed by the residues of the polypeptide segment between the histidine and proline. The pseudo torsion angle *α* characterizing the imidazole heme orientation has values between –25° and +20°.

Mono-Cc6

These heme proteins exhibit a similar globular *α*-helical structure to the previous subgroup, but the sequence similarity with the mono-C*c*124 cytochromes is low. The coordinated histidine forms the hydrogen bond with the backbone CO group of glycine, arginine, or asparagine in this subgroup. The corresponding pseudo torsion angle α has values between –25° and –10°. These structures possess a proline located relatively close to histidine in amino acid sequence but not close enough to form a hydrogen bond with the ligated histidine.

Mono-Cc'

The coordinated histidine of these proteins is accessible to a solvent, so the NδH group of the imidazole ring forms an H-bond with a water molecule. Orientation of the imidazole ring, with the NδH group pointing toward the propionic acid groups, is prevented because of the propionic acid charge reduction caused by three arginines. The usually preferred imidazole ring orientation is also sterically hindered by two other residues. Although angle α values cover the interval from 80° to 120°, most of it is close to 100°.

The group of mono-histidine ligated cytochrome also includes cytochrome *f*, the globular protein with a predominantly β-strand structure. In this heme protein, although the NδH group of ligated histidine forms a hydrogen bond with a water molecule, it is oriented toward the propionic acid groups with $\alpha = 10°$.

H-Bonds of Axially Ligated Imidazoles in Bis-Histidine Ligated Cytochromes

This type of heme protein includes the two groups: cytochrome *b* (C*b*) and bis-histidine ligated cytochrome *c* (bis-C*c*). These groups are very heterogeneous and contain heme proteins with various three-dimensional structures. Consequently, the hydrogen bond patterns can be very different. Nevertheless, the coordinated histidine and the protein backbone oxygen form the H-bond in most cases.

Yet, for some coordinated histidines, no suitable hydrogen bonding partner was found. Since mutants are not available, the conclusion of whether the hydrogen-bonding scheme influences the orientation of the imidazole ring coordinated to the heme was not possible. However, the bis-C*c* group has most of the coordinated histidines oriented toward the propionic acid groups.

Influence of H-Bonds on Imidazole Orientation in Heme Proteins

In myoglobin, the NH group of imidazole participates in the bifurcated hydrogen bond, causing some variations in the orientation of imidazoles axially ligated to heme. However, the orientation of ligated imidazole is more or less pointed toward the propionic group PRA. A significant deviation from this orientation is a consequence of the ligated imidazole forming a strong hydrogen bond with a negatively charged acidic amino acid. It can be a reason for unusually high values of the torsion angle in the C*c*Po family and a few other exceptions.

In C*c*Po heme proteins, the imidazole NH group generally forms a strong hydrogen bond with a negatively charged aspartate. In mutants, where alanine replaces aspartate, the imidazole ring does not change its orientation significantly. In this group of proteins, with very few exceptions, the imidazole orientation relative to the heme exhibits only minor changes, while the hydrogen-bonding pattern can be quite different.

Possible H-Bonds of Imidazoles Axially Ligated to Heme in Heme Proteins

As is explained in the METHOD section, the imidazole ring of axially ligated histidines was rotated around the Nδ–Fe bond and monitored the distance of the Nδ atom to possible hydrogen bond acceptors as a function of the pseudo-torsion angle α. This search established the possibility of the NδH group forming a hydrogen bond that differs from the one found in the native structure.

The study included a large number of different proteins, and two typical examples for each of the six groups of heme proteins are displayed in Table 3. In most of the considered cases, there are several possibilities for making the H-bonds. The criterion for distance in the detection of hydrogen bonds was rather conservative. If the structural relaxation is also considered, the number of possible H-bonds can become quite large.

In Figure 22, the possible hydrogen bonds of imidazole, axially coordinated to heme in heme proteins, are shown. The hydrogen bonds of axially coordinated imidazole are studied by rotating the imidazole ring around its Nδ–Fe bond and monitoring the distance between the Nδ1 atom and possible hydrogen bond acceptor atoms. In the figure, the inverse of this distance, which is a rough measure of the hydrogen bonding strength, is displayed, as a function of the pseudo-torsion angle α, as defined in Figure 1. Only hydrogen bonding partners whose minimum distance to the Nδ atom is smaller than 4.0 Å are considered.

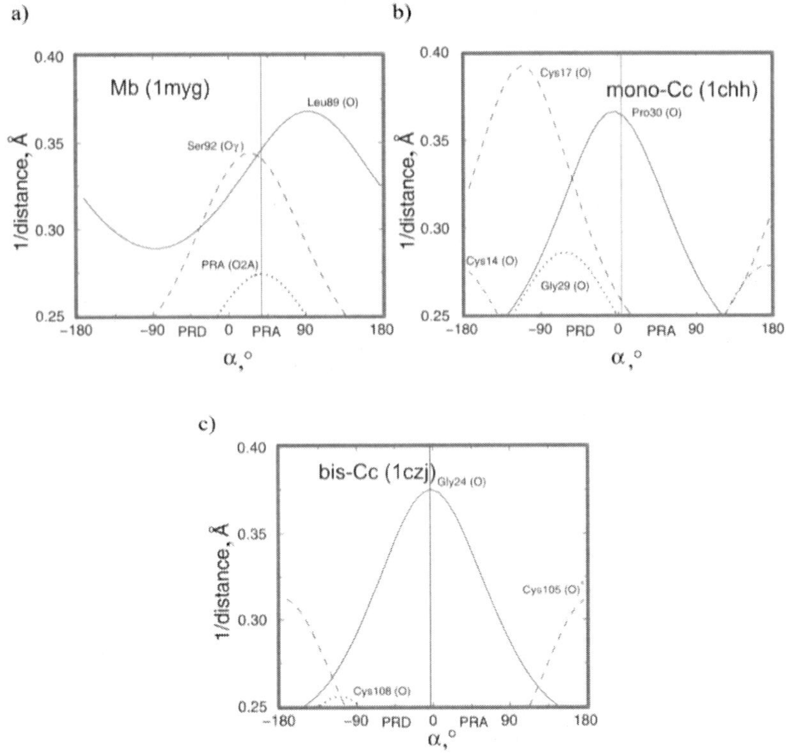

Figure 22. Possible hydrogen bonds of imidazole axially coordinated to heme in heme proteins. The vertical solid line marks the value of the pseudo-torsion angle α of the crystal structure. a) heme protein Mb with PDB code 1myg, chain B (Oldfield et al., 1992). Possible hydrogen bonds are formed with the backbone oxygen of Leu89, with the side-chain oxygen of Ser92, and with the acidic group of propionic acid PRA. For the optimal orientations, the corresponding values of α are 97°, 27°, and 37°, respectively. The orientation assumed in the crystal structure is at $\alpha = 37°$. b) heme protein mono-C*c* with PDB code 1chh. Possible hydrogen bonds are formed with the backbone oxygens of Pro30, Cys17, Gly29, and Cys24. For optimal orientations, the corresponding values of α are –3°, –113°, –63°, and 167°, respectively. The orientation assumed in the crystal structure is at $\alpha = 7°$. c) heme protein bis-C*c* with PDB code 1czj (Czjzek et al., 1996). Possible hydrogen bonds are formed between the axially ligated His109 and the backbone oxygens of Gly24, Cys105, and Cys108. For the optimal orientations, the corresponding values of α are –3°, –173°, and –113°, respectively. The orientation assumed in the crystal structure is at $\alpha = -3°$.

Table 3. Possible H-bonds of imidazole axially ligated to heme

No	heme protein/resolution/ PDB code/chain [a]	residue(atom type)/ ligated histidine [b]	Distance [Å] [c]	angle α [d] [degree]	angle γ [e] [degree]
1	Mb/1.75 Å/1myg /B	Leu89(O)/His93	2.72 (2.89)	97 (37)	–30
		Ser92(Oγ)/His93	2.91 (2.93)	27	
		heme-PRA(O2A)	3.64	37	
2	Mb/1.90 Å/5mba	Phe91(O)/His95	3.07 (3.08)	79 (69)	–10
3	Hb/2.00 Å/1a0x	Leu83(O)/His87	2.71 (2.71)	61 (61)	–25
4	Hb/1.40 Å /1ecn	Ser86(Oγ)/His87	2.80 (2.84)	–30(–50)	24
		Phe83(O)/His87	3.03 (3.19)	–100 (–50)	
		Ser86(O)	3.41	40	
		H_2O63	3.79	–40	
5	*Cc*Po/2.20 Å/1ccp	Asp235(Oδ1)/His175	2.90 (2.90)	124 (124)	–72
		Asp235(Oδ2)	3.58	84	
		Met172(O)	3.20	–156	
		Ala174(O)	3.90	–86	
6	*Cc*Po/1.60 Å/1arv	Asp246(Oδ2)/His184	2.87 (2.89)	85 (95)	–72
		Asp246(Oδ1)/His184	3.20 (3.25)	120 (95)	
7	mono-C*c*/1.97 Å/1chh	Pro30(O)/His18	2.73 (2.75)	–3 (7)	–93
		Cys17(O)	2.55	–113	
		Gly29(O)	3.49	–63	
		Cys14(O)	3.59	167	
8	mono-C*c*/1.80 Å/1cgo	H_2O226/His120	2.91 (2.92)	112 (102)	–92
		Cys116(O)	2.80	–158	
		H_2O270	3.63	62	

Table 3. (Continued)

No	heme protein/resolution/ PDB code/chain [a]	residue(atom type)/ ligated histidine [b]	Distance [Å] [c]	angle α [d] [degree]	angle γ [e] [degree]
9	bis-C*c*/2.16 Å/1czj	Gly24(O)/His109	2.66 (2.66)	–3 (–3)	–96
		Cys105(O)	3.21	–173	
		Cys108(O)	3.91	–113	
10	bis-C*c*/2.16 Å /1czj	Tyr73(O)/His77	2.66 (2.68)	–1 (15)	–86
		Tyr73(Oη)	3.77	105	
		Phe76(O)	2.99	–115	
11	C*b*/1.50 Å /1cyo	Gly42(O)/His39	2.75 (2.75)	–96 (–96)	40
		Gly41(O)	3.78	–46	
12	C*b*/1.50 Å /1cyo	Phe58(O)/His63	2.71 (2.71)	–114(–114)	11
		Val61(O)	3.73	–34	
		heme-PRA(O2A)	3.93	36	
		H_2O509	3.31	–4	

[a] Heme proteins belonging to Mb, Hb, C*c*Po, mono-C*c*, bis-C*c*, and C*b* groups, crystal structure resolution, PDB code, and chain considered.

[b] Listed residues (type and number) and corresponding atom types in brackets involved in an H-bond with the axially coordinated histidine. The first residue listed is the actual hydrogen bonding partner in the crystal structure.

[c] Provides distance of the closest approach of non-hydrogen atom pairs, which could be involved in an H-bond with the NδH group of histidine axially ligated to heme. The distances were varied by rotating the imidazole ring with respect to the pseudo torsion angle α. The distance of the actual hydrogen bond is given in brackets. [d] Provides pseudo torsion angle α of closest approach as defined in footnote c. For a definition of the angle α, see Figure 1. The angle α in the actual crystal structure is given in brackets. [e] For a definition of the pseudo torsion angle γ, see Figure 1.

Figure 22a demonstrates the typical hydrogen-bonding pattern of myoglobin. The bifurcated hydrogen bond between Leu89 and Ser92 is easily recognized. Also, a hydrogen bond with the propionic group PRA can be relatively strong due to the negatively charged PRA group, although its donor-acceptor distance is larger in comparison with the two other hydrogen bonds. Interestingly, in the native structure, the axially ligated imidazole is directed toward this propionic acid.

Figure 22b shows four possible hydrogen bonding partners of the axially ligated histidine in the mono-C*c* heme protein with PDB code 1chh (Lo et al., 1995). The minimal distance to the possible bonding partner Cys17 is even smaller than the one found in the crystal structure. However, this hydrogen bond is not possible due to unfavorable interactions of the imidazole ring with its backbone (see later). Instead, proline Pro30 forms a hydrogen bond with the NδH group of the imidazole so that the NδH group points toward the propionic acid groups.

There are three possible hydrogen bond partners of the coordinated His109 in bis-C*c* heme protein (Figure 22c). Here, unfavorable interactions with the histidine backbone did not prevent a hydrogen bond with Cys105. Nevertheless, in the crystal structure, the coordinated histidine forms a hydrogen bond with Gly24. Again, the resulting orientation has the NδH group of the imidazole pointing toward the propionic acid groups of the heme.

Conclusion

Factors determining conformations of imidazole axially coordinated to heme were investigated by analyzing the crystal structures of proteins from the PDB. Six groups of heme proteins were discriminated: myoglobin (Mb), hemoglobin (Hb), cytochrome *c* with one (mono-C*c*) or two (bis-C*c*) axially coordinated histidines, cytochrome *c* peroxidase (C*c*Po) and bis-histidine ligated cytochrome *b* (C*b*). Data from PDB structures show that in most cases, the preferred orientation of the NδH group of imidazole ligated to heme is toward the propionic groups. Therefore, there is a specific interaction of imidazole with the propionates. Also, the imidazole adopts a preferred orientation concerning its histidine backbone. This interaction prohibits conformations, where the imidazole plane is oriented parallel to the Cα–Cβ bond of its histidine backbone. Consequently, the orientation of the histidine backbone relative to the heme simultaneously determines the imidazole orientation concerning the heme plane.

Considering these two factors, mainly of electrostatic nature, the orientation of axially coordinated imidazoles in all families of heme proteins, except the group C*c*Po, was possible to rationalize. Analyzing the crystal structures of heme proteins, we can conclude that the hydrogen bonding pattern does not determine the overall orientation of imidazole, although it is probably a natural pathway to fine-tune the axial imidazole orientation relative to heme. The H-bond acceptor is most often the CO group of the protein backbone. Since the CO group is abundant everywhere in a protein, it does not impose a serious constraint on possible orientations of the imidazole ring. Results indicated that the NδH group of imidazole ligated to heme could assume many hydrogen bonds that differ from the native one. Besides that, in mutant structures, the orientation of the ligated imidazole often does not change significantly, although the mutant altered the hydrogen-bonding scheme involving the imidazole. In some cases, as in the group of cytochrome c peroxidase, the strong hydrogen bond of the imidazole NδH group with the negatively charged acidic residue of Asp235 enforces the imidazole ring to adopt an unusual orientation.

By molecular force field computations, the interactions of imidazole with propionic groups, histidine backbone, and two cysteines covalently attached to the heme were evaluated and correlated with results from searching in the PDB. The conclusions based on force field computations are in agreement with experimental data. Namely, the results have shown that an energy minimum corresponds to conformer with the NδH group of the imidazole oriented toward the propionic groups, as well as to orientations when the imidazole ring is orthogonal to the plane defined by the Cα–Cβ and Cβ–Cγ bonds of the histidine. The computations also demonstrated that these interactions are mainly of electrostatic origin.

Density functional theory calculations were applied to the heme model system with ionized propionic acid groups and two axially coordinated imidazoles. Different mutual orientations of the imidazole plane were considered in a vacuum, water, and protein environment. The calculated energy minima correspond to the conformations with imidazole NδH groups pointing toward the heme propionic acids, and electrostatic interactions between these groups have a strong influence. Such results are supported by previously collected crystallographic data of heme proteins and force fields calculations. Calculations with simulated water or protein environment revealed the additional impact on imidazole-heme conformation. The energy barriers of the solvated system are smaller than in a vacuum. These energy values are a consequence of the appearance of newly formed interaction of the dipole of the imidazole-heme complex with the reaction field but at the expense of weakening electrostatic interaction between imidazole and propionates. Also, strengthening the interaction between the dipole of the imidazole-heme complex and the reaction field is directly related to increasing the dielectric constant of the environment.

One may wonder that there is so little direct influence coming from the amino acids of the protein environment. Nevertheless, the protein determines the conformations of the propionic acids and the histidine backbone, and thus, indirectly, it significantly affects the conformations of imidazoles axially coordinated to heme.

This interplay of axial ligation, the histidine backbone, and the hydrogen bonding illustrates the multi-conformational landscapes that histidine can access due to its microenvironment. The presented investigation should shed some light on the histidine orientation in proteins and suggest improvements in the design of artificial proteins.

Acknowledgments

This work has been financially supported by the Ministry of Education, Science and Technological Development of the Republic of Serbia (MPNTR): Grant No. 451-03-68/2022-14/200026 and project OI/127035. This article/publication is based upon work from COST Action CA21101, supported by COST (European Cooperation in Science and Technology).

References

ACD/ChemSketch, 1994–2001. Advanced Chemistry Development Inc., Toronto, ON.

Adam, W., Lazarus, M., Saha-Möller, C. R., Weichold, O., Hoch, U., Häring, D. and Schreier, P. (1999). *Biotransformations with Peroxidases*. In *Biotransformations. Advances in Biochemical Engineering/Biotechnology*, edired by Kurt Faber, 63:73-108. Springer, Berlin, Heidelberg.

Anđelković, Lj., Grubišić, S., Đorđević, I., Zlatar, M., Niketić, S. and Gruden-Pavlović, M. (2010) Consistent force field for metalloporphyrins. *Journal of the Serbian Chemical Society*, 75: 1671−1683.

Aubert, C., Giudici-Orticoni, M. T., Czjzek, M., Haser, R., Bruschi, M., and Dolla, A. (1998). Structural and kinetic studies of the Y73E mutant of octaheme cytochrome *c*3

(Mr = 26 000) from Desulfovibrio desulfuricans Norway. *Biochemistry*, 37: 2120–2130.

Bashford, Donald (1997). *An object-oriented programming suite for electrostatic effects in biological molecules: An experience report on the MEAD project.* In *Scientific Computing in Object-Oriented Parallel Environments. ISCOPE 1997. Lecture Notes in Computer Science*, vol 1343, edited by Yutaka Ishikawa, Rodney R. Oldehoeft, John V. W. Reynders and Marydell Tholburn, 8–11. Springer. Berlin, Heidelberg.

Bashford, D., and Gerwert, K. (1992). Electrostatic calculations of the pKa values of ionizable groups in bacteriorhodopsin. *Journal of Molecular Biology*, 224:473–486.

Bayly, C., Cieplak, P., Cornell, W., and Kollman, P. (1993). A well-behaved electrostatic potential based method using charge restraints for deriving atomic charges: the RESP model. *The Journal of Physical Chemistry*, 97:10269–10280.

Becke, A. D. (1988). Density-functional exchange-energy approximation with correct asymptotic behavior. *Physical Review A*, 38:3098–3100.

Berman, H. M., Westbrook, J., Feng, Z., Gilliland, G., Bhat, T. N., Weissig, H., Shindyalov, I. N. and Bourne, P. E. (2000). The Protein Data Bank. *Nucleic Acids Research*, 28:235–242.

Berry, E. A., and Walker, F. A. (2008). Bis-histidine-coordinated hemes in four-helix bundles: how the geometry of the bundle controls the axial imidazole plane orientations in transmembrane cytochromes of mitochondrial Complexes II and III and related proteins. *Journal of Biological Inorganic Chemistry*, 13:481–498.

Bryson, J. W., Betz, S. F., Lu, H. S., Suich, D. J., Zhou, H. X., O'Neil, K. T., and DeGrado, W. F. (1995). Protein Design: A Hierarchic Approach, *Science*, 270:935–941.

Chroma, C. T., Lear, J. D., Nelson, M. J., Dutton, P. L., Robertson, D. E., and DeGrado, W. F. (1994). Design of a heme-binding four-helix bundle. *Journal of the American Chemical Society*, 116: 856–865.

Cornell, W., Cieplak, P., Bayly, C., and Kollman, P. (1993). Application of RESP charges to calculate conformational energies, hydrogen bond energies, and free energies of solvation. *Journal of the American Chemical Society*, 115:9620–9631.

Czjzek, M., Guerlesquin, F., Bruschi, M., and Haser, R. (1996). Crystal structure of a dimeric octaheme cytochrome c3 (M(r) 26,000) from Desulfovibrio desulfuricans Norway. *Structure*, 4: 395-404.

Đorđević, I. S., Grubišić, S., Milčić, M., and Niketić, S. (2015). Derivation of a new set of force field parameters for ammine complexes of chromium(III) containing halogeno ligands: modelling the trans-influence of halogenido ligands, *Journal of the Serbian Chemical Society*, 80:329–342.

Đorđević, I., and Niketić, S. R. (2012). Atomic partial charges for mixed chloroammine chromium(III) complexes fitted to the molecular electrostatic potential, *Computational and Theoretical Chemistry*, 1001:20–25.

Dupradeau, F. Y., Pigache, A., Zaffran, T., Savineau, C., Lelong, R., Grivel, N., Lelong, D., Rosanski, W., and Cieplak, P. (2010). The R.E.D. tools: advances in RESP and ESP charge derivation and force field library building. *Physical chemistry chemical physics*, 12:7821–7839.

Faller M, Matsunaga M, Yin S, Loo JA, Guo F. (2007). Heme is involved in microRNA processing. *Nature Structural & Molecular Biology*, 14:23–29.

Fufezan, C., Zhang, J., and Gunner, M. R. (2008). Ligand preference and orientation in *b*- and *c*-type heme-binding proteins. *Proteins*, 73:690–704.

Gallivani, J. P., and Dougherty, D. A. (1999). Cation-π interactions in structural biology. *Proceedings of the National Academy of Sciences of the United States of America*, 96:9459–9464.

Galstyan, S. G., Zarić, S. D., and Knapp, E.-W. (2005). Computational studies on imidazole heme conformations. Journal of Biological Inorganic Chemistry, 10(4):343-54.

Ghosh, A., Gonzales, E., and Vangberg, T. (1999). Theoretical Studies of Low-Spin Six-Coordinate Iron(III) Porphyrins Relevant to Cytochromes *b*: Variable Electronic Configurations, Ligand Noninnocence, and Macrocycle Ruffling. *The Journal of Physical Chemistry B*, 103:1363–1367.

Gibney, B. R., Isogai, Y., Rabanal, F., Reddy, K. S., Grosset, A. M., Moser, C. C., and Dutton, P. L. (2000). Self-assembly of heme A and heme B in a designed four-helix bundle: Implications for a cytochrome c oxidase maquette. *Biochemistry*, 39:11041–11049.

Goodin, D. B., and McRee, D. E. (1993). The Asp-His-Fe triad of cytochrome c peroxidase controls the reduction potential, electronic structure, and coupling of the tryptophan free radical to the heme. *Biochemistry*, 32:3313-3324.

Gray, H. B., and Winkler, J. R. (1996). Electron transfer in proteins. *Annual Review of Biochemistry*, 65:537–561.

Harris, D., Loew, G., and Waskell, L. (1998). Structure and Spectra of Ferrous Dioxygen and Reduced Ferrous Dioxygen Model Cytochrome P450. *Journal of the American Chemical Society*, 120:4308–4318.

Hay, P. J., and Wadt, W. R. (1985). Ab initio effective core potentials for molecular calculations. Potentials for K to Au including the outermost core orbitals. *The Journal of Chemical Physics*, 82:299–310.

Hon, T., Dodd, A., Dirmeier, R., Gorman, N., Sinclair, P. R., Zhang, L. and Poyton, R. O. (2003) A mechanism of oxygen sensing in yeast. Multiple oxygen-responsive steps in the heme biosynthetic pathway affect Hap1 activity. *Journal of Biological Chemistry*, 278:50771–50780.

Hu, C., Noll, B. C., Schulz, C. E., and Scheidt W. R. (2008). Hydrogen bonding influence of 1,10-phenanthroline on five-coordinate high-spin imidazole-ligated iron(II) porphyrinates. *Inorganic Chemistry*, 47(19):88884–88956.

Hu, C., Noll, B. C., Schulz, C. E., and Scheidt, W. R. (2018). Hydrogen-Bonding Effects in Five-Coordinate High-Spin Imidazole-Ligated Iron(II) Porphyrinates. *Inorganic Chemistry*, 57(2):793 – 803.

JAGUAR 4.2, 1991–2000. Schrödinger, Inc, Portland, OR.

Jenkins, R. M., Singleton, M. L., Leamer, L. A., Reibenspies, J. H., and Darensbourg M. Y. (2010). Orientation and Stereodynamic Paths of Planar Monodentate Ligands in Square Planar Nickel N_2S Complexes. *Inorganic Chemistry*, 49:5503–5514.

Jewsbury, P., Yamamoto, S., Minato, T., Saito, M., and Kitagawa, T. (1994). The Proximal Residue Largely Determines the CO Distortion in Carbon Monoxy Globin Proteins. An ab Initio Study of a Heme Prosthetic Unit. *Journal of the American Chemical Society*. 116:11586–11587.

Karakus, Yonca Y., 2020. *Typical Catalases: Function and Structure*. In *Glutathione System and Oxidative Stress in Health and Disease*, edited by Margarete Bagatini. London: IntechOpen.

Kavanaugh, J. S., Weydert, J. A., Rogers, P. H., and Arnone, A. (1998). High-resolution crystal structures of human hemoglobin with mutations at tryptophan 37beta: structural basis for a high-affinity T-state. *Biochemistry*, 37:4358–4373.

Kavanaugh, J. S., Rogers, P. H., and Arnone, A. (2005). Crystallographic evidence for a new ensemble of ligand-induced allosteric transitions in hemoglobin: the T-to-T(high) quaternary transitions. *Biochemistry*, 44:6101–6121.

Kolatkar, P. R., Ernst, S. R., Hackert, M. L., Ogata, C. M., Hendrickson, W. A., Merritt, E. A., and Phizackerley, R. P. (1992). Structure determination and refinement of homotetrameric hemoglobin from *Urechis caupo* at 2.5 A resolution. *Acta Crystallographica B*, 48:191–199.

Lee, C., Yang, W., and Paar, R. G. (1988). Development of the Colle-Salvetti correlation-energy formula into a functional of the electron density. *Physical Review B*, 46:785–789.

Li, P., and Merz, K. M. Jr. (2017). Metal Ion Modeling Using Classical Mechanics. *Chemical Reviews*, 117:1564−1686.

Liu, J., Chakraborty, S., Hosseinzadeh, P., Yu, Y., Tian, S., Petrik, I., Bhagi, A. and Lu, Y. (2014). Metalloproteins Containing Cytochrome, Iron–Sulfur, or Copper Redox Centers. *Chemical Reviews*,114(8):366–4469.

Lloyd, E., Burk, D. L., Ferrer, J. C., Maurus, R., Doran, J., Carey, P. R., Brayer, G. D., and Mauk, A. G. (1996). Electrostatic modification of the active site of myoglobin: characterization of the proximal Ser92Asp variant. *Biochemistry*, 35:11901–11912.

Lloyd, E., Hildebrand, D. P., Tu, K. M. and Mauk, A. G. (1995). Conversion of Myoglobin into a Reversible Electron Transfer Protein That Maintains Bishistidine Axial Ligation. *Journal of the American Chemical Society*, 117:6434–6438.

Lo, T. P., Guillemette, J. G., Louie, G. V., Smith, M., and Brayer, G. D. (1995). Structural studies of the roles of residues 82 and 85 at the interactive face of cytochrome *c*. *Biochemistry*, 34:163-171.

MacKerell, A. D., Jr., Bashford, D., Bellot, M., Dunbrack, R. L., Jr., Evanseck, J. D., Field, M. J., Fischer, S., Gao, J., Guo, H., Ha, S., Joseph-McCarthy, D., Kuchnir, L., Kuczera, K., Lau, F. T. K., Mattos, C., Michnick, S., Ngo, T., Nguyen, D. T., Prodholm, B., Reiher, W. E., III, Roux, B., Schlenkrich, M., Smith, J. C., Stote, R., Straub, J., Watanabe, M., Wiórkiewicz-Kuczera, J., Yin, D., and Karplus, M. (1998). All-Atom Empirical Potential for Molecular Modeling and Dynamics Studies of Proteins. *The Journal of Physical Chemistry B*, 102:3586–3616.

Marques, H. M., Munro, O. Q., Grimmer, N. E., Levendis, D. C., Marsicano, F., Pattrick, G., and Markoulides, T. (1995). A force field for molecular mechanics studies of iron porphyrins. *Journal of the Chemical Society, Faraday Transactions*, 91:1741–1749.

Marten, B., Kim, K., Cortis, C., Friesner, R. A., Murphy, R. B., Ringnalda, M. N., Sitkoff, D., and Honig, B. (1996). New Model for Calculation of Solvation Free Energies: Correction of Self-Consistent Reaction Field Continuum Dielectric Theory for Short-Range Hydrogen-Bonding Effects. *The Journal of Physical Chemistry*, 100:11775–11788.

Martin, A. C., Orengo, C. A., Hutchinson, E. G., Jones, S., Karmirantzou, M., Laskowski, R. A., Mitchell, J. B., Taroni, C., and Thornton, J. M. (1998). Protein folds and functions. *Structure*, 6:875–884.

Matias, P. M., Morais, J., Coelho, R., Carrondo, M. A., Wilson, K., Dauter, Z., and Sieker, L. (1996). Cytochrome *c*3 from *Desulfovibrio gigas*: crystal structure at 1.8 A resolution and evidence for a specific calcium-binding site. *Protein Science*, 5:1342–1354.

Mauri, F., Omnaas, J., Davidson, L., Whitfill, C., and Kitto, G. B. (1991). Amino acid sequence of a globin from the sea cucumber *Caudina (Molpadia) arenicola*. *Biochimica et Biophysica Acta*, 1078:63–67.

Medaković, V., and Zarić, S. D. (2003). Theoretical study on orientations of axially coordinated imidazoles in model systems of cytochromes. *Inorganica Chimica Acta*, 349:1–5.

Mense, S., and Zhang, L. (2006). Heme: a versatile signaling molecule controlling the activities of diverse regulators ranging from transcription factors to MAP kinases. *Cell Research*, 16:681–692.

Menyhard, D. K., and Keseru, G. M. (1998). Stereoelectronic control on the coordination of substrates to globin proteins. The role of proximal His93 on the NO release from myoglobin. *Journal of the American Chemical Society*, 120:7991–7992.

Merlino, A., Vergara, A., Sica, F., and Mazzarella, L. (2009). The bis-histidyl complex in hemoproteins: A detailed conformational analysis of database protein structures and the case of Antarctic fish hemoglobins. *Marine Genomics*, 2:51–56.

Miehlich, B., Savin, A., Stoll, H., and Pruess, H. (1989). Results obtained with the correlation energy density functionals of Becke and Lee, Yang and Parr. *Chemical Physics Letters*, 157:200–206.

Mitchell, D. T., Kitto, G. B., and Hackert, M. L. (1995). Structural analysis of monomeric hemichrome and dimeric cyanomet hemoglobins from *Caudina arenicola*. *Journal of Molecular Biology*, 251:421–431.

Momot K. I., and Walker, F. A. (1997a). Investigations of Rotation of Axial Ligands in Six-Coordinate Low-Spin Iron(III) Tetraphenylporphyrinates: Measurement of Rate Constants from Saturation Transfer Experiments and Comparison to Molecular Mechanics Calculations. *The Journal of Physical Chemistry A*, 101:2787–2795.

Momot, K. I., and Walker, F. A. (1997b). Investigations of Rotation of Axial Ligands in Six-Coordinate Low-Spin Iron(III) Tetraphenylporphyrinates: Measurement of Rate Constants from Saturation Transfer Experiments and Comparison to Molecular Mechanics Calculations. *The Journal of Physical Chemistry A*, 101:2787–2795

Munro, O. Q., Marques, H. M., Debrunner, P. G., Mohanrao, K., and Scheid, W. R. (1995). Structural and Molecular Mechanics Studies on Highly Ruffled Low-Spin (Porphinato)iron(III) Complexes, *Journal of the American Chemical Society*, 117:935–954.

Munro, O. Q., Serth-Guzzo, J. A., Turkovska-Tyrk, I., Mohanrao, K., Shokhireva, T. Kh., Walker, F. A., Debrunner, P. G., and Scheidt, W. R. (1999). Two Crystalline Forms of Low-Spin [Fe(TMP)(5-MeHIm)$_2$]ClO_4. Relative Parallel and Perpendicular Axial Ligand Orientations. *Journal of the American Chemical Society*, 121:11144–11155.

Mutter, M., Altmann, E., Altmann, K. H., Hersperger, R., Koziej, P., Nebel, K., Tuchscherer, G., Vuilleumier, S., Gremlich, H. U., and Müller, K. (1988). Non-native architectures in protein design and mimicry. *Helvetica Chimica Acta*, 71:835–847.

Nakamura, M., Ikeue, T., Neya, N., Fanasaki, N., and Nakamura, N. (1996). Fixation of the 2-Methylimidazole Ligand and Anomalous Pyrrole Chemical Shifts in Bis(2-methylimidazole)(meso-tetraalkylporphyrinato)iron(III) Chloride Caused by the Nonplanar Porphyrin Ring. *Inorganic Chemistry*, 35:3731–3732.

Nakamura, M., Tajima, K., Tada, K., Ishizu, K., and Nakamura, N. (1994). Rotation and orientation of axially coordinated imidazoles in low-spin ferric porphyrin complexes. *Inorganica Chimica Acta*, 224:113–124.

Narula, J., Pandey, P., Arbustini, E., Haider, N., Narula, N., Kolodgie, F. D., Dal Bello, B., Semigran, M. J., Bielsa-Masdeu, A., Dec, G. W., Israels, S., Ballester, M., Virmani, R., Saxena, S., and Kharbanda, S. (1999). Apoptosis in heart failure: Release of cytochrome c from mitochondria and activation of caspase-3 in human cardiomyopathy. *The Proceedings of the National Academy of Sciences*, 96(14):8144–8149.

Negron, C., Fufezan, C. and Koder, R. L. (2009). Geometric constraints for porphyrin binding in helical protein binding sites. *Proteins*, 74:400–416.

Nesset, M. J. M., Shokhirev, N. V., Enemark, P. D., Jacobson, S. E., and Walker F. A. (1996). Models of the cytochromes. Redox properties and thermodynamic stabilities of complexes of 'hindered' iron(III) and iron(II) tetraphenylporphyrinates with substituted pyridines and imidazoles. *Inorganic Chemistry*, 35:5188–5200.

Olafson, B. D., and Goddard W. A. (1977). Molecular description of dioxygen bonding in hemoglobin. *Proceedings of the National Academy of Sciences*, 74(4):1315–1319.

Oldfield, T. J., Smerdon, S. J., Dauter, Z., Petratos, K., Wilson, K. S., and Wilkinson, A. J. (1992). High-resolution X-ray structures of pig metmyoglobin and two CD3 mutants: Mb(Lys45----Arg) and Mb(Lys45----Ser). *Biochemistry*, 31: 8732-8739.

Padmanaban, G., Venkateswar, V. and Rangarajan P. N. (1989). Haem as a multifunctional regulator. Trends in Biochemical Sciences, 14:492–496.

Polam, J. R., Shokhireva, T. Kh., Raffi, K., Simons, U., and Walker, F. A. (1997). Rates of axial ligand rotation in diamagnetic d^6 Co(III) and Fe(II) porphyrinates. *Inorganica Chimica Acta*, 263:109–117.

Pond, A. E., Roach, M. P., Sono, M., Rux, A. H., Franzen, S., Hu, R., Thomas, M. R., Wilks, A., Dou, Y., Ikeda-Saito, M., de Montellano, P. R. O., Woodruff, W. H., Boxer, S. G., and Dawson, J. H. (1999). Assignment of the Heme Axial Ligand(s) for the Ferric Myoglobin (H93G) and Heme Oxygenase (H25A) Cavity Mutants as Oxygen Donors Using Magnetic Circular Dichroism. *Biochemistry*, 38:7601–7608.

Popović, D. M., and Đorđević, I. S. (2020). Catalytic center of cytochrome c oxidase: Effects of protein environment on the pKa values of CuB histidine ligands. *Journal of the Serbian Chemical Society*, 85(11):1429–1444.

Popović, D. M., and Stuchebrukhov, A. A. (2005). Proton Exit Channels in Bovine Cytochrome *c* Oxidase. *The Journal of Physical Chemistry B*, 109:1999–2006.

Popović, D. M., and Stuchebrukhov, A. A. (2012). Coupled electron and proton transfer reactions during the O→E transition in bovine cytochrome *c* oxidase. *Biochimica et Biophysica Acta*, 1817(4): 506–517.

Popović, D. M., Zarić, S. D., Rabenstein, B., and Knapp, E. W. (2001). Artificial Cytochrome *b*: Computer Modeling and Evaluation of Redox Potentials. *Journal of the American Chemical Society*, 123:6040–6053.

Poulos, T. L., Finzel, B. C., and Howard, A. J. (1987). High-resolution crystal structure of cytochrome P450cam. *Journal of Molecular Biology*, 195:687–700.

Quenneville, J., Popović, D. M., and Stuchebrukhov, A. A. (2004). Redox-Dependent pKa of CuB Histidine Ligand in Cytochrome *c* Oxidase. *The Journal of Physical Chemistry B*, 108:18383–18389.

Quenneville, J., Popović, D. M., and Stuchebrukhov, A. A. (2006). Combined DFT and electrostatics study of the proton pumping mechanism in cytochrome *c* oxidase. *Biochimica et Biophysica Acta*, 1757:1035–1046. Review

Rakić, A. A., Medaković, V. B., and Zarić, S. D. (2006). Orientations of axially coordinated imidazoles and pyridines in crystal structures of model systems of cytochromes. *Journal of Inorganic Biochemistry*, 100:133–142.

Rau, H. K., and Haehnel, W. (1998). Design, Synthesis, and Properties of a Novel Cytochrome b Model. Journal of the American Chemical Society, 120:468–476.

Reedy C. J., and Gibney B. R. (2004). Heme Protein Assemblies. *Chemical Reviews*, 104: 617–649.

Rovira, C., and Parrinello, M. (1999). Factors Influencing Ligand-Binding Properties of Heme Models: A First Principles Study of Picket-Fence and Protoheme Complexes. *Chemistry–A European Journal*, 5:250–262.

Safo, M. K., Nesset, M. J. M., Walker, F. A., Debrunner, P. G. and Scheidt, W. R. (1997). Models of the Cytochromes. Axial Ligand Orientation and Complex Stability in Iron(II) Porphyrinates: The Case of the Noninteracting d_{π} Orbitals. *Journal of the American Chemical Society*, 119:9438–9448.

Safo, M. K., Walker, F. A., Raitsimring, A. M., Walters, W. P., Dolata, D. P., Debrunner, P. G., and Scheidt, W. R. (1994). Axial Ligand Orientation in Iron(III) Porphyrinates: Effect of Axial π-Acceptors. Characterization of the Low-Spin Complex [Fe(TPP)(4-CNPy)$_2$]ClO_4. *Journal of the American Chemical Society*, 116:7760–7770.

Shelnutt, J. A., Franko, R., Ma, J. G., Lu, Y., and Ferreira, G. C. (2000). Porphyrin Interactions with Wild-type and Mutant Mouse Ferrochelatase. *Biochemistry*, 39:2517–2529.

Shelnutt, J. A., Song, X.-Z., Ma,, J.-G., Jia, S.-L., Jentzen, W., Medforth, C. J., and Medforth. C. J. (1998). Nonplanar porphyrins and their significance in proteins. *Chemical Society Reviews*, 27:31–42.

Shokhirev, N. V. and Walker, F. A. (1995). Analysis of the Temperature Dependence of the ^{1}H Contact Shifts in Low-Spin Fe(III) Model Hemes and Heme Proteins: Explanation of "Curie" and "Anti-Curie" Behavior within the Same Molecule. *The Journal of Physical Chemistry A*, 99:17795–17804.

Shokhirev, N. V., Shokhireva, T. K., Polam, J. R., Watson, C. T., Raffii K., Simons, U., and Walker, F. A. (1997). ^{2}D NMR Investigations of the Rotation of Axial Ligands in Six-Coordinate Low-Spin Iron(III) and Cobalt(III) Tetraphenylporphyrinates Having 2,6-Disubstituted Phenyl Rings: Quantitation of Rate Constants from ^{1}H EXSY Cross-Peak Intensities. *The Journal of Physical Chemistry A*, 101:2778–2786.

Simoes, P., Matias, P. M., Morais, J., Wilson, K., Dauter, Z., and Carrondo, M. A. (1998). Refinement of the Three-Dimensional Structures of Cytochrome C3 from *Desulfovibrio Vulgaris* Hildenborough at 1.67 Angstroms Resolution and from *Desulfovibrio Desulfuricans* Atcc 27774 at 1.6 Angstroms Resolution. *Inorganica Chimica Acta*, 273:213–224.

Slater, John C. (1974). *Quantum Theory of Molecules and Solids*, Vol. 4: *The Self-Consistent Field for Molecules and Solids*, edited by McGraw-Hill Education. New York.

Smerdon, S. J., Krzywda, S., and Wilkinson, A. J. (1993). Serine92 (F7) contributes to the control of heme reactivity and stability in myoglobin. *Biochemistry*, 32:5132–5138.

Song, X-Z., Jentzen, W., Jaquinod, L., Khoury, R. G., Medforth, C. J., Jia, S-L., Ma, J-G., Smith, K. M., and Shelnutt, J. A. (1998). Substituent-Induced Perturbation Symmetries and Distortions of meso-tert-Butylporphyrins. *Inorganic Chemistry*, 37:2117 –2128.

Sowole, M. A., Vuong, S. and Konermann, L. (2015). Interactions of Hemoglobin and Myoglobin with Their Ligands CN–, CO, and O2 Monitored by Electrospray Ionization–Mass Spectrometry. Analytical Chemistry, 87(19):9538–9545.

Steigemann, W., and Weber, E. (1979). Structure of erythrocruorin in different ligand states refined at 1.4 Å resolution. *Journal of Molecular Biology*, 127:309-338.

Tannor, D. J., Marten, B., Murphy, R., Friesner, R. A., Sitkoff, D., Nicholls, A., Ringnalda. M., Goddard III, W. A., and Honig B (1994). Accurate First Principles Calculation of Molecular Charge Distributions and Solvation Energies from Ab Initio Quantum Mechanics and Continuum Dielectric Theory. *Journal of the American Chemical Society*, 116:11875–11882.

Tarricone, C., Galizzi, A., Coda, A., Ascenzi, P., and Bolognesi, M. (1997). Unusual structure of the oxygen-binding site in the dimeric bacterial hemoglobin from Vitreoscilla sp. *Structure*, 5:497-507.

Vangberg, T., and Ghosh, A. (1999). A First-Principles Quantum Chemical Analysis of the Factors Controlling Ruffling Deformations of Porphyrins: Insights from the Molecular Structures and Potential Energy Surfaces of Silicon, Phosphorus, Germanium, and Arsenic Porphyrins and of a Peroxidase Compound I Model. *Journal of the American Chemical Society*, 121:12154–12160.

Vosko, S. H., Wilk, L., and Nusair, M. (1980). Accurate spin-dependent electron liquid correlation energies for local spin density calculations: a critical analysis. *Canadian Journal of Physics*, 58:1200–1211.

Walker, F. A. (1999). Magnetic spectroscopic (EPR, ESEEM, Mössbauer, MCD and NMR) studies of low-spin ferriheme centers and their corresponding heme proteins. *Coordination Chemistry Reviews*, 185–186:471–534.

Walker, F. A., Huynh, B. H., Schedt, W. R. and Osvath, S. R. (1986). Models of the cytochromes *b*. Effect of axial ligand plane orientation on the EPR and Moessbauer spectra of low-spin ferrihemes. *Journal of the American Chemical Society*, 108:5288–5297.

Walker, F. A., Nasri, H., Turowska-Tyrk, I., Mohanrao, K., Watson, C. T., Shokhirev, N. V., Debrunner, P. G., and Scheidt, W. R. (1996). π-Acid Ligands in Iron(III) Porphyrinates. Characterization of Low-Spin Bis(tert-butylisocyanide)(porphyrinato) iron(III) Complexes Having $(d_{xz},d_{yz})^4(d_{xy})^1$ Ground States. *Journal of the American Chemical Society*. 118:12109–12118.

Wallace, C. J. A., and Clark–Lewis, I. (1992). Functional role of heme ligation in cytochrome c. Effects of replacement of methionine 80 with natural and non-natural residues by semisynthesis. *Journal of Biological Chemistry*, 267:3852–3861.

Wang, J. M., Mauro, M., Edwards, S. L., Oatley, S. J., Fishel, L. A., Ashford, V. A., Xuong, N. H., and Kraut, J. (1990). X-ray structures of recombinant yeast cytochrome c peroxidase and three heme-cleft mutants prepared by site-directed mutagenesis. *Biochemistry*, 29:7160–7173.

Willett, P. (2011). Chemoinformatics: a history. *Wiley Interdisciplinary Reviews: Computational Molecular Science*, 1:46–56.

Yu, F., Cangelosi, V. M., Zastrow, M. L., Tegoni, M., Plegaria, J. S., Tebo, A. G., Mocny, C. S., Ruckthong, L., Qayyum, H., and Vincent L. Pecoraro, V. L. (2014). Protein Design: Toward Functional Metalloenzymes. *Chemical Reviews*, 114(7):3495–3578.

Zarić, S. D., Popović, D. M., and Knapp, E.-W. (2000). Metal Ligand-Aromatic Cation-pi Interactions in metallo-proteins: ligands coordinated to metal interact with aromatic residues. *Chemistry–A European Journal*, 6:3935–3942.

Zarić, S. D., Popović, D. M., and Knapp, E.-W. (2001). Factors Determining the Orientation of Axially Coordinated Imidazoles in Heme Proteins. *Biochemistry*, 40:7914–7928.

Zhuang, J., Amoroso, J. H., Kinloch, R., Dawson, J. H., Baldwin, M. J., and Gibney, B. R. (2006). Evaluation of electron-withdrawing group effects on heme binding in designed proteins: implications for heme a in cytochrome *c* oxidase. *Inorganic Chemistry*, 45:4685–4694.

Biographical Sketches

Ivana S. Đorđević, PhD

Affiliation: University of Belgrade, Institute of Chemistry, Technology and Metallurgy, National Institute of the Republic of Serbia

Education:

2001-2009 Graduated with honors from the Faculty of Chemistry of University of Belgrade

2009-2017 University of Belgrade, a PhD degree in Chemistry

Business Address:

12 Njegoševa, Belgrade, Republic of Serbia

Research and Professional Experience:
Assistant Research Professor at University of Belgrade, Institute of Chemistry, Technology and Metallurgy, National Institute of the Republic of Serbia since 2018.

Professional Appointments:
Member of Group of Molecular Modeling and Dynamics, Laboratory of Theoretical Chemistry and Molecular Modeling, Department of Chemistry, University of Belgrade, Institute of Chemistry, Technology and Metallurgy, National Institute of the Republic of Serbia.

Publications from the Last 3 Years:

1. Bondžić, A. M., Žakula, J. J., Korićanac, L. B., Keta, O. D., Janjić, G. V., Đorđević, I. S. and Rajković, S. U. (2022). Cytotoxic activity and influence on acetylcholinesterase of series dinuclear platinum(II) complexes with aromatic nitrogen-containing heterocyclic bridging ligands: Insights in the mechanisms of action. *Chemico-Biological Interactions*, 351:109708−109719.
2. Franich, A. A., Đorđević, I. S., Živković, M. D., Rajković, S. U., Janjić, G. V. and Đuran, M. (2022). Dinuclear platinum(II) complexes as the pattern for phosphate backbone binding: a new perspective for recognition of binding modes to DNA. *Journal of Biological Inorganic Chemistry*, 27:65−75.
3. Gak-Simić, K., Đorđević, I. S., Janjić, G. V., Datz, D. V., Toth-Katona, T. and Trišović, N. (2021). On the photophysical properties of a liquid crystal dimer based on 4-nitrostilbene: A combined experimental and theoretical study. *Journal of Molecular Liquids*, 339:116969−116974.
4. Popadić, M. G., Marinović, S. R., Mudrinić, T. M., Milutinović-Nikolić, A. D., Banković, P. T., Đorđević, I. S. and Janjić, G. V. (2021). A novel approach in revealing mechanisms and particular step predictors of pH dependent tartrazine catalytic degradation in presence of Oxone®. *Chemosphere*, 281:130806−130814.
5. Dahmani, R., Grubišić, S., Djordjević, I., Yaghlane, S. B., Boughdiri, S., Chambaud, G. and Hochlaf, M. (2021). In silico design of a new Zn-triazole based metal-organic framework for CO_2 and H_2O adsorption. *The Journal of Chemical Physics*, 154(2):024303−024303-10.
6. Simić, K. Gak, Đorđević, I. S., Lazić, A. M., Radovanović, L. D., Petković-Benazzouz, M. M., Rogan, J. R., Trišović, N. P. and Janjić, G. V. (2021). On the supramolecular outcomes of fluorination of cyclohexane-5-spirohydantoin derivatives. *Cryst Eng Comm*, 23:2606−2622.
7. Janjić, G. V., Jelić, S. T., Trišović, N. P., Popović, D. M., Đorđević, I. S. and Milčić, M. K. (2020). New Theoretical Insight into Fluorination and Fluorine-Fluorine Interactions as a Driving Force in Crystal Structures. *Crystal Growth & Design*, 20(5):2943–2951.
8. Franich, A. A., Živković, M. D., Ilić-Tomić, T. R., Đorđević, I. S., Nikodinović-Runić, J., Pavić, A. B. and Janjić, G. V., Rajković, S. U. (2020). New minor groove covering

DNA binding mode of dinuclear Pt(II) complexes with various pyridine-linked bridging ligands and dual anticancer-antiangiogenic activities. *Journal of Biological Inorganic Chemistry*, 25(3):395–409.
9. Đorđević, I. S., Popadić, M. G., Sarvan, M. Z., Petković-Benazzouz, M. M. and Janjić, G. V. (2020). Supramolecular insight into the substitution of sulfur by selenium, based on crystal structures, quantum-chemical calculations and biosystem recognition. *Acta Crystallographica Section B: Structural Science, Crystal Engineering and Materials*, 76:122–136.

Dragan M. Popović, PhD

Affiliation: Department of Chemistry, University of Belgrade, Institute of Chemistry, Technology and Metallurgy, National Institute of the Republic of Serbia, Belgrade, Serbia

Education:

2003-2010, University of California, Davis, USA, postdoctoral scholar and Assistant research professor in the field of biochemistry, biophysics, and theoretical/computational chemistry

1999-2002, Free University of Berlin, Germany, a PhD degree in Chemistry (*magna cum laude*); dissertation: "Modeling of conformation and redox potentials of hemes and other cofactors in proteins"

1988-1993, Graduated with a BSc in Chemistry with honors (*GPA: 9.87/10.0*) from the Faculty of Chemistry, University of Belgrade, Serbia

Business Address:

Group for Molecular Modeling and Dynamics, Studentski Trg 12, 11000 Belgrade, Serbia

Research and Professional Experience:

Associate research professor at the Institute of Chemistry, Technology and Metallurgy, University of Belgrade, since 2015. He was visiting professor teaching postgraduate students "Computational methods in biochemistry" at University of the Philippines Diliman, Manila, in 2017.

Professional Appointments:

Member of the Group for Molecular Modeling and Dynamics. Associate editor of the journal Frontiers in Chemistry. Author of several chapters in international books.

Honors:

Award of Serbian Chemical Society for the best-graduated student. Scholarship of honor to academic excellence. Professional recognition: research fellowships from the German Research Society, DFG (SFB 498 A5 and GRK 268).

Publications from the Last 3 Years:

1. Protić-Rosić, I., Nešić, A., Lukić, I., Miljković, R., Popović, D.M., Atanasković-Marković, M., Stojanović, M. and Gavrović-Jankulović M. (2021). Recombinant Bet v 1-BanLec chimera modulates functional characteristics of peritoneal murine macrophages by promoting IL-10 secretion, *Molecular Immunology* 138: 58-67.
2. Popović, D.M. (2021). Photolyase – molecular mechanism for repair of UV-damaged DNA. *Hemijski Pregled* 62: 50-62.
3. Lopandić, Z., Dragačević, L., Popović, D. M., Andjelković, U., Minić, R., Gavrović-Jankulović, M. D. (2021). BanLec-eGFP chimera as a tool for evaluation of lectin binding to high-mannose glycans on microorganisms, *Biomolecules* 11: 180, pp 1-13.
4. Popović, D. M. and Djordjević, I. S. (2020). Catalytic center of cytochrome *c* oxidase: Effects of protein environment on p*Ka* values of CuB histidine ligands. *Journal of Serbian Chemical Society* 85: 1429-1444.
5. Lazić, A. M., Djordjević, I. S., Radovanović, L. D., Popović, D. M., Rogan, J. R., Janjić, G. V. and Trisović, N. P. (2020). Self-assembly and biorecognition of a spirohydantoin derived from α-tetralone: Interplay between chirality and intermolecular interactions. *Chem Plus Chem* 85: 1220-1232.
6. Janjić, G. V., Jelić, S. T., Trisović, N. P., Popović, D. M., Djordjević, I.S. and Milčić, M. K. (2020). New theoretical insight into fluorination and fluorine-fluorine interactions as a driving force in crystal structures. *Crystal Growth & Design* 20: 2943-2951.

Chapter 3

Comparative Study on Structure-Function Correlation of Native and Synthetic *Cytochrome b*: Protonation and Redox Properties

Dragan M. Popović*

Department of Chemistry, University of Belgrade, Institute of Chemistry, Technology and Metallurgy, National Institute of the Republic of Serbia, Belgrade, Serbia

Abstract

In the previous chapter, we have generated atomic coordinates of a synthetic protein recently synthesized (SynCyt*b*) to model the central part of the native cytochrome *b* (Cyt*b*) subunit consisting of a four-helix bundle with two hemes as cofactors. Since the X-ray structure is not solved, the structural elements of SynCyt*b* were assembled from scratch using all known chemical and structural information available, thereby avoiding strain and atomic clashes. The molecular mechanics was subsequently employed to fully relax the model structure of SynCyt*b* by applying the stepwise constraint energy minimization. Molecular dynamics simulations are applied to the computer-generated synthetic protein structure and the crystal structure of native Cyt*b*, revealing very similar root-mean-square deviation (RMSD) profiles over time and, thus, demonstrating a relative rigidness, stability, and strain-free of the finally modeled structure.

In this chapter, to further test and validate the computer-generated structure of SynCyt*b*, characteristic observables have to be computed and

* Corresponding Author's Email: dragan.popovic@ihtm.bg.ac.rs.

In: Structure and Function of Cytochromes
Editor: Elaine J. Mathis
ISBN: 979-8-88697-102-6

compared with measured quantities. One of these is the redox potential, a key parameter of redox-active proteins. The results of the redox potentials calculation are discussed here. For comparison, protonation patterns and heme redox potentials in the SynCyt*b* model structure and native Cyt*b* from the respiratory cytochrome bc_1 (Cytbc_1) complex are calculated from the electrostatic energies by solving the Poisson-Boltzmann equation. The computed redox potentials generally agree within 20 mV with the experimental measurements. The factors of protein environment and different energy terms determining a shift in the heme redox potentials are elucidated and analyzed in detail. In addition, the pH dependence of the redox potential of all hemes is evaluated in the physiologically relevant pH range. This chapter also rationalizes the effects of the applied ionic strength, external redox potential of bulk solvent (aqueous solution), the coupling of protonation and reduction reactions, and the distribution and population of the microscopic redox states in the studied biomolecular systems. The redox-Bohr effect in native Cyt*b* is modeled successfully, whereby the essential protonatable groups responsible for this phenomenon are identified. A good agreement between calculated and measured redox potentials offers the possibility to analyze the factors of the protein environment determining the shift in the redox potentials of hemes, which are cofactors in the synthetic and native cytochromes. That opens new avenues to understand how nature tunes the redox potentials of cofactors in proteins to perform their biological functions and how one can construct synthetic proteins whose cofactors have desired redox potentials. The applied approach is valuable in studying structure–function correlation and understanding protein–cofactor interactions in atomic detail.

Keywords: computational chemistry, molecular modeling, Poisson-Boltzmann equation, electrostatic energies, artificial protein, cyt bc_1 complex, cyt *b*, heme, redox potential, acid-base equilibrium, protein-cofactor interactions, redox-Bohr effect

Abbreviations

Cyt*b*	cytochrome *b*
Cytbc_1	cytochrome bc_1
Cytb_6f	cytochrome b_6f
Cytc_1	cytochrome c_1
E^0	standard redox potential
E_m, $E_{1/2}$	redox midpoint potential

E_{sol}	solution (external, ambient) redox potential
ET	electron transfer
ISP	iron-sulfur protein
LPBE	linear Poisson-Boltzmann equation
MD	molecular dynamics
PR (PRA, PRD)	heme propionic groups
PT	proton transfer
QM	quantum-mechanics
RMSD	root-mean-square deviation
SynCyt*b*	synthetic (artificial) cytochrome *b*

Introduction

The redox proteins contain redox-active centers, which can be organic or metal compounds and complexes essential for their function. The redox potential of the cofactors represents the key parameter of these proteins. The redox-active cofactors can be involved in inter- and intra-molecular electron transfer (ET) and light-induced charge separation processes. These ET proteins conduct, transport, or donate electrons. Besides the more basic assignment of a protein matrix to lock the cofactors in a preferred position and orientation to enable or facilitate function, the protein also manages to tune the redox potentials of its cofactors. In this way, an enzyme optimizes the requirements of its function. The same redox-active cofactor can have different redox potential values, depending on details of protein environment and solvent exposure, and thus fulfill different functions in different proteins.

In the companion chapter, "*Comparative Study of Native and Synthetic Cytochrome b: Modeling, Dynamics, and Bioinformatics Study*," we described a protocol for generating the structure of the artificial Cyt*b* from scratch (Rau and Haehnel 1998; Popović et al. 2001). The attention was concerted on protein design, molecular modeling, and characterization of the synthetic Cyt*b* protein.

In this chapter, we use the optimized structure of SynCyt*b* to evaluate the protonation pattern(s) and the redox potentials of its two hemes. Furthermore, we discuss the results of the redox potential calculations on the SynCyt*b* and analyze the factors of the protein environment that determine the heme redox potentials. Subsequently, we have analyzed the results of the redox potential calculations on the native Cyt*b* subunit of the mitochondrial Cyt*bc*$_1$ complex from the bovine heart (Iwata et al. 1998) to validate the procedure of calculating the redox potential of the hemes in proteins of known X-ray structure, to test the set of charges that we used in these applications and to compare these results with available experimental data. We have also provided a comprehensive and comparative study on the protonation properties and redox potential magnitudes of the native and synthetic Cyt*b*.

This chapter's primary research interest focuses on protein electrostatics, molecular bioenergetics, and the ET/PT reactions occurring in redox proteins. Accordingly, the native and synthetic Cyt*b* are the central subjects of our considerations. The effort is directed to calculating the protonation and oxidation patterns, evaluating their cofactors' redox potentials, and protein energetics. This chapter also rationalizes the effects of the applied ionic strength, external redox potential of solvent, coupling of the ET and PT reactions, the redox-Bohr effect, and the distribution and population of the microscopic redox states in the studied systems.

Section "*Material and Methods*" introduces the essential details of electrostatic theory applied to investigate the titration behavior of proteins. Section "*Respiratory Electron Transport Chain*" (previous chapter) provides necessary information about the structure, function, and working environment of the native Cytbc_1 complex in the mitochondrial respiratory electron transport chain, which is relevant for understanding its biochemical role and redox heterogeneity of the two Cyt*b* hemes.

The results presented here consist of the continuum electrostatic calculations by solving the linear Poisson-Boltzmann equation (LPBE) with a subsequent Monte Carlo titration of proteins. This approach is applied to the synthetic and native Cyt*b* to examine the titration behavior of ionizable residues and evaluate the redox potential of heme groups. The factors determining the redox potential of the two hemes in the SynCyt*b* are analyzed regarding the influence of different structural elements and amino acid residues.

Material and Method

Titration Curves of Ionizable Groups

The theoretical background of a method that calculates the titration curves of proteins is provided here. After a brief introduction to the acid-base and redox equilibria, the electrostatic calculations of the protonation and oxidation probabilities by solving the Poisson-Boltzmann equation numerically on a grid with a subsequent Monte Carlo titration are explained. Among the 20 essential amino acids in proteins, about half contain side chains that are protonatable groups. Also, some redox-active cofactors have a role in electron transfer processes. Since protonation and redox reactions are coupled, their equilibria have to be considered simultaneously. It requires redefining the standard definitions of the pK_a and E^0 values, which become dependent on pH and the ambient redox potential (E_{sol}). Only by using these values can the titration behavior and energetics in proteins be determined appropriately.

Acid-Base Equilibrium of a Single Titratable Group

Proton transfer reactions represent one of the most fundamental reactions in chemistry. The Lewis theory gives the most general definition of acid-base reactions, which defines the electron-pair acceptor group as an acid and the electron-pair donor group as a base. However, the proton transfer reactions in organic and inorganic chemistry as well as in biochemistry are ubiquitous, which explains why acid-base reactions are still identified with the Brönsted–Lowry definition, in which the acidic and basic character of a chemical species is defined by its ability to donate or accept protons, as a standalone type of reactions.

Among amino acids, two types of titratable residues can take part in the protonation equilibrium – acidic and basic amino acids. The protonation equilibrium of a single titratable group can be described for acids by Eq. 1 and bases by Eq. 2.

$$HA \xrightleftharpoons{K_a} A^- + H^+ \tag{1}$$

$$BH^+ \xrightleftharpoons{K_a} B + H^+ \tag{2}$$

The most important quantity that measures the thermodynamic acidity of a chemical species (AH, BH^+) is the acidity constant K_a, which is defined by the mass action law:

$$K_a = \frac{[A^-][H^+]}{[HA]} \tag{3}$$

The solution pH and pK_a of acid are the negative decadic logarithms of the hydrogen ion concentration and the K_a value.

$$\mathrm{pH} = -\log[\mathrm{H}^+] \tag{4}$$

$$\mathrm{p}K_a = -\log K_a \tag{5}$$

Using these definitions, the Henderson-Hasselbach equation (Eq. 6) is obtained from Eq. 3.

$$pH = \mathrm{p}K_a + \log\frac{[\mathrm{A}^-]}{[\mathrm{HA}]} \tag{6}$$

The pK_a value is related to the standard reaction (protonation) free energy $\Delta G^0{}_{prot}$ of the acid-base equilibrium, Eq. 7, where R is the gas constant, and T is the temperature.

$$\Delta G^0_{prot} = -RT\ln 10 pK_a \tag{7}$$

The probability <x> that the acid HA is protonated is given as, <x> = [HA] / ([HA] + [A$^-$]). With this definition and equations 3 and 6, one obtains the protonation probability of an acid HA as:

$$\langle x\rangle = \frac{10^{pK_a - \mathrm{pH}}}{1 + 10^{pK_a - \mathrm{pH}}} \tag{8}$$

This equation defines the pH dependence of the protonation probability of acid, and its graphical presentation is a sigmoidal curve, the so-called acid-base titration curve, in which <x> is plotted against pH. At pH equal to pK_a, protonation probability becomes <x> = 1/2, e.g., acid is in equilibrium where one-half of all molecules are deprotonated. The free energy required to protonate a single titratable group at a given pH and temperature is expressed by Eq. 9:

$$\Delta G_{prot} = RT\ln 10(pH - pK_a) = -RT\ln\frac{\langle x\rangle}{1-\langle x\rangle} \tag{9}$$

Redox Equilibrium of a Single Redox-Active Group

The redox-active group and its redox equilibrium can be treated similarly. Redox-active groups often found in proteins are specific cofactors like hemes, chlorophylls, copper centers, quinones, flavins (FAD), nicotinamide adenine dinucleotides (NAD), iron-sulfur clusters, and some aromatic amino acid residues, which can take part in the ET through a protein matrix using a hopping mechanism. For a single, isolated redox-active group, the thermodynamic equilibrium between the redox couple A^+_{oxd}/A_{red} is given as:

$$A^+_{oxd} + e^- \underset{}{\overset{K_{ET}}{\rightleftarrows}} A_{red} \tag{10}$$

and the K_{ET} equilibrium constant is defined with equation 11.

$$K_{ET} = \frac{[A_{red}]}{[A^+_{oxd}][e^-]} \tag{11}$$

In analogy to the derivation of the Henderson-Hasselbalch equation (Eq. 6), one defines the solution (ambient) redox potential (or the electromotive force) E_{sol} and the standard redox potential E^0 of the redox couple A^+_{oxd}/A_{red}, respectively as:

$$E_{sol} = -\frac{RT}{F}\ln[e^-] \tag{12}$$

$$E^0 = \frac{RT}{F}\ln K_{ET} \tag{13}$$

Here, the natural logarithm rather than the decadic logarithm is used. F is the Faraday constant, R is the universal gas constant, and T is the absolute temperature in K. The quantities E_{sol} and E^0 are given in the units of Volt. Taking the logarithm of Eq. 11, multiplying by a factor RT/F, and using the definition of E_{sol} and E^0 as given in Eqs. 12 and 13, one obtains the Nernst law defined by Eq. 14. The similarity with the Henderson-Hasselbalch equation (Eq. 6) is obvious.

$$E_{sol} = E^0 + \frac{RT}{F}\ln\frac{[A^+_{oxd}]}{[A_{red}]} \tag{14}$$

The E^0 value relates to the standard reaction free energy $\Delta G^0{}_{redox}$ given by Eq. 15.

$$\Delta G^0_{redox} = FE^0 \tag{15}$$

The probability $<x>$ that a redox-active group is in its oxidized state is defined as, $<x> = [A_{oxd}] / ([A^+{}_{oxd}] + [A_{red}])$. While the protonation probability of an acid depends on the solution pH value (see Eq. 8), the oxidation probability of an isolated redox couple depends on the solution's redox potential.

$$\langle x \rangle = \frac{10^{\frac{F}{RT}(E_{sol}-E^0)}}{1+10^{\frac{F}{RT}(E_{sol}-E^0)}} \tag{16}$$

The graphical presentation of this equation gives the sigmoidal redox titration curve, where $<x>$ is plotted against the solution redox potential E_{sol}. The concentration of the reduced and oxidized forms becomes equal at $E_{sol} = E^0 = E_m$ (redox midpoint potential).

From Eq. 16, one can read that the free energy required to oxidize a redox-active group at a given solution redox potential and temperature is:

$$\Delta G_{redox} = -F(E_{sol} - E^0) = -RT\ln\frac{\langle x \rangle}{1-\langle x \rangle} \tag{17}$$

The Model of a Protein in Solution

The pK_a (E^0) value of a titratable group in a protein can significantly differ from one in an aqueous solution. To understand the shift of the pK_a (E^0) value, one should look at the energetics of the group protonation (oxidation) equilibrium in different mediums. The titration behavior of a single acid, base, or redox-active molecular group in water is entirely determined by its pK_a (E^0) value in an aqueous solution. It displays standard sigmoidal curve that obeys the Henderson-Hasselbach (or Nernst) law. Nevertheless, it is well known that titration curves of natural proteins differ substantially from the sums of unperturbated titrations of the constituent acidic and basic groups. The dominant factor responsible for the difference arises from electrostatic interactions between titratable groups, e.g., (Tanford and Roxby 1972; Olson, Williams, and Allen 2013). If a titratable group is buried in a low dielectric hydrophobic region, it can also show non-standard titration behavior due to the protein environment. Hydrogen bonds could affect pK_a values, too.

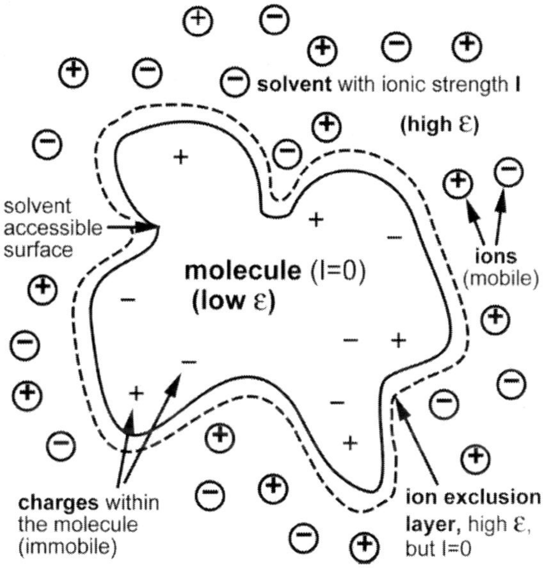

Figure 1. A protein molecule in a heterogeneous dielectric medium, using the continuum electrostatic model for a solvent. A protein can be modeled as a low dielectric cavity with fixed charges in a high dielectric environment with mobile ions.

Thus, in proteins, the local environment of titratable groups is much more complex due to the interactions between titratable groups, the atomic charges of polar groups, and ions in the surrounding solvent. Finally, the dielectric environment changes when the group is *transferred* from the aqueous solution into the protein. Figure 1 depicts a protein molecule in a heterogeneous dielectric medium and simplifies the continuum electrostatics model used for the calculations in this application. In this approach, a protein is described by the atomic partial charges and positions of all atoms placed in a low dielectric medium (typically, $\varepsilon = 2$ or 4) and ionic strength $I = 0$ M. The protein interior is separated from the solvent by the solvent accessible surface (the Connolly surface), indicated by a solid line. In contrast, a dashed line marks the border of the ion exclusion layer. The dielectric constant ε has a high value within the ion exclusion layer, but the ionic strength (I) is still equal to 0. The solvent is not represented by explicit molecules but rather implicitly by a medium of high dielectric value (for water, $\varepsilon = 80$) and ionic strength $I \neq 0$ (typically, $I = 0.1$ M). Thus, the mobile ions are also represented implicitly through ionic strength. The solvent-accessible surface is defined by rolling a sphere of radius, $r = 1.4$ Å (so-called probe radius) along the protein surface. This way, all holes and cavities inside a protein that can adopt and host the sphere of probe radius will be assigned a dielectric constant of $\varepsilon = 80$.

Continuum Electrostatic Method

The electrostatic calculations have been performed using the program suite MEAD (Bashford and Gerwert 1992; Bashford 1997), as described and reviewed in detail elsewhere (Gunner and Honig 1991; Bashford and Gerwert 1992; Yang et al. 1993; Bashford et al. 1993; Beroza and Fredkin 1996; Antosiewicz et al. 1996; Li et al. 1998; Warshel and Papazyan 1998; Martel et al. 1999). Here, we summarize the approach and describe the modifications employed in our calculations.

The methodology to calculate the protonation probability of titratable groups in proteins for a single and later for multiple protein conformations have been developed during the last three decades (Bashford and Karplus 1990, 1991; Beroza et al. 1991; Yang et al. 1993; Antosiewicz, McCammon, and Gilson 1994; You and Bashford 1995; Beroza and Fredkin 1996; Beroza and Case 1996; Alexov and Gunner 1997; Sham, Chu, and Warshel 1997; Beroza and Case 1998; Rabenstein, Ullmann, and Knapp 2000). These calculations are performed in two steps. In the first step, the intrinsic pK_a values and interaction energies between charged titratable groups are computed, solving the linear Poisson-Boltzmann equation (LPBE) numerically on a grid for a protein in an inhomogeneous dielectric medium and using a continuum electrostatic model for solvent. Suppose we assume that electrostatic interactions are the major contribution responsible for the difference between the protonation energies of titratable groups in a protein and an aqueous solution. It is possible to evaluate these energy differences with high accuracy using the LPBE (Honig and Nicholls 1995). Using the additivity of the equation solution, it becomes possible to separate it into several independent energy contributions. There are also applications where the non-linear Poisson-Boltzmann equation (NPBE) has been used, for instance, (Oberoi and Allewell 1993; Vila et al. 1998) and the references therein. However, the additivity of the linear PB equation gives the advantage to the linearized form compared with the non-linear methods.

In the second step, the obtained energy terms (the intrinsic pK_a values and W matrix elements with interaction energies between titratable groups) are used to compute titration curves for all titratable sites within a protein. Several approximation methods have been developed to compute protonation probabilities because the exact Boltzmann sum for a protein that often contains several tens to hundreds of titratable sites is impossible.

The pK_a and E^0 Calculations in Proteins

The Poisson-Boltzmann Equation

The electrostatic interactions and energies within the protein and with its solvent environment, as depicted in Figure 1, can be determined by the Poisson-Boltzmann equation in terms of the electrostatic potential $\phi(\mathbf{r}, \rho)$ at the position **r**, which arises from the charge distribution (density) $\rho(\mathbf{r'})$ at the position **r'**. Its derivation originates from the Poisson equation (the fundamental equation of electrostatics), which for an inhomogeneous dielectric medium ensures the form:

$$\nabla\left[\varepsilon(\vec{r})\,\nabla\phi(\vec{r})\right] = -4\pi\rho(\vec{r}) \tag{18}$$

All quantities, the electrostatic potential $\phi(\mathbf{r})$, the charge density $\rho(\mathbf{r})$ inside a protein, and the dielectric constant $\varepsilon(\mathbf{r})$ depend on the spatial coordinates and position **r**. The simple Coulomb law is no longer valid for the system with spatially variable dielectric properties.

The Poisson-Boltzmann equation can be derived by applying the Debye-Hückel theory of ionic solutions. If the mobile ions in the solvent are also considered, they arrange themselves in the electric field according to the Boltzmann distribution. Thus, an additional term should be added to the Poisson equation to describe the charge density of all ions in an aqueous solution.

$$\nabla\left[\varepsilon(\vec{r})\,\nabla\phi(\vec{r})\right]=-4\pi\left(\rho(\vec{r})+\rho_{ion}(\vec{r})\right)=-4\pi\left(\rho(\vec{r})+\sum_{s}c_{s}^{bulk}(\vec{r})q_{s}\exp\left(-\beta q_{s}\phi(\vec{r})\right)\right) \tag{19}$$

The sum runs over all kinds of ions *s*, where q_s and c_s^{bulk} are the charges, and the initial concentration of ions *s* and $\beta = 1/(k_BT)$. Note that the bulk ion concentrations depend upon position **r**, possessing zero value inside the solute (protein) and having the bulk solution values outside.

Introducing the ionic strength $I(\mathbf{r})$ and the inverse Debye length $\kappa(\mathbf{r})$ simplifies the second term of the equation:

$$I(\vec{r})=\frac{1}{2}\sum_{s}c_{s}^{bulk}(\vec{r})q_{s}^{2},\quad \sum_{s}c_{s}^{bulk}(\vec{r})q_{s}=0, \qquad \kappa(\vec{r})=\sqrt{8\pi\beta\, I(\vec{r})}\;. \tag{20}$$

Finally, with the additional term, the Poisson equation is transformed into the linear Poisson-Boltzmann equation (LPBE):

$$\nabla\left[\varepsilon(\vec{r})\,\nabla\phi(\vec{r})\right]=-4\pi\rho(\vec{r})+\kappa^{2}(\vec{r})\phi(\vec{r}) \tag{21}$$

The charge distribution $\rho(\mathbf{r})$ is derived from the atomic partial charges of a given molecular system. The ionic strength is defined through the parameter κ, the inverse Debye length. A dielectric constant of 80 is used for the solvent and 4 within the protein to obtain the electrostatic potential $\phi(\mathbf{r})$; for a complete discussion on the choice of dielectric constants, see the corresponding subsection below.

The electrostatic interaction $G_{i,j}$ of two molecular components *i* and *j* whose atoms possess partial charges q^{i}_{m} and q^{i}_{n} can be computed as

$$G_{ij}=\sum_{m}q^{i}_{m}\,\phi^{j}(\vec{r}^{\,i}_{m}) \tag{22}$$

where $\phi^{j}(\mathbf{r}^{i}_{m})$ is the electrostatic potential at the position of atom *m* belonging to molecular group *i*. Its value is obtained by solving the LPBE (Eq. 21) with the charge distribution from molecular group *j* as given below,

$$\rho^j(\vec{r}) = \sum_n q_n^j \, \delta(\vec{r} - \vec{r}_n^{\,j}) \tag{23}$$

If $I(\mathbf{r}) = 0$, we get to the Poisson equation back. The advantage of the LPBE is the additivity of electrostatic potentials and charge densities, which makes calculations faster and easier. However, electrostatic potentials computed from the LPBE are additive as long as the dielectric boundaries of a macromolecule do not change. Thus, the conformation of titratable groups (or, more generally, a protein conformation) cannot be changed during a proton exchange. Due to the additivity of the potentials obtained from LPBE, it is possible to separate the individual contributions of the titratable group protonation energy and treat them independently.

For a detailed discussion on the LPBE and the Debye-Hückel theory, see the textbook of McQuarrie (1976). This equation enables computing the electrostatic potential of a macromolecule of known structure in solution. However, an analytical solution is possible only for elementary geometries, such as planes or spheres (Kirkwood 1934; Daune 1997). It can be solved only numerically for proteins with complex dielectric boundaries at the protein surfaces. The most common method implemented in the electrostatics simulation programs, such as **D**elPhi (Nicholls and Honig 1991; Honig and Nicholls 1995) and **M**EAD (Bashford and Gerwert 1992; Bashford 1997), and used in this study is a finite difference method (Warwicker and Watson 1982; Nicholls and Honig 1991). Alternatively, but not so often, other numerical methods were applied, for instance, boundary element methods (Zauhar and Varnek 1996) or multigrid-based methods (Holst et al. 1994).

The Boltzmann Sum and Protonation/Redox State Energy

The linear Poisson-Boltzmann equation (LPBE) was solved for the appropriate molecular systems to obtain the electrostatic energies of the different protonation and redox states. Protonation and redox states of the individual ionizable groups depend on each other. The probability $\langle \mathbf{x}_\mu \rangle$ of a variably charged group μ to be protonated or oxidized can be expressed by a thermodynamic average over all 2^N possible charge patterns of the N variably charged groups (protonatable and redox-active) (Yang et al. 1993; Beroza and Case 1998; Ullmann and Knapp 1999) of the molecular system according to the Boltzmann sum:

$$\langle \mathbf{x}_{\mu} \rangle = \sum_{n=1}^{2^N} x_{\mu}^{(n)} \cdot \exp\left(-\frac{G^{(n)}}{RT}\right) \Big/ \sum_{n=1}^{2^N} \exp\left(-\frac{G^{(n)}}{RT}\right) \tag{24}$$

where T is the absolute temperature and R is the universal gas constant. For simplicity, we assumed that only two different charge states are available for each variably charged molecular group. However, a generalization is easily possible. The charge state of group μ for the charge pattern (n) is characterized by the integer $x_{\mu}^{(n)}$, which adopts the value of 0 or 1 depending on whether group μ is deprotonated (reduced) or protonated (oxidized).

The free energy $G(\boldsymbol{n})$ of a specific protonation/redox state of the enzyme $(\boldsymbol{n})$ is given as,

$$G(\vec{n}) = \sum_{\mu=1}^{P} x_{\mu}^{(n)} RT\ln 10\left(pH - pK_{a,\mu}^{intr}\right) - \sum_{\eta=1}^{K} x_{\eta}^{(n)} F\left(E_{ext} - E_{intr,\eta}^{0}\right) + \frac{1}{2}\sum_{\mu=1}^{N}\sum_{\nu=1}^{N} x_{\mu}^{(n)} x_{\nu}^{(n)} W_{\mu\nu} \tag{25}$$

or in a more compact form as,

$$G(\vec{n}) = \sum_{\mu=1}^{N}\left[x_{\mu}^{(n)} G_{\mu}^{\text{int}} + \sum_{\nu>\mu}^{N} x_{\mu}^{(n)} x_{\nu}^{(n)} W_{\mu\nu}\right] \tag{26}$$

where $(\boldsymbol{n})$ is the N-dimensional protonation/redox state vector, P is the number of protonatable groups, K is the number of redox sites, and $N = P + K$ is the total number of ionizable groups in a protein.

The mutual interaction energy $W_{\mu\nu}$ of the titratable groups, μ and ν, refer to their charge pattern, where all other ionizable groups of the molecular system are in the neutral (uncharged or reduced) reference state. The strength of the electrostatic coupling between the two titratable groups, μ and ν, is expressed by the $W_{\mu\nu}$ matrix elements. The intrinsic free energy, ΔG_{μ}^{intr}, accounts for the three additional energy terms:

$$\Delta G_{\mu}^{intr}(pH) = RT\ln 10\left(pH - pK_{a,\mu}^{\text{model}}\right) + \left(\Delta G_{Born,\mu}^{protein} - \Delta G_{Born,\mu}^{\text{model}}\right) + \left(\Delta G_{back,\mu}^{protein} - \Delta G_{back,\mu}^{\text{model}}\right) \tag{27}$$

The first term in Eq. 27 accounts for the electrostatic free energy of the corresponding titratable group μ, which is used as a model compound in an aqueous solution with a known pK_a (E^0) value. For titratable amino acids and redox-cofactors, the experimental values of the model compounds in an aqueous solution are taken from the literature (Nozaki and Tanford 1967; Tanokura 1983; Moore and Pettigrew 1990; Wilson 1983) and listed in reference (Popović et al. 2002). Transferring the titratable group μ from an aqueous solution into a protein in which all other titratable groups are in their uncharged reference state will cause an energy shift that can be further divided into two contributions. The second term, the Born energy (desolvation penalty), accounts for the difference in self-energy to protonate (oxidize) the titratable group μ in the protein and aqueous solution. The last term provides the corresponding energy difference resulting from the interactions of group μ with the background charges, including backbone dipoles, i.e., the charges of non-titratable residues and the charges of all other titratable sites in their neutral reference state. In case the group μ is redox-active, the first term in Eq. 27 must be replaced by $F(E_\mu^{model} - E_{sol})$, where F is the Faraday constant, E_{sol} is the external redox potential of the solvent, and E_μ^{model} is the redox potential of the model compound in aqueous solution.

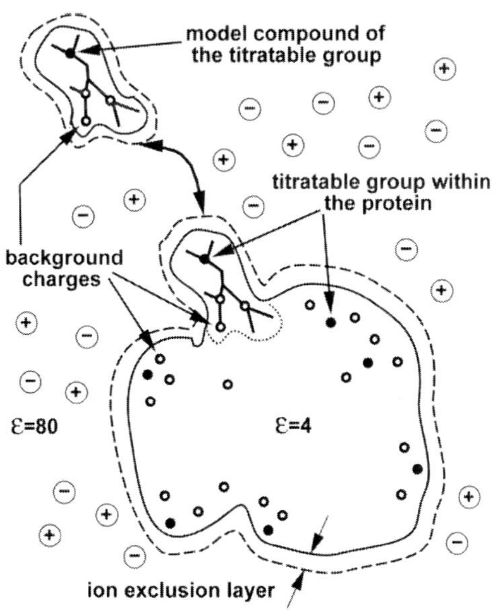

Figure 2. A titratable group within a protein and the corresponding model compound (in the same conformation) in an ionic aqueous solution. The dielectric environment and the electrostatic interactions are changed by transferring the titratable group from the aqueous phase into the protein, and this causes a shift in the protonation energy, which can be divided into four energy contributions.

All three energy terms relate to the situation where the protonation of the titratable sites in a protein is considered independently of each other in the sense that all other sites are uncharged and neutral. However, the protonation behavior of the titratable sites depends firmly on neighboring charges, which could be very strongly coupled. It means that one more energy term is needed to describe the interactions between the titratable sites when they are charged. These energies are given in the form of the W-matrix elements, assuming that $W_{\mu\mu} = 0$. See the references (Beroza and Fredkin 1996; Bashford 1997) for a detailed description of these energy terms.

Thus, the protonation energy of a titratable group μ in a protein can be considered to consist of four contributions. These contributions for group μ are the pK_a value of the model compound (p$K_{a,\mu}^{\text{model}}$), the Born-solvation energy term ($\Delta\Delta G_\mu^{\text{Born}}$), the influence of background charges ($\Delta\Delta G_\mu^{\text{back}}$), and the energy of coupling two charged sites ($W_{\mu\nu}$). These energy terms are depicted in the self-explanatory Figure 2.

The free energy $G(\boldsymbol{n})$, Eq. 25, of the charge pattern ($\boldsymbol{n}$) of the considered molecular system is determined by calculating the electrostatic energies from the solution of the LPBE using Bashford's program **M**EAD (MULTIFLEX) (Bashford and Gerwert 1992; Bashford 1997). With the electrostatic energy $G(\boldsymbol{n})$ for each configuration ($\boldsymbol{n}$) and Eq. 24, one can compute the thermodynamically averaged charged state of all titratable groups in the protein and evaluate their pH dependence. However, the exact Boltzmann averaged sums over all 2^N configurations for a protein containing over 1,000 titratable sites is not feasible. Therefore, several approximation methods have been developed: the Tanford-Roxby approximation, which is a mean-field approximation to the exact theory (Tanford and Roxby 1972), the reduced site approximation (Bashford and Karplus 1991), a hybrid statistical mechanical/Tanford-Roxby approximation (Yang et al. 1993), the cluster method (Gilson 1993) and the Monte Carlo (MC) method (Beroza et al. 1991; Alexov and Gunner 1997; Rabenstein and Knapp 2001). A detailed review of all these different methods can be found in (Gilson 1995). In this work, to evaluate protonation (oxidation) probabilities $\langle x_\mu \rangle$, the Monte Carlo (MC) method with the Metropolis criterion (Metropolis et al. 1953) implemented in the program **K**ARLSBERG (Rabenstein 1999) has been employed.

Definition of pK_a and E^0 of Ionizable Groups in a Protein Environment

The free energy difference between the protonated (oxidized) and deprotonated (reduced) states of the ionizable group μ depending on pH value and redox potential of the solvent E_{sol} is defined as the ratio of probabilities that this group is charged or uncharged at given ambient conditions according to:

$$\Delta G_{\mu}\left(pH, E_{sol}\right) = -RT\ln\frac{\langle x_{\mu}\rangle}{1-\langle x_{\mu}\rangle} \quad . \tag{28}$$

ΔG_{μ} represents the free energy of group μ thermodynamically averaged over all protein protonation and redox states at a given solvent pH and E_{sol}. With this free energy difference, the pK_a value of a protonatable group μ in the protein environment can be written as,

$$pK_{a,\mu}^{protein}\left(pH, E_{sol}\right) = pH - \frac{\Delta G_{\mu}\left(pH, E_{sol}\right)}{RT\ln 10} \quad . \tag{29}$$

This expression is related to the sigmoidal Henderson-Hasselbalch titration curve, which describes the protonation probability $<x_{\mu}>$ dependence on pH, as explained in detail elsewhere (Mao, Hauser, and Gunner 2003; Popović and Djordjević 2020). For a redox-active group η, the standard redox potential of the group in the protein environment can be written as,

$$E_{\eta}^{protein}\left(pH, E_{sol}\right) = E_{sol} + \frac{\Delta G_{\eta}\left(pH, E_{sol}\right)}{F} \tag{30}$$

It is related to Nernst law, which is analogous to the Henderson-Hasselbalch titration curve. Note that for an isolated variably charged group μ, the expressions 29 and 30 are equivalent to the conventional definitions of the p$K_{a,\mu}$ value, and the standard redox potential $E^0{}_{\eta}$. For several mutually interacting variably charged groups, the free energy difference ΔG_{μ} in Eqs. 29 and 30 possess an implicit pH and E_{sol} dependence besides the trivial dependence for independent variably charged groups, RT·ln10[pH – p$K_a{}^{\text{model}}$] for protonatable groups, and F[$E_{\mu}{}^{\text{model}}$ – E_{sol}] for redox-active groups. This trivial dependence ensures that the pK_a and E^0 values are constant for isolated variably charged groups. However, with the generalized definitions 29 and 30, the pK_a value and the standard redox potential E^0 become pH and E_{sol}–dependent (Baptista, Martel, and Soares 1999; Popović et al. 2001). Formally, one can get rid of these dependencies by defining p$K_{1/2,\mu}{}^{\text{protein}}$ and $E_{1/2,\eta}{}^{\text{protein}}$ values that are set equal to the pH (E_{sol}) value of the solvent where the probabilities of the charged and uncharged state of the considered variably charged group μ are equal, i.e., at $<x_\mu> = 1/2$ and consequently $\Delta G_\mu = 0$. These values are equivalent to the conventional pK_a and E^0 values for an isolated variably charged group. The above expressions for pK_a^{protein}(pH, E_{sol}) and the standard redox potential E_η^{protein}(pH, E_{sol}) may be more useful if one prefers to characterize the energetics of protonation and redox reactions in proteins in all detail (Gunner and Alexov 2000; Ullmann 2000; Popović et al. 2001). We consider the later definitions in this contribution, but for the synthetic Cyt*b*, the p$K_{1/2,\mu}^{\text{protein}}$ and $E_{1/2,\eta}^{\text{protein}}$ values are also used.

Dielectric Constant

In this approach, the solvation effects are modeled by a dielectric constant ε that depends on the spatial position **r**. The effect of the medium on the electrostatic interactions is described with an appropriate ε value. $\varepsilon(\mathbf{r})$ varies within the protein-solvent system, and a choice of the dielectric constant value for a protein is a controversial topic of the phenomenological nature (Warshel and Russel 1984; Gilson and Honig 1986; Harvey 1989; Warshel and Aqvist 1991; Gilson 1995; Warshel and Papazyan 1998).

Experimental and theoretical investigations suggest that if a "poor charge model" is used, better results are obtained with a dielectric constant of 20 for a protein (Antosiewicz, McCammon, and Gilson 1994). However, with a very "detailed charge model," a dielectric constant of 4 is more appropriate (Antosiewicz et al. 1996). In the former model, the unit charge is placed only on one atom of a charged titratable group, while in the latter one, the charge is distributed over all atoms of the titratable group. Also, a large dielectric constant (ε =10–20) gives better agreement with experiments for the residues on the protein surface, while a small dielectric constant (ε = 2–4) explains better the behavior of the residues buried in a protein (Simonson, Perahia, and Bricogne 1992; Simonson and Perahia 1995a, b; Demchuk and Wade 1996; Simonson and Brooks 1996; Lancaster et al. 1996; Alexov and Gunner 1997). That is not surprising since residues on the protein surface are partially exposed to or in direct contact with solvent molecules, while the residues inside a protein are usually surrounded by other inner residues.

Further, a large value of ε compensates for an unrealistic charge distribution used in a "poor model," but even with a "detailed charge model," a dielectric constant of 2 or 4 has to be used instead of 1. The explanation is that a factor of 2 should compensate for the effects of electronic polarization. Another factor of 2 (or more) is used for the effects of the nuclear polarization, i.e., for the reorientation of permanent dipoles and displacement of atoms carrying atomic partial charges due to thermal fluctuations within the protein structure.

Computational Models

The protonation states of the protonatable residues and the oxidation states of the redox-active groups (hemes) were determined simultaneously by calculating the electrostatic energies, solving the linearized Poisson-Boltzmann equation (LPBE) numerically on a grid with a finite difference method, and using a subsequent Monte Carlo (MC) titration technique. The redox potentials of the hemes in the native and synthetic Cyt*b* were evaluated by averaging over all sampled protonation and redox states of the titratable groups. All four considered hemes from the native and synthetic Cyt are *b*-type hemes (protoporphyrin IX) with two histidines axially coordinated to the iron. Although all hemes are of the same chemical nature, their redox potentials differ since they are placed in different protein environments. The fundamental theoretical problem in calculating the pK_a and redox potential values is to predict how the protein environment shifts these values of the variably charged groups in the protein from that of the isolated amino acid or heme in solution. Assuming that electrostatic interactions play the central role in tuning pK_as and redox potentials in the proteins, the differences in the redox potentials are caused by different interactions with other charged and polar groups of the protein and by different solvent accessibility of the hemes. Because the protonatable groups also contribute to the redox potential difference due to the mutual interactions with hemes, the redox potential of the hemes also depends on the ambient pH.

The electrostatic energy of titratable groups in the protein environment is calculated relative to suitable model systems, which serve as a reference. The model compounds of titratable amino acid side chains are the corresponding isolated amino acids in an aqueous solution where the N- and C-termini are acetylated and amidated. For the redox potential of the heme, we used as a reference system the *bis*-histidine heme model compound in an aqueous solution, where the experimental value of the redox potential is –220 mV (Wilson 1983). The electrostatic method assumes that the difference in titration behavior of an ionizable group in protein and a model compound can be accounted for by calculating the difference in the electrostatic work of altering the charges from the unprotonated to the protonated state in the protein and the work of making the same alteration in the model compound. The electrostatic potential is given by the LPBE in which the protein interior has a low dielectric value (ε = 4); for the choice of this value, see: (Honig and Nicholls 1995; Warshel, Papazyan, and Muegge 1997; Warshel and Papazyan 1998; Rabenstein, Ullmann, and Knapp 1998b). The solvent has a high dielectric value (ε = 80), and the Boltzmann counterion terms are excluded from a region near the protein surface. Using the finite difference method to solve the LPBE, we can consider the detailed shape of the protein surface as defined by the atomic coordinates of the protein and the charge distribution given by the atomic partial charges. The PB calculations were carried out using the program **M**ULTIFLEX from the **M**EAD (Macroscopic Electrostatics with Atomic Detail) suite of programs (Bashford and Gerwert 1992; Bashford et al. 1993; Bashford 1997).

First, it computes the electrostatic potentials for each titratable group in an aqueous solution and within the protein for protonated (oxidized) and deprotonated (reduced) forms. Further from the electrostatic potentials, the Born solvation energy term, the energy term of the interactions with background charges, and the coupling energy between each pair of variably charged sites are calculated. The first two terms, together with the $pK_{mod,\mu}$ value of the proper model compound (Popović et al. 2001), contribute to the intrinsic $pK_{int,\mu}$ value, while the interaction energies between the charged titratable sites are elements of the so-called W-matrix. The mobile ions in a solvent were modeled by the ionic strength of 100 mM. An ion exclusion layer of 2 Å and a solvent probe radius of 1.4 Å was used. To solve the LPBE, a grid focusing procedure was used with decreasing grid spacing.

The computation of the electrostatic energies $G(n)$ (Eq. 25) for the different charge patterns n of a protein and a subsequent Monte Carlo titration was done with the program **K**ARLSBERG (Rabenstein 1999; Rabenstein and Knapp 2001). The number of titratable groups in synthetic and native Cyt*b* is too large to evaluate the Boltzmann averaged sums (Eq. 24) directly. Therefore, we used a Monte Carlo (MC) titration method, incorporated in the program **K**ARLSBERG, to sample the protein's protonation and redox states and evaluate the protonation and oxidation probabilities of titratable groups at different pH and E_{sol} values. To improve the MC titration's sampling efficiency, two (three) titratable groups, which coupling is stronger than 2.5 (5.0) pK_a units, are included to change their protonation state in one MC move simultaneously. Such double and triple moves were made in addition to simple moves. Each MC scan consists of at least N simple MC moves, where it is attempted to change the protonation state of one titratable group. For each MC titration, we performed 5,000 full MC scans and 50,000 reduced scans, where all titratable groups that did not change their protonation state after the first 5,000 MC scans were fixed in their respective protonation states and excluded from further sampling. This MC titration procedure led to a standard deviation of fewer than 0.01 protons in each titratable group. We applied a complete sampling procedure for each pH value varying from 0 to 14 in steps of 0.1 pH units at 300 K and a solution redox potential of $E_{sol} = 0$ mV, corresponding to an acid-base titration. For the SynCyt*b*, we performed redox titrations at pH range from 5 to 9 and 300 K, changing the solution redox potential E_{sol} with a 20 mV increment in the range from -400 to +300 mV. A redox titration was done for each step of 0.2 pH units in the pH range between 5 and 9.

Accordingly, in this application, we considered for SynCyt*b* the definitions of the $pK_{a,\mu}(pH, E_{sol} = 0)$ and $E_{\eta}^{0}(pH, E_{sol})$ (see Eqs. 29 and 30), as well as the $pK_{1/2,\mu}$ and $E_{1/2,\mu}$ if not otherwise stated.

Titratable Groups

The C-termini, N-termini, heme propionates, aspartates, glutamates, arginines, lysines, cysteines, tyrosines, and histidines, except for the histidines coordinated to hemes, are considered as titratable groups, regardless whether they are involved in anionic bridge or not. The two heme porphyrins with the corresponding axially ligated imidazoles are redox-active sites (H_l and H_h) in the native and artificial Cyt*b*. Therefore, each heme contributes with three variably charged groups (one redox-active site and two PR groups). Each non-ligating histidine possesses two titratable sites. Specifically, two hydrogen atoms (δ-H and ε-H) at the imidazole ring can get deprotonated. Our calculation considered both forms (Bashford et al. 1993; Beroza and Case 1998) by representing them as two strongly coupled titratable sites, which cannot simultaneously be deprotonated. The Cytbc_1 complex contains two additional cofactors, a heme c_1 of the Cytc_1 subunit and the iron-sulfur cluster from the Rieske (ISP) subunit. Both are taken to be in the reduced state because their redox potentials have high positive E^0 values. Furthermore, since the heme of the Cytc_1 subunit is more than 35 Å away from the closest heme (H_l) of the Cyt*b* subunit, the solvent-exposed PR groups of heme c_1 are considered as non-titratable and are fixed in their deprotonated state.

Atomic Partial Charges

Most of the charges and atomic radii that we used in this application have been used successfully before to calculate the redox potentials of cofactors (Popović et al. 2001; Popović et al. 2002; Medvedev et al. 2008; Couch, Popović, and

Stuchebrukhov 2011), the protonation patterns in proteins (Rabenstein, Ullmann, and Knapp 1998a; Rabenstein and Knapp 2001; Popović and Stuchebrukhov 2005, 2006; Popović et al. 2010; Popović and Djordjević 2020), and bioenergetics of ET and PT reactions in different enzymes and proteins (Rabenstein, Ullmann, and Knapp 1998b; Popović and Stuchebrukhov 2004; Quenneville, Popović, and Stuchebrukhov 2006; Popović and Stuchebrukhov 2012).

All additional charges needed in this application were computed with previous procedures we used in similar applications. The atomic partial charges of the amino acids, including the protonated and deprotonated states of the titratable amino acids, were taken from the CHARMM22 (MacKerell et al. 1998) parameter set. However, for some titratable residues (Arg, Cys, Lys, Tyr, C-terminus, N-terminus), charges are available for the standard protonation state only. For these residues, the atomic partial charges were calculated with the program GAUSSIAN98 (Frisch et al. 1998; Frisch, Pople, and Binkley 1984) using the semi-empirical PM3 method and the CHELPG procedure (Breneman and Wiberg 1990). The obtained charges appropriately reproduce the electrostatic potential outside the considered molecular group. Yet, the charges for the iron-sulfur cluster were taken from Izrailev et al. (1999).

The heme was divided into three parts, where the porphyrin system and the PR groups were treated as independent titratable groups that interact electrostatically only. Thereby, the porphyrin moiety of the heme with the two axially ligated imidazoles is considered to be the redox-active site. For the atomic partial charges of the propionic acid, the same charges were used as for the side chain of glutamate taken from the CHARMM22 parameter set. The electronic wave functions of different redox states of the *bis*-imidazole heme model compounds were calculated with the Hartree-Fock method from GAUSSIAN98 using the 6-311+G* basis set for the iron atom and 6-31+G* basis set for all other atoms. The atomic partial charges were obtained from the wave functions of the quantum chemical computations by adjusting the electrostatic potential in the neighborhood of the heme with the Merz-Kollman approach (Besler, Merz, and Kollman 1990).

To reduce the number of independent atomic partial charges, which have to be determined, we constrained the charges of hydrogen atoms bound at CH_2 or CH_3 groups to be equivalent, except for the hydrogens at the CAA and CAD carbon atoms, where we fixed the charges of two hydrogens at 0.09 and of one at 0.00. In the complete heme, the neutral hydrogen is replaced by the charge-neutral CH_2 group of the propionic acid. Also, the hydrogen atom charge at the axially ligated imidazole rings that will be later replaced by the $C_{\beta}H_2$ group was constrained to a value of 0.09 since it is the total charge of that group in CHARMM force-field. In similar applications, one can get the negative charge for a metal atom/ion. Even the charge set with a negative charge at the iron atom gives virtually the same results for the electrostatic energies and the redox potentials (Rabenstein, Ullmann, and Knapp 1998a, b, 2000). Nevertheless, we preferred to avoid a negative charge at the heme iron and fixed its charge to be +0.48 and +0.25 for oxidized and reduced *bis*-histidine heme, respectively. The above-mentioned restrained charges and the dipole and quadrupole moments were constrained using the program RESP (Restrained ElectroStatic Potential) from Kollman's group. The RESP fits the quantum mechanically calculated electrostatic potential at molecular surfaces using an atom-centered point charge model (Bayly et al. 1993; Cornell et al. 1993; Cornell et al. 1995). A particularly convenient feature of this program is the possibility of constraining partial atomic charges of some atoms and then calculating a molecule's charge distribution, keeping the dipole and quadrupole moment constant. The partial atomic charges and calculated redox potentials depend on the orientation of the ligated imidazole rings and the heme conformation (Popović and Stuchebrukhov 2004, 2012).

Therefore, we performed the quantum chemical computation for all four conformations of the heme model compounds (two for the native and two for the synthetic Cyt*b*). With this method, the atomic partial charges were calculated to faithfully represent the electrostatic potential in the neighborhood of the heme model compound. The partial atomic charges of the *bis*-histidine heme model compound and for all other titratable residues are given in the original publication by Popović et al. (2001).

Preparation of Structures for Electrostatic Computations

The electrostatic calculations of the native Cyt*b* were done using the X-ray structure (PDB code 1BE3, Iwata et al. 1998) of the bovine heart Cytbc_1 complex, consisting of 11 subunits. The three central subunits of the complex III include Cyt*b*, Cytc_1, and Rieske iron-sulfur protein. The hydrogen atom coordinates were generated with the HBUILD command of CHARMM22, leading to an all-atom representation of the Cytbc_1 complex. Subsequently, the hydrogen atom positions were energetically minimized, keeping all other atoms fixed.

For calculations using only the Cyt*b* subunit, we took the C chain of the Cytbc_1 complex. It consists of 379 residues and two hemes with a total number of 6242 atoms. It possesses 79 titratable groups, including the hemes, propionates, and histidines. To solve the LPBE for that system, a two-step grid focusing procedure was applied, starting with a cubic lattice of 151 Å length and 1.0 Å grid spacing, followed by a 22.75 Å cube with 0.25 Å grid spacing. The Cytbc_1 complex contains 2030 residues, three hemes, and one iron-sulfur cluster with a total of 31996 atoms. We considered 661 variably charged sites as protonatable or redox-active. The LPBE for the whole Cytbc_1 complex was solved using a three-step grid focusing, starting from a 302 Å cube lattice with a grid spacing of 2.0 Å, followed by a 101 Å cube with 1.0 Å grid spacing, and finally a 25.25 Å cube with a grid spacing of 0.25 Å. The lattices of the lower resolution were centered on the geometric center of the protein. The lattice with the highest resolution was centered at the considered titratable groups.

In comparison, the SynCyt*b* contains 122 residues and two hemes, with 2218 atoms. The 52 titratable amino acids, two redox-active hemes, and four PR groups give 58 variably charged groups in the system. To solve the LPBE for the SynCyt*b*, a two-step grid focusing was performed, starting with a cubic lattice of 105 Åwith 1.0 Å grid spacing, followed by a 22.75Å cube with 0.25 Å grid spacing.

Results and Discussion

Calculations on the Synthetic Cytochrome *b*

Protonation Pattern of Titratable Groups

Since the protonation and redox states of the titratable groups depend on each other, for a given pH and solution redox potential E_{sol}, the ensemble of protonation and redox patterns had to be calculated simultaneously by a statistical averaging method employed in an MC program. Most titratable groups, at pH = 7 and a solution redox potential of E_{sol} = 0 mV, are in the standard protonation state. Only Arg5(F_1) and the propionate PRA(H_h) exhibit deviations from standard protonation behavior, see Figure 3. Their protonation probabilities at pH = 7 and E_{sol} = 0 mV, are 0.916 and 0.654, respectively. The propionate PRA(H_h) has a higher p$K_{a,1/2}$ value than PRA(H_l) and starts to get deprotonated just below pH = 6. That may be due to the salt bridge of PRA(H_h) with Arg5(F_1), which is slightly "weaker" (more distant) than the corresponding salt bridge involving PRA(H_l), and due to the interaction of the positively charged Lys23(G_1) and Lys23(G_2), which are close to PRA(H_h) and may weaken its salt bridge. These two Lys could be responsible for what Arg5(F_1) starts to deprotonate already at pH = 6.5. In the pH range of 7 to 12, Arg5(F_1) exhibits a non-monotonous titration behavior. In the same pH range, the titration curve of PRA(H_h) exhibits a plateau. Both are probably due to a strong electrostatic interaction between Arg5(F_1) and PRA(H_h), which are involved in a salt bridge. They are situated in the more hydrophobic core of the four-helix bundle, which also may influence their mutual interactions.

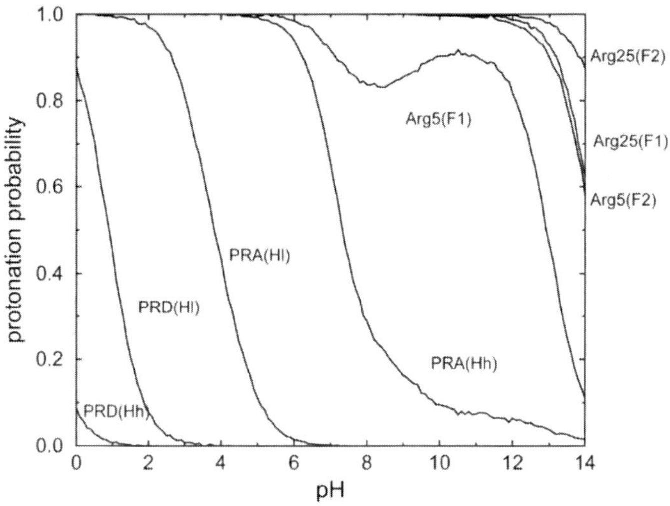

Figure 3. Titration curves of arginines and heme propionates in the SynCyt*b* at E_{sol}= 0 mV. The PR groups are denoted according to the pyrrole ring of porphyrin attached to (A or D) and the heme identity (high- H_h or low- H_l potential heme). The arginines Arg5(F_1), Arg5(F_2), Arg25(F_1), and Arg25(F_2) form salt bridges with PRA(H_h), PRD(H_h), PRA(H_l), and PRD(H_l), respectively. The corresponding O···H distances between the PR groups and the arginines are 2.4 Å, 1.7 Å, 2.1 Å, and 1.6 Å.

At neutral pH, the charges of four arginines and four propionates and the charges of lysines and glutamates cancel each other. Nevertheless, the total charge of the model protein is +2 due to the charges of the two oxidized hemes. Therefore, the pI value (isoelectric point), which corresponds to the zero charge of a protein, is shifted towards higher pH values (pI = 9.3). Monitoring the pH dependence of the total charge of the SynCyt*b* (Figure 4), one can clearly distinguish titration in two steps, where the acidic groups titrate at a pH of around 4 and basic groups at a pH of about 11.

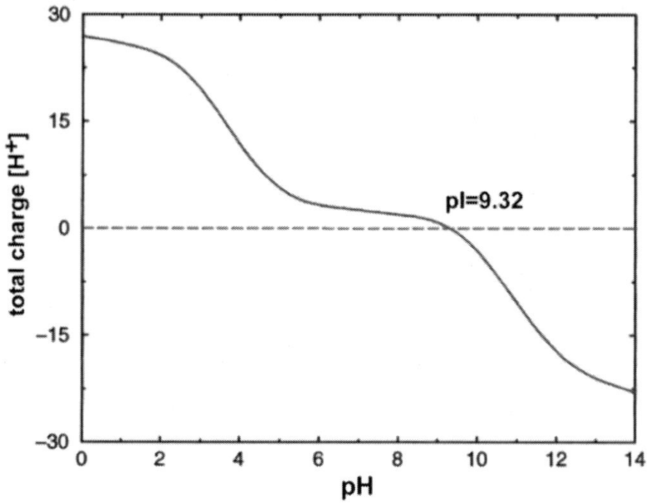

Figure 4. Titration curve of the SynCyt*b* at E_{sol}= 0 mV. The total charge of SynCyt*b* is given as a function of pH.

Experimental Values of Redox Potentials

The redox potential of the hemes in the SynCyt*b* at T = 293 K, pH = 7, and I = 100 mM was determined by redox potentiometry. The redox titration curve was obtained by plotting the absorbance difference at the α-band (560nm) of the reduced minus that of the fully oxidized species against the ambient redox potential in the range between +50 to –300 mV. The data for the titration in reductive and oxidative directions exhibit two distinct midpoint potentials of –106 mV and –170 mV with a standard deviation of ± 8 mV (Rau and Haehnel 1998). These values are significantly higher than the midpoint potential of –220 mV of a synthetic four-helix bundle which accommodates one *bis*-histidine ligated heme (Choma et al. 1994), and the midpoint potential of –215 mV of an imidazole-histidine ligated single heme octapeptide (Harbury et al. 1965).

In line with these values are also the potentials of –205 and –215 mV of the two non-interacting hemes of the four-helix bundle cytochromes of Robertson et al. (1994). These synthetic four-helix bundles consist of a single polypeptide chain, which cannot efficiently protect the hemes from the solvent. Hence, the values around –220 mV seem to be typical for *bis*-histidine ligated hemes in a hydrophilic protein environment and without arginines to form salt bridges with the PR groups. The redox potential of the isolated heme with axially ligated imidazoles in solution was also measured to be –220 mV

(Wilson 1983). The structure of Haehnel's synthetic Cyt*b* is characterized by close packing of the four-helix bundle around the hemes, better shielding against the solvent, and strong electrostatic interactions of the heme PR groups with four arginines. All of that results in higher values of the redox potential of the hemes.

Compared with the native complexes, the closest values of the heme redox potential are –50 mV and –170 mV (Kramer and Crofts 1994; Kramer et al. 1994) found for the two hemes in the cytochrome b_6f (Cytb_6f) complex. The two hemes in the Cyt*b* subunit of the native Cytbc_1 complex from the bovine heart show more positive values of the redox potential, +93 mV and -34 mV at pH = 7 (Nelson and Gellerfors 1974, 1975). Although the design of the heme-binding core was adopted, being very close to the Cyt*b*-subunit of the Cytbc_1 complex from *Rhodobacter captulatus*, there are significant differences in the wavelength of maximum absorbance and the redox potential of the hemes in the native and synthetic Cyt*b*.

In the native Cyt*b*, H_h and H_l refer to the hemes with high and low redox potentials, which correspond to the positions near the open end and the template, respectively, when the orientation of the helices is used as a reference (Rau & Haehnel 1998). However, the heme near the template is better shielded from the solvent and should, therefore, possess a higher redox potential value than the one at the open end. A reliable assignment of the high and low potential heme in the SynCyt*b* was possible by orienting the four-helix bundle with the open ends of the two *G* helices fixed at a gold electrode and observing the voltage dependence of the current of the ET from the gold electrode to the SynCyt*b* hemes in a time-resolved experiment (Willner et al. 1999). This method measured the redox potentials of the two hemes close to the open end and the template as –182 mV and –115 mV, respectively. Hence, concerning the helix orientation, the high and low potential hemes in the SynCyt*b* are exchanged relative to the native Cyt*b*. Our computations will show to what extent a distinct heme environment and the electrostatic interaction between two almost identical hemes cause the difference in the redox potentials.

Calculated pH-Dependent Heme Redox Potentials

Using the calculated oxidation probabilities and expressions 28 and 30, we evaluated the redox potential of the hemes. The experimental and theoretically calculated redox potentials at pH = 7 and E_{sol} = 0 mV are given in Table 1.

The obtained values of –91 mV for the heme near the template and –167 mV for the heme near the open end of the four-helix bundle agree very well with the measured values of the SynCyt*b* in solution, which are –106 mV and –170 mV (Rau & Haehnel 1998). However, they differ from the values –115 mV and –182 mV obtained for the SynCyt*b* attached to a gold electrode (Willner et al. 1999). These discrepancies are presumably due to the different dielectric boundaries.

The different arrangement of the substituents ($–CH_3$ and $–CH=CH_2$) at the pyrrole rings B and C breaks a mirror symmetry of the heme (protoporphyrin IX). Consequently, there are two possibilities for the mutual orientation of the hemes in the SynCyt*b*. For most of our computations, we consider the orientation of the hemes, which corresponds to that found in the native Cyt*b* ("native" conformation). Since there is no experimental evidence that only this conformation is assumed in the SynCyt*b*, we also computed the redox potentials for the alternative conformation obtained by rotating one of the hemes by 180° around the CHA–CHC axis. We refer to that orientation as the "non-native" conformation of the hemes. The corresponding redox potentials (–90 and –162 mV) exhibit only minor changes, as shown in Table 1. That is not surprising since both substituents are non-polar, and there are only minor changes in the structure by rotating the heme.

The pH dependence of the heme redox potentials in the SynCyt*b* is displayed in Figure 5A. The redox potential of the heme is pH-dependent due to the electrostatic coupling with the protonation reactions at the titratable residues of the protein. The stronger the coupling, the more the redox potential decreases with increased pH. From Figure 5A, one can see that heme H_h exhibits a stronger pH dependence than heme H_l. The residues that are strongly coupled with the redox-active part of the hemes and that at the same time change their charge state are responsible for the different behavior of the two hemes. Those residues are the four arginines and the related PR groups involved in salt bridges in the model structure. As discussed later, the buried salt bridges show a more substantial influence on the redox potential than solvent-exposed ones, where the high dielectric constant of water screens the interactions. Accordingly, the redox potential of H_h heme decreases by 18 mV/pH, while it remains practically constant for H_l heme at pH values between 6 and 9. At lower pH values, the redox potential of that heme also becomes pH-dependent since its PR groups start to get deprotonated and change their charge state.

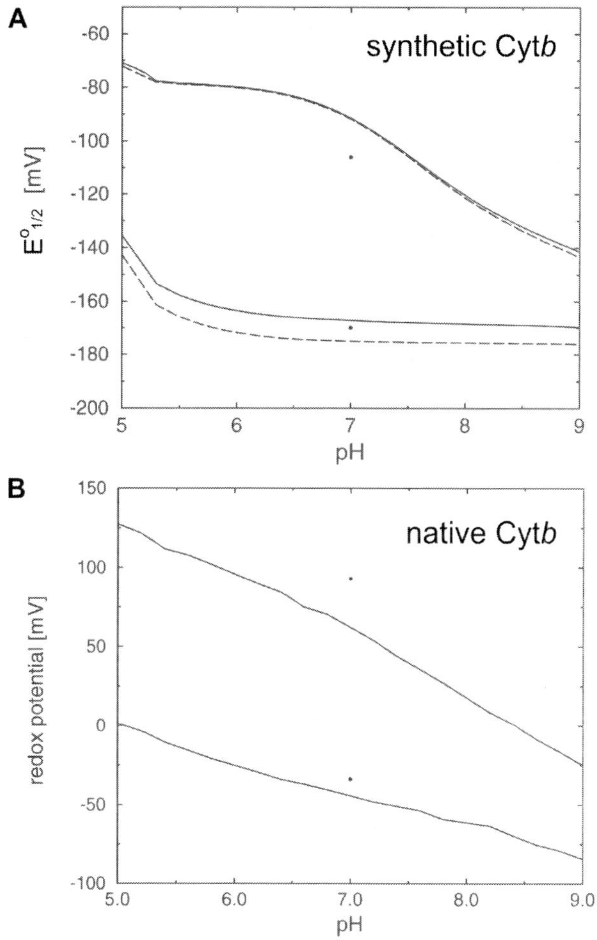

Figure 5. (A) pH dependence of the redox potential of the two hemes in the SynCyt*b*. The dots show the experimental values. The pH dependence of the redox potential of the two hemes is displayed in the physiologically relevant pH regime from 5 to 9. The redox potentials were evaluated $E_\eta{}^0$(pH, E_{sol} = 0 mV) (solid lines) according to Eq. 30 and as $E_{1/2,\eta}$ value (dashed lines), see "*Material and Methods*." Upper curves provide data for the high potential heme H_h close to the template, and lower curves for the low potential heme H_l near the open end of the four-helix bundle. (B) The pH dependence of the redox potential of the two hemes in the native Cyt*b* subunit. The dots show the experimental values. The pH dependence of the redox potentials of the two hemes is displayed in the physiologically relevant pH regime from 5 to 9. The redox potentials were evaluated as $E_\eta{}^0$(pH, E_{sol} = 0 mV).

Table 1. Redox potentials of hemes in SynCyt*b* in mV at pH = 7 and E_{sol} = 0 mV

Experimental Values	heme H_h near template	heme H_l near open end	Average shift [a]	Difference
bis-histidine heme model in aqueous solution	-220	-220	0	0
Willner et al. (1999) [b]	-115	-182	+72	+67
Rau & Haehnel (1998) [b]	-106	-170	+82	+64
computed values for full model of SynCyt*b*:				
native heme conformation [c]	-91	-167	+91	+76
non-native heme conformation [c]	-90	-162	+94	+72
uncoupled hemes [d]	-100	-176	+82	+76
contributions from different parts by setting atomic charges to zero for:				
all residues	-178	-182	+40.0 (44) [e]	+4 (5) [f]
backbone	-142	-211	+43.5 (48)	+69 (90)
template with links [g]	-95	-167	+89.0 (98)	+72 (95)
inner hydrophobic [h] side chains	-101	-168	+85.5 (94)	+67 (88)
outer hydrophilic [h] side chains	-92	-129	+109.5 (120)	+37 (49)
all four Arg and PR [i]	-140	-178	+61 (67)	+38 (50)
removing coordinates of:				
template with links	-151	-167	+61	+16

[a]The average shift of the redox potentials relative to the solution value according to Eq. 42.

[b]The assignment of the measured high and low redox potentials to the two hemes was not uniquely determined from the original experiment (Rau & Haehnel 1998). In the table, the assignment was made in agreement with more recent experiments (Willner et al. 1999) and the present computational results.

[c]The native heme conformation corresponds to the relative orientation of the hemes as it is in the native Cyt*b*, where the hemes can be matched after two 180° rotations, one around the virtual bond CHA-CHC and a second around CHB-CHD. Omitting the second rotation operation corresponds to the non-native heme conformation. If not otherwise stated, the native heme conformation is used.

[d]The redox states of the two hemes are uncoupled by calculating the electrostatic energies for the redox state of one heme by fixing the redox state of the other heme in the uncharged reduced state.

[e] Fraction in % of the calculated average shift of redox potentials referring to the calculated value of 91 mV.

[f] Fraction in % of the calculated difference of redox potentials referring to the calculated value of 76 mV.

[g] The links to the template involve the residues Gly1(F_1), Gly1(F_2), Lys27(G_1), and Lys27(G_2).

[h] The inner hydrophobic and outer hydrophilic side chains are defined in Figure 5 (in the companion chapter), excluding arginines and histidines.

[i]The four arginines, which form salt bridges with the PR groups of the hemes, are Arg5(F_1), Arg5(F_2) and Arg25(F_1), and Arg25(F_2).

Role of the Phe and Trp and the Electrostatic Coupling of the Hemes

The two conserved phenylalanines, Phe15(F_1) and Phe15(F_2), are placed in the middle of the four-helix bundle between the two hemes. Their role in shielding the heme redox potentials can be tested by replacing them with two alanines. The resulting shifts in the redox potentials +4 and –8 mV for H_h and H_l, respectively, are pretty small. Hence, the cavities with a dielectric constant of $\varepsilon = 80$ created by removing the phenyl rings have a similar effect on the redox potentials as the phenyl rings. However, the phenyl rings may serve to mediate the ET between the two hemes. The influence of the phenylalanine side chain charges on the redox potential of the two hemes was investigated by setting them to zero. The obtained shifts in redox potentials are +1 and –5 mV for high and low potential heme, respectively. It demonstrates that the phenylalanine side chain charges practically have no influence. Since the phenylalanine is a non-polar and hydrophobic residue, its low dielectric constant can influence the redox potential rather than the atomic charges themselves.

The two tryptophans in the structure of the SynCyt*b* are supposed to shield the low potential heme on the open end of the four helix-bundle from the solvent. However, the tryptophans cannot protect the heme H_l near the open end as efficiently as the template does with the heme H_h. Hence, the solvent accessibility of the two hemes differs, which is one of the main reasons for the difference in their redox potentials. Similar to the phenylalanines, the influence of the tryptophan side chain charges on the redox potential of the hemes is relatively small (0 and +11 mV) since tryptophan is a non-polar residue with relatively small atomic charges. Thus, they act instead by changing the dielectric boundaries of the protein system.

It is also interesting to investigate to what extent the couplings of the redox states of the two hemes can influence each other. For that purpose, we considered only one of the two hemes is redox-active, while the other remains in its reduced uncharged redox state. However, the calculated difference in the redox potentials between the coupled and uncoupled hemes is less than 10 mV, see Table 1. The shortest distance between the two hemes is 14.9 Å (Fe–Fe distance is 21.6 Å), and most importantly, their PR groups are oriented opposite to each other. Experimental data also suggest a small mutual coupling of the two hemes (Fahnenschmidt et al. 1999; Iakovleva et al. 2002; Albrecht et al. 2005).

Effect of Dielectric Medium and Specific Charge Distribution

In Table 1, the third column gives the average shift of the redox potential of the two hemes relative to the value of –220 mV, which is the redox potential of an isolated *bis*-histidine heme in an aqueous solution. The average shift of the redox potential of the two hemes relative to the value in the aqueous solution is:

$$\Delta E_{sol} = \frac{1}{2}\left[E^0(H_l) + E^0(H_h)\right] - E_{sol} \tag{42}$$

The fourth column displays the influence of setting specific atomic charges of the SynCyt*b* to zero on the difference in the redox potentials of the two hemes. Both relative contributions are also given as a percentage in brackets.

To get a rough idea about the influence of the atomic charges on the redox potential of the hemes, we set up the charges of all residues in the SynCyt*b* to zero. The obtained potentials of the two hemes become very similar, and their values are shifted toward the value of –220 mV in an aqueous solution. However, even after removing all atomic charges, the redox potentials of the H_h and H_l hemes are 42 and 38 mV, more positive than the solution redox potential of the heme. The redox potential of the heme H_l near the open end is slightly closer to the solution value than the redox potential of the heme H_h, which is close to the template and therefore better shielded from the solvent.

The low dielectric medium is still left inside the four-helix bundle, and by using Eq. 42, we estimated this contribution to be about 44% of the total shift of ΔE_{sol}(total) = 91 mV. The remaining 56% of the overall shift is due to the specific charge distribution in the SynCyt*b*.

Removing the atomic charges of the protein backbone, the difference in the redox potentials of the two hemes remains almost the same. However, the redox potentials of both hemes become more negative, such that the backbone charges contribute about 52% (the complement of 48%) to the total average shift relative to the solution value of the redox potential. Going in the opposite direction from the zero backbone charges toward the charges that the backbone atoms assume in the structure of the SynCyt*b*, the heme redox potentials increase by about 48 mV on average. Hence, the backbone dipoles generate a positive potential in this model protein. Because the backbone has zero net charge, the non-zero potential is unexpected. In a recent study, Gunner et al. (2000) have shown that the asymmetry in packing the peptide amide dipoles results, on average, in more positive than negative electrostatic potential inside proteins of all folding motifs. An explanation is that the backbone dipoles with their carbonyl (C=O) groups are more likely to be surface exposed, and amide protons (N–H) prefer to be buried. Hence, the reduced form of redox-active groups is stabilized by a positive potential of the backbone dipoles, increasing the cofactors' redox potential.

The effect of atomic charges of the template on the hemes redox potentials can be probed by setting these charges to zero. No significant change in redox potentials is observed (Table 1) since the template does not contain polar or titratable groups. The template charges have only a weak (4 mV) direct influence on the redox potential of the H_h heme, but indirectly it shields particularly strongly the H_h heme from the impact of the bulk solvent (water). An additional electrostatic computation shows that the difference of redox potentials disappears significantly (–151 vs. –168 mV for the H_h and H_l hemes, respectively) by removing the coordinates of all atoms from the template and linking residues. The redox potential of the H_h heme shifts by 60 mV toward the corresponding value of –220 mV for the heme in an aqueous solution. Hence, the template with its dielectric boundaries and dielectric constant of 4 makes the protein environment around the H_h heme more hydrophobic and the heme H_h less solvent accessible. Buried hemes prefer to be in the reduced state and generally have higher redox potential values than water-exposed ones.

Influence of Different Residues on the Redox Potentials

Next, we considered the effect of different residue types on the shift of redox potentials. Turning off all atomic charges of the hydrophobic side chains in the interior of the four-helix bundle has only a minor effect on the shift of the redox potentials. That is not surprising since the corresponding charges are small. Practically, only H_h heme is affected by these charge changes, indicating that H_h heme is somewhat better packed in the hydrophobic pocket of the helix bundle than H_l heme. The influence of charges of the hydrophobic residues on the difference in the redox potential of the two hemes is also small (12%).

On the other hand, turning off the atomic charges of the charged and polar side chains on the outer solvent-exposed surface of the four-helix bundle (excluding the arginines and the axially ligated histidines, which are already inside of the binding pocket) causes a significant effect. The charges of the hydrophilic side chains practically do not affect the high potential heme H_h at the better shielded closed end of the helix-bundle, whereas the redox potential of the low potential heme shifts by 38 mV to higher values. For the H_l heme, negatively charged glutamates exhibit a more substantial effect on the redox potential than positively charged lysines. The effects of glutamates and lysines seem to cancel each other for the H_h heme. Consequently, the charges of the hydrophilic side chains contribute more than 50% to the difference in the redox potentials. At the same time, these charges exhibit a mild but negative contribution of 20% to the shift of the average redox potential.

Finally, we consider the influence of the salt bridges between the PR groups of the hemes and the arginines Arg5 and Arg25 of the two *F* helices. Setting the corresponding atomic charges to zero, we obtained the redox potentials of –140 and –178 mV for the heme near the template and the open end, respectively (Table 1). Accordingly, the charges of the four salt bridges with their corresponding molecular groups contribute about 50% to the difference in the redox potentials of the two hemes and 33% to the overall shift of the redox potentials relative to the solution value. The salt bridges of the PR groups with the arginines give rise to a more substantial shift of the redox potential for heme near the template (H_h) than for heme close to the open end (H_l). That is not surprising since buried salt bridges are generally stronger than solvent-exposed ones. Notice that the different contributions do not add exactly up to 100% since these electrostatic energies are not strictly additive.

Calculations on the Native Cytochrome *b*

The accuracy of the atomic coordinates and the reliable set of the atomic charges are critical for electrostatic computations. Since no crystal structure is available for the SynCyt*b*, the atomic coordinates were obtained by a modeling procedure whose quality cannot directly be assessed. In conventional applications, the atomic coordinates are taken from a crystal structure. Good agreement between the calculated and experimental redox potentials for the computer-generated structure of the synthetic Cyt*b* has motivated us to test our computational procedure and set of partial atomic charges that we used, also on the crystal structure of a native *bis*-histidine heme protein. Therefore, we performed the same computations on the Cytbc_1 complex of the bovine heart (Iwata et al. 1998), obtained at pH = 6.6 at 3.0 Å resolution.

Protonation Pattern of Titratable Groups

In the native Cyt*b* subunit, at pH = 7 and E_{sol} = 0 mV, the following residues are in non-standard protonation state (protonation probability is given in brackets): Asp254 (0.39), Asp331 (1.00), Glu162 (0.24), Glu202 (0.06), Glu271 (0.99), Lys227 (0.68), N-terminus (0.045), PRA(H_l) (0.045), PRD(H_l) (0.96), PRA(H_h) (0.64) and PRD(H_h) (0.32). From eight histidine residues in the Cyt*b* subunit only for three (His68, His201, and His345), the fraction of the double protonated form is somewhat larger, 0.75, 0.10, and 0.10, respectively. All other histidines are neutral, with a proton residing on the δ- or ε-nitrogen. The number of titratable groups in the whole bc_1 complex, which possess non-standard protonation, is considerably larger. Interestingly, Glu271 is very close to the H_l heme (the shortest distance is 3.4 Å) and near the propionic group PRA of the H_l heme (5.4 Å away from the O1A atom). Since, Glu271 is situated in a hydrophobic pocket formed by Pro270, Phe128, and Phe274, despite the negatively charged PRA(H_l) group in the neighborhood, it is entirely protonated.

In the physiological pH range between 6 to 8 and solvent redox potential of E_{sol} = 0 mV, the high potential heme (H_h) is reduced, and the low potential heme (H_l) is in the oxidized state. At the same time, looking at the titration curves of the heme propionates (Figure 6), we noticed that the propionic groups of the H_h heme are symmetrically protonated and exhibit only mild pH dependence. In contrast, the protonation pattern of the propionates of the H_l heme is asymmetric. Namely, PRD(H_l) is almost entirely protonated, while PRA(H_l) is nearly deprotonated. This protonation pattern can be explained by the salt bridges formed between the PR groups and arginines and by additional interactions with the local protein environment. Fully protonated Arg80 ($\langle x \rangle$ = 1.0) forms a strong salt bridge with PRA(H_l) (O···H distance 2.08 Å), and only a weaker salt bridge (O···H distance 3.8 Å) with the second PR group (PRD(H_l)) at the same heme. Furthermore, the later PR group is surrounded by several hydrophobic residues (Leu51, Tyr55, Val66, and Pro134) and does not form an H-bond. The PR group PRA(H_h) from heme H_h also forms a strong salt bridge with Arg100 ($\langle x \rangle$ = 0.98) (O···H distance 1.99 Å) and a weaker salt bridge (O···H distance 3.2 Å) with the other propionate PRD(H_h). In contrast to the corresponding PRD(H_l) group, the PRD group of heme H_h is, in addition to the salt bridge also, involved in two H-bonds with the amide hydrogen of Asn206 (O···H distance 1.90 Å) and the hydroxyl hydrogen of Ser106 (O···H distance 2.57 Å).

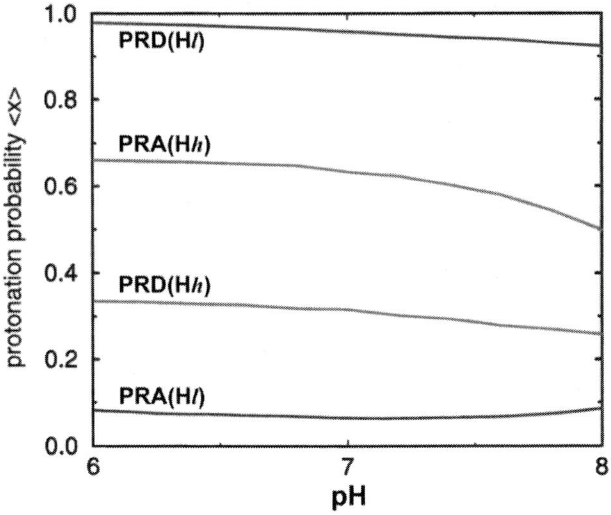

Figure 6. Titration curves of arginines and heme propionates in the native Cyt*b* at $E_{sol} = 0$ mV. The PR groups are denoted according to the heme pyrrole ring attached to (A or D) and the heme identity (low- H_l or high- H_h potential heme). The arginines Arg80 and Arg100 form salt bridges with PRA(H_l), PRD(H_l), and PRA(H_h), PRD(H_h), respectively. The corresponding O···H distances between the PR groups and the arginines are 2.1 Å, 3.8 Å, 2.0 Å, and 3.2 Å.

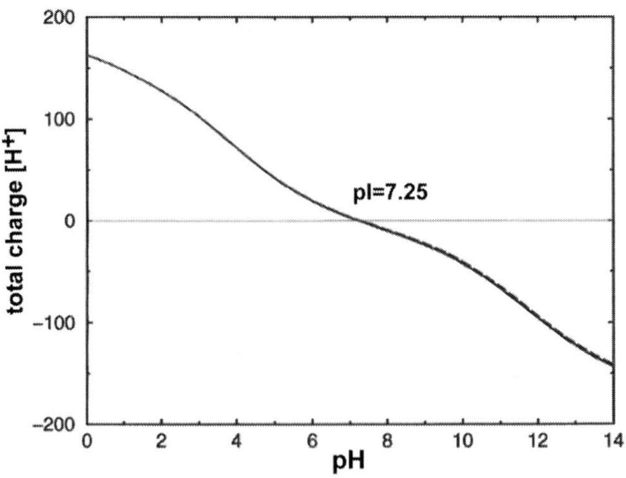

Figure 7. Titration curve of the native Cyt*bc₁* complex at $E_{sol} = 0$ mV. The total charge of the Cyt*bc₁* complex is given as a function of pH. Apoprotein charge changes are shown by a black line, while the red line also includes the charges of the two Cyt*b* hemes.

In summary, the electrostatic interactions of the PR groups of heme H_h are more symmetric than that of heme H_l. It is not accidental that each PR pair belonging to the identical heme at neutral pH carries one negative charge since only one arginine per heme is involved in a salt bridge. In that way, the propionic and the corresponding arginine charges neutralize each other in a low dielectric medium of the binding pocket. In fact, this trend is actually valid for a whole physiological pH range of 6 to 8.

The calculated isoelectric point of the native Cytbc_1 complex is pI = 7.25. Since Cytbc_1 is a transmembrane protein complex consisting of predominantly hydrophobic residues, it is not surprising that its total charge is close to zero at a neutral pH value. In contrast to the SynCyt*b* (Figure 4), the titration curve of the whole Cytbc_1 complex (Figure 7) does not show a clear separation between acidic and basic residues. Since many protonatable groups are buried in the low dielectric medium of the protein, their pK_a values can be considerably shifted from the solution pK_a values. There are also 57 histidines, which are mostly titrated at neutral pH. Thus, the combined contribution of all titratable residues together leads to a titration curve of the whole protein complex that appears relatively smoothly.

The calculated pI value of the isolated Cyt*b* subunit (including both hemes) of 7.68 is just slightly lower than the one experimentally determined (pI = 7.8) by Schilling et al. (2006). Although from the experimental isolation procedure, it is unclear whether the reported value is related to the apoenzyme or whole protein, including two hemes.

Experimental Values of the Heme Redox Potentials

The complex III from respiratory and the photosynthetic electron-transport chains have the universal redox center composition of two cytochrome *b* hemes, one cytochrome *c* heme (c_1 or *f*), one Fe_2S_2 cluster, and up to one bound ubiquinone or plastoquinone (Hurt and Hauska 1981). Substoichiometric amounts of the Rieske protein and quinone (Weiss and Kolb 1979; Engel, Schaegger, and von Jagow 1980) are caused by using Triton X-100 during the isolation procedure and might be responsible for their partial loss. The redox potentials of hemes in native Cyt*b* differ considerably from the value in aqueous solution, and with the preparation from the bovine heart by Hatefi (1978), two midpoint potentials were found (Table 2). The higher value corresponds to a component with an α-peak at 562 nm, the lower to a component with an α-peak at 565 nm plus a shoulder at 558 nm (Nelson and Gellerfors 1974, 1975).

Table 2. Redox potentials of hemes in native Cyt*b* at different pH in mV

Experimental values	High potential heme	Low potential Heme	Difference
bovine heart mitochondrial cytochrome bc1 complex[a]	+93	-34	+127
Cytb isolated from bovine heart cytochrome bc1[b]	-5	-85	+80
Rhodobacter sphaeroides cytochrome bc1 complex[c]	+50	-60	+110
cytochrome b6f from chloroplasts[d]	-50 -45	-170 -150	+120 +105
computed values from cytochrome bc1 complex	pH 6.6, 7.0, 8.0	pH 6.6, 7.0, 8.0	pH 6.6, 7.0, 8.0
whole complex in solution ($\varepsilon = 80$)	+75, +61, +19	-37, -45, -62	+112, +106, +81
Cytb from C chain only in solution ($\varepsilon = 80$)	+46, +35, -1	-42, -52, -67	+88, +87, +66
Cytb from C chain only embedded in membrane[e]	+106, +91, +42	+22, 0, -52	+84, +91, +94

[a] Measured at pH = 7.0, (Nelson and Gellerfors 1974, 1975).
[b] Measured at pH = 7.0, (von Jagow et al. 1978; von Jagow and Sebald 1978).
[c] Measured at pH = 7.0, (Gabellini et al. 1982; Gabellini and Hauska 1983).
[d] Measured at pH = 7.0; upper value (Hurt and Hauska 1982); lower value (Kramer and Crofts 1994; Kramer et al. 1994).
[e] A membrane of 50 Å thickness was assumed. Within the membrane, the dielectric constant was set to $\varepsilon = 4$.

The difference in redox potentials between low and high potential hemes is 127 mV. Both potentials are pH-dependent in the physiological range (von Jagow and Sebald, 1978; von Jagow et al., 1981), being about 40 mV more negative at pH 8 than pH 7. After isolating the Cyt*b* from the complex, the spectral differences disappear, leaving only the α-peak at 560 nm, but the redox heterogeneity is retained with a considerable shift to lower midpoint potentials (–5 and –85 mV), which are still pH-dependent (von Jagow et al., 1978). The redox potential of the H_h heme in isolated Cyt*b* becomes 98 mV more negative, while the shift of redox potential of H_l heme is 51 mV, compared with their potentials within the Cytbc_1 complex. This suggests that some conformational changes occur in the structure of Cyt*b*. The individual values of the redox potentials of the Cyt*b* hemes vary for different species, as seen in Table 2, for bovine heart and *Rhodobacter sphaeroides*. Due to the differences in the preparation procedure, the measured values of the redox potential can differ by 20 mV or even more. In comparison with the SynCyt*b* (Table 1), the closest redox potential values were found in the cytochrome b_6f

complex from chloroplasts of plants (–50 and –170 mV), which is the functional analog but non-homolog of the Cyt*bc*$_1$ complex (Hurt and Hauska 1982).

Calculated Redox Potentials in the Whole Cytbc$_1$ Complex

The calculated redox potentials of the hemes in the native Cyt*b* are displayed in Table 2 at three different pH values. They are considerably higher than the redox potential of heme in solution. The pH value of 6.6 was also chosen for our computations since the crystal structure of the Cyt*bc*$_1$ complex was solved at this pH value (PDB code 1BE3, determined by Iwata et al. (1998)). We calculated the redox potentials of the Cyt*b* hemes within the whole Cyt*bc*$_1$ complex, assuming a solvent environment with a dielectric constant $\varepsilon = 80$ outside of the protein complex. The calculated values of the redox potentials deviate from the corresponding experimental values at pH 7 by 32 mV for H_h and 11 mV for H_l heme (32 mV is equivalent to about 1/2 p*K*-unit). The calculated values of the redox potentials for both hemes are more negative than the measured values; however, they are still in good agreement with experimental values, which are +93 and –34 mV for the high and low potential heme, respectively.

Moreover, the calculated difference between the redox potentials of two hemes correlates well with the experimental value (106 vs. 127 mV at pH 7). The pH dependence of the redox potential is also in good agreement with experiments (von Jagow et al. 1978; von Jagow and Sebald 1978; von Jagow et al. 1981), decreasing by 42 mV and 17 mV for the high and low potential hemes at a pH between 7 to 8 (see Figure 5B). The strong pH dependence of the redox potential of the two Cyt*b* hemes arises from the strong coupling of the redox-active part of the hemes with the corresponding PR groups and arginines (Arg80, Arg100) involved in the salt bridges. Other neighboring titratable residues (His68, Lys227, and Asp254) also show a moderate influence. Since both hemes are well buried in the low dielectric medium of the hydrophobic pocket formed by the four-helix bundle, their electrostatic interactions with other charged groups in the protein are not screened by nearby aqueous solution, as is partially the case for the SynCyt*b*. Consequently, their redox potentials are higher, and their pH dependences are more pronounced than for the SynCyt*b*.

Calculated Redox Potentials of the Hemes in the Cytb Subunit

Comparing the experimental data for the whole Cyt*bc*$_1$ complex and the Cyt*b* subunit, one can see a considerable shift of the heme redox potentials to lower values. That shift is -98 mV for the H_h heme and -51 mV for the H_l heme. The difference in the behavior of the two hemes can be understood from the structure of the isolated Cyt*b* subunit (see Figure 4C in the companion chapter), where the low potential heme remains well buried, whereas the high potential heme becomes a more solvent-exposed. Our calculated values show the same trend, where the isolation of the Cyt*b* from the protein complex causes a more substantial decrease in the redox potential of the H_h heme. Namely, the computed redox potential of the H_h heme decreases by 26 mV, but the redox potential of the H_l heme is changed only by -7 mV. However, the obtained shifts toward the more negative values of the redox potentials are much smaller than in experiments.

Moreover, the calculated redox potentials are about 35 mV higher than the experimental values for the native Cyt*b* subunit isolated from the Cytbc_1 complex. This discrepancy is significant and may be due to differences between the conformation that the Cyt*b* subunit adopts in the crystal structure of the whole Cytbc_1 complex used for the computations and the conformation that the native Cyt*b* isolated from the Cytbc_1 complex assumes. On the other hand, the calculated difference of the redox potentials (87 mV) agrees very well with the corresponding experimental value for Cyt*b* isolated from the Cytbc_1 complex (80 mV). In the structure of isolated Cyt*b* in a solution, both hemes are probably more solvent-exposed than in the corresponding crystal structure of the Cyt*b* subunit. Hence, it is likely that some conformational changes occur in the structure of the Cyt*b* if prepared in isolation. In agreement with experiments (von Jagow et al. 1978; von Jagow and Sebald 1978), both redox potentials remain pH-dependent, decreasing by 36 and 15 mV for the H_h and H_l heme, respectively, at pH between 7 and 8.

We also investigated the influence of the dielectric medium of the membrane on the heme redox potentials in the native Cyt*b* subunit. The membrane was modeled as a dielectric layer of 50 Å thickness with $\varepsilon = 4$. The Cyt*b* subunit was embedded in the center of a membrane layer, such that the normal of a layer was oriented parallel to the axis of the four-helix bundle. The calculated redox potentials increased by about 55 mV at pH 7 compared to the computed values in an aqueous solution, with $\varepsilon = 80$. However, we expect that the membrane influence is much smaller for the whole Cytbc_1 complex since both hemes are then well buried in a low dielectric of the protein complex. Interestingly, in our computation, the membrane does not affect the difference in the redox potentials (91 mV), which practically remains the same as in an aqueous solution (87 mV).

The Cytbc_1 complex and the dielectric medium of the membrane can be considered to have a similar influence on the isolated Cyt$b$ subunit. Therefore, we tried comparing the redox potentials calculated using the membrane with the experimental values for the whole Cytbc_1 complex. Interestingly, the agreement between the calculated redox potential of the H_h heme within the membrane and the experimentally measured value within the whole Cytbc_1 complex at pH 7.0 is perfect (+91 *vs.* +93 mV). At the same time, the redox potential of the H_l heme for the Cyt$b$ subunit in the membrane is 34 mV higher than the corresponding experimental value in the whole Cytbc_1 complex. Hence, this agreement of the calculated and measured redox potential for the H_h heme should not be taken with full weight.

Redox Titrations

The Coupling of Protonation and Redox Reactions

The protonation and redox equilibria in a protein have to be considered simultaneously to evaluate any relevant energy parameter. The coupling between protonation and redox reactions plays a significant role in many essential biochemical charge transfer processes, and it is often caused by electrostatic interactions between protonatable and redox-active groups. Specifically, the positive charge of a proton can neutralize the negative charge of an electron. Thus, reducing a redox-active moiety can go together with the protonation of neighboring acid-base groups, increasing their pK_a values. In contrast, the oxidation of a redox group can lead to the deprotonation of adjacent protonatable groups, decreasing their pK_a values (Louro et al. 1997; Popović and Stuchebrukhov 2012). Thereby, the redox and protonatable groups may belong to the same or different molecular groups. The electrochemical coupling phenomenon is, also known as the redox-Bohr effect described in experimental (Wyman 1968; Urban and Klingenberg 1969; Dutton, Wilson, and Lee 1970; Kilmartin and Rossi-Bernardi 1973; Dutton and Wilson 1974; Papa, Guerrieri, and Izzo 1979; Papa et al. 1998; Papa 2005) and theoretical/computational studies (Soares, Martel, and Carrondo 1997; Martel et al. 1999; Teixeira, Soares, and Baptista 2002; Fonseca et al. 2009).

The applied theoretical approach can explore the coupling between protonation and redox reactions. The first sum in Eq. 25 runs over P protonatable groups, and it is the only term where the ambient pH value appears. The second sum relates to the K redox-active groups, and only there the solvent redox potential E_{sol} appears as a variable. That can lead to a trivial and wrong conclusion that only protonation probabilities and titration curves of acidic and basic groups depend on pH, and only oxidation probabilities of redox-active groups depend on the solution (ambient) redox potential E_{sol}.

Increasing pH causes deprotonation of protonatable groups, and increasing the solution redox potential E_{sol} oxidizes the redox-active groups. However, due to the mutual coupling between protonatable and redox groups ($W_{\mu\eta}$ terms in Eq. 25), oxidation probabilities are also pH-dependent, and protonation probabilities can depend on the external redox potential E_{sol}. Specifically, increasing pH causes the deprotonation of neighboring protonatable groups, and a total protein charge becomes more negative, stabilizing the oxidized form and decreasing the E^0 potential of the redox-active group η. Many protonatable groups usually surround the redox group in a protein, and even if the mutual coupling is not strong, the total effect of changing the charges of protonatable groups will cause a change in redox potential with pH.

In contrast, a few redox-active groups, changing the oxidation states due to the change of solution redox potential, can affect the protonation probabilities of the titratable groups μ that are in their vicinity and strongly coupled with them. Therefore, the number of protonatable groups that depend on the solution's redox potential is usually not so large. The pK_a values of the protonatable groups commonly drop by increasing the solution potential E_{sol}, but that is not a general rule. In principle, the more redox-active group is oxidized, the harder titratable groups in the vicinity can get protonated. However, these protonatable groups are also mutually coupled, rendering the interaction pattern even more intricate.

Hence, the mutual coupling of protonatable and redox-active groups is responsible for the pH-dependence of the redox potentials and the redox-Bohr effect, i.e., the dependence of protonation probabilities (and pK_as) of protonatable groups on the external redox potential. Strictly speaking, the redox-Bohr effect is related to the deprotonation of acidic/basic groups with the oxidation of the redox centers in protein.

The Effects of Ionic Strength, External Redox Potential, and the Redox-Bohr Effect

The pH dependence of the redox potentials of all hemes is evaluated in the physiologically relevant pH range and presented in the subsection "*Calculated pH-dependent Heme Redox Potentials.*" In this chapter, we also rationalize the effects of the applied ionic strength, external redox potential of bulk solvent (aqueous solution), the coupling of protonation and reduction reactions, and the distribution and population of the redox microstates in the studied Cyt*b* systems. The redox-Bohr effect in native Cyt*b* is modeled successfully, whereby the essential protonatable groups responsible for this phenomenon are identified.

The acid-base titrations of ionizable groups in proteins have been widely investigated theoretically using the continuum electrostatic method (Yang et al. 1993; Beroza and Case 1996; Lancaster et al. 1996; Antosiewicz, McCammon, and Gilson 1994). In some work, the redox potentials of the redox-active cofactors were considered (Gunner and Honig 1991; Gunner, Nicholls, and Honig 1996; Muegge et al. 1996; Rabenstein, Ullmann, and Knapp 1998b; Zheng and Gunner 2009) or the pH dependence of the redox potential was studied (Beroza et al. 1995; Baptista, Martel, and Soares 1999; Popović et al. 2002; Popović and Stuchebrukhov 2004; Song, Michonova-Alexova, and Gunner 2006). In most of these studies, the influence of the redox state on the protonation pattern is considered. However, the effect of the solvent (ambient) redox potential on the oxidation and protonation pattern of variably charged groups in the proteins was frequently neglected. An essential aspect of the electrostatic computations, which we included in the consideration, is the possibility of performing a redox titration of a protein by varying the external (ambient) redox potential E_{sol} in Eqs. 29 and 30 and evaluating the corresponding protonation and oxidation probabilities of titratable groups by a Monte Carlo method. We changed the solution potential in 20 mV increments ranging from -400 to +300 mV. A similar approach was recently applied to cytochrome c_3 (Ullmann 2000; Teixeira, Soares, and Baptista 2002; Soares and Baptista 2012); however, the probabilities were calculated using less accurate hybrid statistical mechanics with the Tanford-Roxby approximation (Yang et al. 1993). This section shows a few selected examples demonstrating the benefits and a wide range of electrostatic applications, which include redox titrations. The plots in Figure 8 illustrate which information can be obtained from the electrochemical redox titration simulations.

Figure 8A displays redox titration curves of the two hemes in the SynCyt*b* at pH = 7 and I = 0 and 0.1M, where the oxidation probabilities of the redox-active groups are given as a function of the solution redox potential.

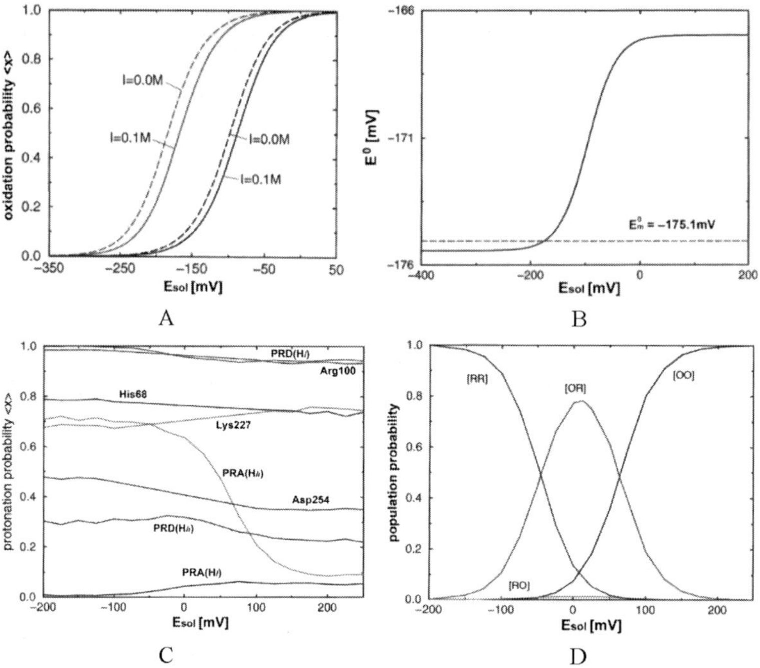

Figure 8. Selected examples illustrate how the results of the redox titrations can be used to obtain additional information. (A) Redox titration curves of the two hemes in the *SynCytb* at pH = 7 and I = 0.0 and 0.1M. Redox titration curves of redox-active hemes in proteins may deviate more or less from the Nernst law because of the coupling with other charged groups. From this plot, one can read the midpoint potential E_m values of the two hemes at different ionic strengths. (B) Comparison of the solution (external) redox potential dependent E^0 value (solid line) and E_m value (dashed line) for the H_l heme of the *SynCytb* at pH 7. This plot shows how much the actual value of redox potential deviates from the ideal Nernst behavior ($E_{1/2}$) due to the coupling with other titratable sites. (C) Dependence of the protonation probabilities of some titratable groups in the *native* Cyt*b* on the solution redox potential at pH 7. The displayed residues exhibit the most significant changes in the protonation state during the redox titration. It arises from a strong coupling between these ionizable residues and two redox-active hemes, whose oxidation probabilities depend directly on the ambient redox potential. However, the protonation probabilities of some ionizable residues can also rely indirectly on the E_{sol} value. (D) Population probabilities of all four redox states of the native Cyt*b* from the bovine heart at pH 7. Both hemes are in the reduced state [RR] at very negative values of the solution redox potential. The low potential heme gets first oxidized by increasing the external redox potential, populating the microscopic [OR] redox state. At $E_{sol} = 0$ mV, the [OR] state is at the peak, becoming the most populated state with a probability of 0.80. Finally, the high potential heme also gets oxidized at higher values of solvent redox potential.

The midpoint potentials ($E_m = E_{1/2}$) at pH = 7 and I = 0.1M, which one can read from the plot, are –92 and –175 mV for the high and low potential heme, respectively. Due to the electrostatic coupling of the two hemes, these curves slightly deviate from the standard behavior of an isolated uncoupled heme, whose titration behavior is defined by the Nernst law. However, since the coupling of the two hemes in the SynCyt*b* is small (< 10 mV), this deviation is only minor. Figure 8A also shows the effect of the ionic strength on the redox potentials of the two SynCyt*b* hemes. These effects are small and only slightly stronger for the low potential H_l heme, which is more solvent-exposed and therefore more sensitive to changes in ionic strength. Increasing the ionic strength causes an increase in the redox potential of both hemes. A possible explanation for such behavior is stabilizing and neutralizing effect of the mobile counterions from aqueous solution on the negatively charged solvent-exposed glutamate residues.

Figure 8B displays the external redox potential dependent E^0 value (defined by Eq. 30) as a solid line and, in addition, the midpoint potential E_m value (dashed line) for the same H_l heme of SynCyt*b* at pH 7. The midpoint potential ($E_m = E_{1/2}$) corresponds to the value of the solution redox potential at which the concentrations of the reduced and oxidized forms are equal. Interestingly, the redox potential of the H_l heme varies between –175.5 and –166.5 mV in the inspected range of the external (solvent) potential, indicating a small coupling (less than 10 mV) with the charged state of the other heme. The figure also indicates that the midpoint potential only gives a strictly correct redox potential value for one point at a given pH. The redox potential is much better defined by Eq. 30 in the whole external potential (E_{sol}) range. Consequently, the standard redox potential is not a constant in the redox proteins.

The influence of solvent redox potential on the titration behavior of protonatable groups in SynCyt*b* is negligible. Since the two hemes of native Cyt*b* are buried in a low dielectric environment and better shielded against the aqueous solution, their electrostatic coupling with the protonation reactions of titratable amino acid side chains is much stronger. Figure 8C displays the influence of solvent redox potential on changes in the protonation probabilities of selected titratable groups (with the most significant shifts) in the native Cyt*b* at pH 7. As one can see, the protonation probability of the propionate PRA(H_h) attached to the high potential heme depends strongly on the external redox potential, and it decreases by 0.62 protons with oxidation of the corresponding heme. Other residues, such as PRD(H_l), Arg100, His68, Asp254, and PRD(H_h), exhibit qualitatively similar but much weaker dependences at pH 7.

The so-called redox-Bohr effect in the native Cyt*b* (Papa et al. 1979) seems to be associated with changes in the protonation states of charged residues with the oxidation of hemes, although the participating residues have not been experimentally identified so far. This plot may indicate that the coupling between hemes and the above-underlined residues is responsible for the redox-Bohr effect in the native Cyt*b* subunit of complex III. The experimental measurements (Papa et al. 1979) suggest that a release of 0.5–0.6 H^+ occurs with the oxidation of the native Cyt*b* hemes. The best candidate is the propionate PRA(H_h), which deprotonation curve follows the oxidation of the high potential heme and is therefore strongly related to its redox state.

Finally, Figure 8D displays the calculated population probabilities of the four different redox microstates of the native Cyt*b* at pH 7 as a function of solution (ambient) redox potential. The two-letter code describes the particular redox microstate of the H_l and H_h hemes. For energetic reasons, only three redox microstates are significantly populated at the different values of the external potential.

A similar dependence is obtained for the SynCyt*b* protein structure for the distribution and population of the redox microstates. The distribution curves are shifted according to the corresponding redox potentials in SynCyt*b,* while the central peak of the population probability for the [OR] microstate is shifted to a somewhat lower value. At $E_{sol} = -130$ mV, the [OR] state is at its peak, becoming the most populated state with a probability of 0.68.

In the previous subsection, the reader can find additional discussion on the coupling of protonation and redox reactions in proteins, the pH-dependent pK_a values, and the external (ambient) redox potential-dependent E^0 values.

Conclusion

A recently synthesized *de novo* SynCyt*b* protein was modeled from scratch using advanced modeling techniques and studied in the previous companion chapter. Its structure consists of a four helix-bundle, two hemes, and a cyclic decapeptide as a template and mimics the central part of the native Cyt*b*. The generation of atomic coordinates, energy minimization, and molecular dynamics was done using the program CHARMM22. The molecular dynamics (MD) simulations have aimed to test the stability and validate the generated structure. They proved that the SynCytb structure has similar stability and robustness as the native Cyt*b*.

The second critically important validation test for our computer-generated structure was the computation of the redox potential of its two hemes. We studied the factors on which the ET process depends and explored the elements of the protein environment that shift the redox potential of heme from the value in an aqueous solution. The successful computation of the redox potentials of the two hemes in the SynCyt*b* also suggests that the atomic coordinates generated by a modeling procedure are reliable.

The same methodology is applied to the natural Cytbc_1 complex from the bovine heart to test the atomic charges and computational procedure to evaluate redox potential in heme proteins. The obtained results agree well with the experiments. The agreement was even better for SynCyt*b*, but that may be by chance. Nevertheless, the isolation and preparation conditions for the Cytbc_1 complex can also affect redox potential measurements. Notice that the content of the quinones and Rieske protein depends on these conditions, and the measurements from different groups can vary by 20 mV or more. Such problems do not appear for SynCyt*b*. Although the atomic coordinates of the SynCyt*b* are subject to considerable uncertainties due to the modeling procedure, the coordinates of the crystal structure of the Cytbc_1 complex obtained at 3.0 Å resolution may also be uncertain to some extent.

Analysis of the factors determining the shift and difference in the heme redox potentials in SynCyt*b* gave the following results. (1) The electrostatic coupling of the redox potential of the hemes is relatively small (<10 mV). (2) A change of dielectric medium from aqueous solution ($\varepsilon = 80$) to the protein ($\varepsilon = 4$) causes about 45% of the total shift of the heme redox potentials; the rest of 55% is due to the specific charge distribution in SynCyt*b*. (3) The backbone charges contribute 50% to the total shift of the redox potentials but do not affect the difference of the redox potentials. (4) The charges of hydrophilic residues contribute about 50% to the difference in the redox potential of two hemes and, at the same time, give rise to –20% of the total shift. (5) The charges of the salt bridges between arginines and propionic acids shift the redox potential of the high potential heme (situated near the template) more than the low potential heme. Therefore, they participate about 50% of the difference and 33% of the overall shift in the redox potential. (6) The template and the inner hydrophobic side chain charges have only a minor influence on the difference and the shift of the heme redox potentials, but they render the hemes less solvent accessible. In that way, the template and inner hydrophobic side chains significantly affect the hemes' redox potentials.

We have demonstrated that a careful, stepwise modeling procedure can generate the reliable atomic coordinates of an artificial protein and that our modeling protocol can provide a guideline on how to construct similar synthetic proteins. A good agreement between the calculated and measured redox potentials has enabled us to analyze the factors of the protein environment determining the shift in the redox potentials of the hemes. It opens up new horizons to understanding how the protein environment tunes the redox potential of cofactors, which is essential for their function, and how one can create synthetic proteins with predefined redox properties.

The following two quotes might best illustrate the synergy of experiments and computations, which we so often meet in scientific research: "*Without structure, one is simply guessing*" (*Paul Sigler*) and "*Without energetics, there is a risk of believing everything one sees*" (*B. Jayaram*). The underlined quotations show that both protein structure and energetics of the ET and PT reactions computed by theoretical methods or obtained from the experiments are equally relevant for making accurate conclusions. Our work combines them to gain new theoretical knowledge about structure–function correlation in proteins, a complete insight into studied biological systems, and a full understanding of how proteins work on the molecular level and in atomic detail.

Acknowledgments

This work has been financially supported by the Ministry of Education, Science and Technological Development of the Republic of Serbia (MPNTR): Grant No. 451-03-68/2022-14/200026 and project OI/127035. This article/publication is based upon work from COST Action CA21101, supported by COST (European Cooperation in Science and Technology).

References

Albrecht, T., W. Li, J. Ulstrup, W. Haehnel, and P. Hildebrandt. 2005. "Electrochemical and spectroscopic investigations of immobilized de novo designed heme proteins on metal electrodes." *Chem Phys Chem* 6:961–970.

Alexov, E. G., and M. R. Gunner. 1997. "Incorporating protein conformational flexibility into pH-titration calculations: Results on T4 Lysozyme." *Biophys. J.* 74:2075–2093.

Antosiewicz, J., J. M. Briggs, A. H. Elcock, M. K. Gilson, and J. A. McCammon. 1996. "Computing ionization states of proteins with a detailed charge model." *J. Comput. Chem.* 17:1633–1644.

Antosiewicz, J., J. A. McCammon, and M. K. Gilson. 1994. "Prediction of pH-dependent properties of proteins." *J. Mol. Biol.* 238:415–436.

Baptista, A. M., P. J. Martel, and C. M. Soares. 1999. "Simulation of electron-proton coupling with a Monte Carlo method: Application to cytochrome c3 using continuum electrostatics." *Biophys. J.* 76:2978–2998.

Bashford, D. 1997. "An object-oriented programming suite for electrostatic field effects in biological molecules." In *Scientific Computing in Object-Oriented Parallel Environments*, edited by Y. Ishikawa, R. R. Oldehoeft, J. V. W. Reynders and M. Tholburn, 233–240. Berlin: Springer.

Bashford, D., D. A. Case, C. Dalvit, L. Tennant, and P. E. Wright. 1993. "Electrostatic calculations of side-chain pKa values in myoglobin and comparison with NMR data for histidines." *Biochemistry* 32:8045–8056.

Bashford, D., and K. Gerwert. 1992. "Electrostatic calculations of the pKa values of ionizable groups in bacteriorhodopsin." *J. Mol. Biol.* 224:473–486.

Bashford, D., and M. Karplus. 1990. "pKas of ionizable groups in proteins: Atomic detail from a continuum electrostatic model." *Biochemistry* 29:10219–10225.

Bashford, D., and M. Karplus. 1991. "Multiple-site titration curves of proteins: An analysis of exact and approximate methods for their calculations." *J. Phys. Chem.* 95:9557–9561.

Bayly, C. I., P. Cieplak, W. D. Cornell, and P. A. Kollman. 1993. " A well-behaved electrostatic potential based method using charge restraints for deriving atomic charges: The RESP model." *J. Phys. Chem.* 97:10269–10280.

Beroza, P., and D. A. Case. 1996. "Including side chain flexibility in continuum calculations of protein titration." *J. Phys. Chem.* 100:20156–20163.

Beroza, P., and D. A. Case. 1998. "Calculations of proton-binding thermodynamics in proteins." *Methods Enzymol.* 295:170–189.

Beroza, P., and D. R. Fredkin. 1996. "Calculation of amino acid pKas in a protein from a continuum electrostatic model: Model and sensitivity analysis." *J. Comput. Chem.* 17:1229–1244.

Beroza, P., D. R. Fredkin, M. Y. Okamura, and G. Feher. 1991. "Protonation of interesting residues in a protein by a Monte Carlo method: Application to lysozyme and the photosynthetic reaction center." *Proc. Natl. Acad. Sci. USA* 88:5804–5808.

Beroza, P., D. R. Fredkin, M. Y. Okamura, and G. Feher. 1995. "Electrostatic calculation of amino acid titration and electron transfer Qa–Qb → QaQb– in the reaction center." *Biophys. J.* 68:2233–2250.

Besler, B. H., K. M. Merz, and P. A. Kollman. 1990. *J. Comp. Chem.* 11:431–439.

Betz, S. F., P. A. Liebman, and W. F. DeGrado. 1997. "De novo design of native proteins: Characterization of proteins intended to fold into antiparallel, rop-like, four-helix bundles." *Biochemistry* 36:2450–2458.

Betz, S. F., J. L. Marmorino, A. J. Saunders, D. F. Doyle, G. B. Young, and G. J. Pielak. 1996. "Unusual effects of an engineered disulfide on global and local protein stability." *Biochemistry* 35:7422–7428.

Breneman, C. M., and K. B. Wiberg. 1990. "Determining atom-centered monopoles from molecular electrostatic potentials. The need for high sampling density in formamide conformational analysis." *J. Comput. Chem.* 11 (3):361–373.

Choma, C. T., J. D. Lear, M. J. Nelson, P. L. Dutton, D. E. Robertson, and W. F. DeGrado. 1994. "Design of a heme-binding four-helix bundle." *J. Am. Chem. Soc.* 116:856–865.

Cornell, W., P. Cieplak, C. Bayly, and P. Kollman. 1993. *J. Am. Chem. Soc.* 115:9620–9631.

Cornell, W. D., P. Cieplak, C. I. Bayly, I. R. Gould, K. M. Merz, D. M. Ferguson, D. C. Spellmeyer, T. Fox, J. W. Caldwell, and P. A. Kollman. 1995. *J. Am. Chem. Soc.* 117:5179–5197.

Couch, V., D. Popović, and A. Stuchebrukhov. 2011. "Redox-Coupled Protonation of Respiratory Complex I: The Hydrophilic Domain." *Biophys. J.* 101 (2):431–438.

Daune, M. 1997. *Molekulare Biophysik*. Braunschweig, Germany: Vieweg Verlag.

Demchuk, E., and R. C. Wade. 1996. "Improving the continuum dielectric approach to calculating pKas of ionizable groups in proteins." *J. Phys. Chem.* 100:17373–17387.

Dutton, P. L., and D. F. Wilson. 1974. "Redox potentiometry in mitochondrial and photosynthetic bioenergetics." *Biochim. Biophys. Acta* 346:165–212.

Dutton, P. L., D. F. Wilson, and C. P. Lee. 1970. "Oxidation-reduction potentials of cytochromes in mitochondria." *Biochemistry* 9:5077–5082.

Engel, W. D., H. Schaegger, and G. von Jagow. 1980. "Ubiquinol-cytochrome c reductase (EC 1.10.2.2). Isolation in Triton X-100 by hydroxyapatite and gel chromatography." *Biochim. Biophys. Acta* 592:211–222.

Fahnenschmidt, M., H. K. Rau, R. Bittl, W. Haehnel, and W. Lubitz. 1999. "Characterization of a de novo designed heme protein by EPR and ENDOR spectroscopy." *Chem. Eur. J.* 5:2327–2334.

Fonseca, B. M., I. H. Saraiva, C. M. Paquete, C. M. Soares, I. Pacheco, C. A. Salgueiro, and R. O. Louro. 2009. "The tetraheme cytochrome from Shewanella oneidensis MR-1 shows thermodynamic bias for functional specificity of the hemes." *J. Biol. Inorg. Chem.* 14:375–385.

Frisch, M. J., J. A. Pople, and J. S. Binkley. 1984. "Self-consistent molecular orbital methods 25. Supplementary functions for Gaussian basis sets." *J. Chem. Phys.* 80 (7):3265–3269.

Gaussian 98 Revision A.6. Gaussian, Inc., Pittsburgh, PA.

Gabellini, N., J. R. Bowyer, E. Hurt, B. A. Melandri, and G. Hauska. 1982. "A cytochrome bc1 complex with ubiquinol-cytochrome c2 oxidoreductase activity from Rhodopseudomonas sphaeroides GA." *Eur. J. Biochem.* 126:105–111.

Gabellini, N., and G. Hauska. 1983. "Characterization of cytochrome b in the isolated ubiquinol-cytochrome c2 oxidoreductase from Rhodopseudomonas sphaeroides GA." *FEBS Letters* 153:146–150.

Gilson, M. K. 1993. "Multiple-site titration and molecular modeling: Two rapid methods for computing energies and forces for ionizable groups in proteins." *Proteins* 15:266–282.

Gilson, M. K. 1995. "Theory of electrostatic interactions in macromolecules." *Curr. Opin. Struct. Biol.* 5:216–223.

Gilson, M. K., and B. Honig. 1986. "The dielectric constant of a folded proteins." *Biopolymers* 25:2097–2119.

Gunner, M. R., and E. Alexov. 2000. "A pragmatic approach to structure based calculation of coupled proton and electron transfer in proteins." *Biochim. Biophys. Acta* 1458:63–87.

Gunner, M. R., and B. Honig. 1991. "Electrostatic control of midpoint potentials in the cytochrome subunit of the Rhodopseudomonas viridis reaction center." *Proc. Natl. Acad. Sci. USA* 88:9151–9155.

Gunner, M. R., A. Nicholls, and B. Honig. 1996. "Electrostatic potentials in Rhodopseudomonas viridis reaction centers: Implications for the driving force and directionality of electron transfer." *J. Phys. Chem.* 100:4277–4291.

Harbury, H. A., J. R. Cronin, M. W. Fauger, T. P. Hettinger, A. J. Murphy, Y. P. Myer, and S. N. Vinogradov. 1965. "Complex formation between methionine and a heme peptide from cytochrome c." *Proc. Natl. Acad. Sci. USA* 54:1658–1664.

Harvey, S. C. 1989. "Treatment of electrostatic effects in macromolecular modeling." *Proteins* 5:78–92.

Hatefi, Y. 1978. "Preparation and properties of dihydroubiquinone: cytochrome c oxidoreductase (complex III)." *Methods Enzymol.* 53:35-40.

Holst, M. J., R. E. Kozack, F. Saied, and S. Subramiam. 1994. "Treatment of electrostatic effects in proteins: Multigrid-based Newton iterative method for solution of the full nonlinear Poisson-Boltzmann equation." *Prot. Struct. Funct. Genet.* 18:231–245.

Honig, B., and A. Nicholls. 1995. "Classical electrostatics in biology and chemistry." *Science* 268:1144–1149.

Hurt, E., and G. Hauska. 1981. "A cytochrome f/b6 complex of five polypeptides with plastoquinol-plastocyanin-oxidoreductase activity from spinach chloroplasts." *Eur. J. Biochem.* 117:591–599.

Hurt, E., and G. Hauska. 1982. "Identification of the polypeptides in the cytochrome b6/f complex from spinach chloroplasts with redox-center-carrying subunits." *J. Bioenerg. Biomembr.* 14:405–424.

Iakovleva, O., M. Reiner, H. Rau, W. Haehnel, and F. Parak. 2002. "Mössbauer and EPR study of a cytochrome b model." *Phys. Chem. Chem. Phys.* 4:655–660.

Iwata, S., J. W. Lee, K. Okada, J. K. Lee, M. Iwata, B. Rasmussen, T. A. Link, S. Ramaswamy, and B. K. Jap. 1998. "Complete structure of the 11-subunit bovine mitochondrial cytochrome bc1 complex." *Science* 281:64–71.

Izrailev, S., A. R. Crofts, E. A. Berry, and K. Schulten. 1999. "Steered molecular dynamics simulation of the Rieske subunit motion in the cytochrome bc1 complex." *Biophys. J.* 77:1753–1768.

Kilmartin, J. V., and L. Rossi-Bernardi. 1973. "Interaction of hemoglobin with hydrogen ions, carbon dioxide, and organic phosphates." *Physiol. Rev.* 53:836–889.

Kirkwood, J. G. 1934. "Theory of solution of molecules containing widely separated charges with special application to zwitterions." *J. Chem. Phys.* 2:351–361.

Kramer, D. M., and A. R. Crofts. 1994. "Reexamination of the properties and function of the b-cytochromes of the thylakoid cytochrome bf complex." *Biochim. Biophys. Acta-Bioenerg.* 1184:193–201.

Kramer, D. M., A. Joliot, P. Joliot, and A. R. Crofts. 1994. "Competition among plastoquinol and artificial quinone/quinol couples at the quinol oxidizing site of the cytochrome bf complex." *Biochim. Biophys. Acta-Bioenerg.* 1184:251–262.

Lancaster, C. R., H. Michel, B. Honig, and M. R. Gunner. 1996. "Calculated coupling of electron and proton transfer in the photosynthetic reaction center of Rhodopseudomonas viridis." *Biophys. J.* 70:2469–2492.

Li, J., M. R. Nelson, C. Y. Peng, D. Bashford, and L. Noodleman. 1998. "Incorporating protein environments in density functional theory: A self-consistent reaction field calculation of redox potentials of [2Fe2S] clusters in ferredoxin and phthalate dioxygenase reductase." *J. Phys. Chem. A* 102:6311–6324.

Louro, R. O., T. Catarino, J. LeGall, and A. V. Xavier. 1997. "Redox-Bohr effect in electron/proton energy transduction: cytochrome c 3 coupled to hydrogenase works as a 'proton thruster' in Desulfovibrio vulgaris." *J. Biol. Inorg. Chem.* 2:488–491.

Mao, J., K. Hauser, and M. R. Gunner. 2003. "Multi-conformation continuum electrostatic analysis of heme electrochemistry in cytochromes with different structures." *Biochemistry* 42:829–840.

Martel, P. J., C. M. Soares, A. M. Baptista, M. Fuxreiter, G. Náray-Szabó, R. O. Louro, and M. A. Carrondo. 1999. "Comparative redox and pKa calculations on cytochrome c3 from several Desulfovibrio species using continuum electrostatic methods." *J. Biol. Inorg. Chem.* 4:73–86.

McQuarrie, D. A. 1976. *Statistical mechanics*. New York: Harper & Row Publishers, Inc.

Medvedev, E. S., A. Kotelnikov, A. V. Barinov, B. L. Psikha, J. M. Ortega, D. M. Popović, and A. A. Stuchebrukhov. 2008. "Protein dynamics control of electron transfer in photosynthetic reaction centers from Rps. sulfoviridis." *J. Phys. Chem. B* 112 (10):3208–3216.

Metropolis, N., A. W. Rosenbluth, M. N. Rosenbluth, and A. H. Teller. 1953. "Equation of state calculations by fast computing machines." *Journal of Chemical Physics* 21:1087–1092.

Moore, G. R., and G. W. Pettigrew. 1990. "Cytochromes c: Evolutionary, structural and physicochemical aspects." In, 309–362. Berlin: Springer.

Muegge, I., J. Apostolakis, U. Ermler, G. Fritzsch, W. Lubitz, and E. W. Knapp. 1996. "Shift of the special pair redox potential: Electrostatic energy computations of the mutants of the reaction center of Rhodobacter sphaeroides." *Biochemistry* 35:8359–8370.

Nelson, B. D., and P. Gellerfors. 1974. "The redox properties of the cytochromes of purified complex III." *Biochim. Biophys. Acta* 357:358–364.

Nelson, B. D., and P. Gellerfors. 1975. "Alkali-induced reduction of b-cytochromes in purified complex III from beef heart-mitochondria." *Biochim. Biophys. Acta* 396:202–209.

Nicholls, A., and B. Honig. 1991. "A rapid finite difference algorithm, utilizing successive overrelaxation to solve the Poisson-Boltzmann equation." *J. Comp. Chem.* 12:435–445.

Nozaki, Y., and C. Tanford. 1967. "Intrinsic dissociation constants of aspartyl and glutamyl carboxyl groups." *J. Biol. Chem.* 242:4731.

Oberoi, H., and N. M. Allewell. 1993. "Multigrid solution of the nonlinear Poisson-Boltzmann equation of titration curves." *Biophys. J.* 65:48-55.

Olson, T. L., J. C. Williams, and J. P. Allen. 2013. "Influence of protein interactions on oxidation/reduction midpoint potentials of cofactors in natural and de novo metalloproteins." *Biochim. Biophys. Acta* 1827:914–922.

Papa, S. 2005. "Role of cooperative H(+)/e(-) linkage (redox Bohr effect) at heme a/Cu(A) and heme a(3)/Cu(B) in the proton pump of cytochrome c oxidase." *Biochemistry (Mosc).* 70 (2):178–186.

Papa, S., N. Capitanio, G. Villani, G. Capitanio, A. Bizzoca, L. L. Palese, V. Carlino, and E. De Nitto. 1998. "Cooperative coupling and role of heme a in the proton pump of heme-copper oxidases." *Biochimie* 80 (10):821–836.

Papa, S., F. Guerrieri, and G. Izzo. 1979. "Redox Bohr-effects in the cytochrome system of mitochondria." *FEBS Letters* 105:213–216.

Popović, D. M., and I. S. Djordjević. 2020. "Catalytic center of cytochrome c oxidase: Effects of protein environment on pKa values of CuB histidine ligands." *Journal of Serbian Chemical Society* 85:1429–1444.

Popović, D. M., I. V. Leontyev, D. G. Beech, and A. A. Stuchebrukhov. 2010. "Similarity of cytochrome c oxidases in different organisms." *Proteins Struct. Funct. Bioinform.* 78:2691–2698.

Popović, D. M., and A. A. Stuchebrukhov. 2004. "Electrostatic study of proton pumping mechanism of bovine heart cytochrome c oxidase." *J. Am. Chem. Soc.* 126:1858–1871.

Popović, D. M., and A. A. Stuchebrukhov. 2005. "Proton exit channels in bovine cytochrome c oxidase." *J. Phys. Chem. B* 109:1999–2006.

Popović, D. M., and A. A. Stuchebrukhov. 2006. "Two conformational states of Glu242 and pKas in bovine cytochrome c oxidase." *Photochem. Photobiol. Sci.* 5:611–620.

Popović, D. M., and A. A. Stuchebrukhov. 2012. "Coupled electron and proton transfer reactions during the O→E transition in bovine cytochrome c oxidase." *Biochim. Biophys. Acta* 1817 (4):506–517.

Popović, D. M., S. D. Zarić, B. Rabenstein, and E. W. Knapp. 2001. "Artificial cytochrome b: Computer modeling and evaluation of redox potentials." *J. Am. Chem. Soc.* 123:6040–6053.

Popović, D. M., A. Zmiric, S. D. Zarić, and E. W. Knapp. 2002. "Energetics of radical transfer in DNA photolyase." *J. Am. Chem. Soc.* 124:3775–3782.

Quenneville, J., D. M. Popović, and A. A. Stuchebrukhov. 2006. "Combined DFT and electrostatics study of the proton pumping mechanism in cytochrome c oxidase." *Biochim. Biophys. Acta* 1757:1035–1046.

Karlsberg online manual. http://lie.chemie.fu-berlin/karlsberg/.

Rabenstein, B., and E. W. Knapp. 2001. "Calculated pH-dependent population of CO-myoglobin conformers." *Biophys. J.* 80:1141–1150.

Rabenstein, B., G. M. Ullmann, and E. W. Knapp. 1998a. "Calculation of protonation patterns in proteins with structural relaxation and molecular ensembles - application to the photosynthetic reaction center." *Eur. Biophys. J.* 27:626–637.

Rabenstein, B., G. M. Ullmann, and E. W. Knapp. 1998b. "Energetics of electron transfer and protonation reactions of the quinones in the photosynthetic reaction center of Rhodopseudomonas viridis." *Biochemistry* 37:2488–2495.

Rabenstein, B., G. M. Ullmann, and E. W. Knapp. 2000. "Electron transfer between the quinones in the photosynthetic reaction center and its coupling to conformational changes." *Biochemistry* 39:10487–10496.

Rau, H. K., and W. Haehnel. 1998. "Design, synthesis, and properties of a novel cytochrome b model." *J. Am. Chem. Soc.* 120:468–476.

Sham, Y. Y., Z. T. Chu, and A. Warshel. 1997. "Consistent calculation of pKa's of ionizable residues in proteins: Semi-microscopic and microscopic approaches." *J. Phys. Chem. B* 101:4458–4472.

Simonson, T., and C. L. Brooks. 1996. "Charge separation and the dielectric constant of proteins: Insights from molecular dynamics." *J. Am. Chem. Soc.* 118:8452–8458.

Simonson, T., and D. Perahia. 1995a. "Internal and interfacial dielectric properties of cytochrome c from molecular dynamics in aqueous solution." *Proc. Natl. Acad. Sci. USA* 92:1082–1086.

Simonson, T., and D. Perahia. 1995b. "Microscopic dielectric properties of cytochrome c from molecular dynamics simulation in aqueous solution." *J. Am. Chem. Soc.* 117:7987–8000.

Simonson, T., D. Perahia, and G. Bricogne. 1992. "Intramolecular dielectric screening in proteins." *J. Mol. Biol.* 218:859–886.

Soares, C. M., and A. M. Baptista. 2012. "Simulation of multihaem cytochromes." *FEBS Lett.* 586:510–518.

Soares, C. M., P. J. Martel, and M. A. Carrondo. 1997. "Theoretical studies on the redox-Bohr effect in cytochrome c3 from Desulfovibrio vulgaris Hildenborough." *J. Biol. Inorg. Chem.* 2:714–727.

Song, Y., E. Michonova-Alexova, and M. R. Gunner. 2006. "Calculated Proton Uptake on Anaerobic Reduction of Cytochrome c Oxidase: Is the Reaction Electroneutral?" *Biochemistry* 45:7959–7975.

Tanford, C., and R. Roxby. 1972. "Interpretation of protein titration curves. Application to lysozyme." *Biochemistry* 11:2192–2198.

Tanokura, M. 1983. "1H-NMR study on the tautomerism of the imidazole ring of histidine residues: I. Microscopic pK values and molar ratios of tautomers in histidine-containing peptides." *Biochim. Biophys. Acta* 742:576–585.

Teixeira, V. H., C. M. Soares, and A. M. Baptista. 2002. "Studies of the reduction and protonation behavior of tetraheme cytochromes using atomic detail." *J. Biol. Inorg. Chem.* 7:200–216.

Ullmann, M. G. 2000. "The coupling of protonation and reduction in proteins with multiple centers: theory, computational method, and application to cytochrome c3." *J. Phys. Chem. B* 104:6293–6301.

Ullmann, M. G., and E. W. Knapp. 1999. "Electrostatic models for computing protonation and redox equilibria in proteins." *Eur. Biophys. J.* 28:533–551.

Urban, P. F., and M. Klingenberg. 1969. "On the redox potentials of ubiquinone and cytochrome b in the respiratory chain." *Eur. J. Biochem.* 9:519–525.

Vila, J. A., D. R. Ripoll, Y. N. Vorobjev, and H. A. Scheraga. 1998. "Computation of the structure-dependent pKa shifts in a polypentapeptide of the poly[fV(IPGVG),fE(IPGEG)] family." *J. Phys. Chem. B* 102:3065–3067.

von Jagow, G., W. D. Engle, H. Schaegger, W. Machleidt, and I. Machleidt. 1981. *Vectorial reactions in electron and ion transport in mitochondria and bacteria*, edited by F. Palmieri, Quagliarello, E., Siliprandi, N. and Slater, E. C., 149-161. Amsterdam, Holland: Elsevier/North.

von Jagow, G., H. Schaegger, W. D. Engle, W. Machleidt, and I. Machleidt. 1978. "Beef heart complex III: Isolation and characterization of cytochrome b." *FEBS Letters* 91:121-125.

von Jagow, G., and W. Sebald. 1978. "B-type cytochromes." *Ann. Rev. Biochem.* 49:281–314.

Warshel, A., and J. Aqvist. 1991. "Electrostatic energy and macromolecular function." *Ann. Rev. Biophys. Biophys. Chem.* 20:267–298.

Warshel, A., and A. Papazyan. 1998. "Electrostatic effects in macromolecules: Fundamental concepts and practical modeling." *Curr. Op. Struct. Biol.* 8:211–217.

Warshel, A., A. Papazyan, and I. Muegge. 1997. "Microscopic and semimicroscopic redox calculations: What can and cannot be learned from continuum models." *J. Biol. Inorg. Chem.* 2:143–152.

Warshel, A., and S. T. Russel. 1984. "Calculations of electrostatic interaction in biological systems and in solution." *Quart. Rev. Biophys.* 17:283–422.

Warwicker, J., and H. C. Watson. 1982. "Calculation of the electrostatic potential in the active site cleft due to α-helix dipoles." *J. Mol. Biol.* 157:671–679.

Weiss, H., and H. J. Kolb. 1979. "Isolation of mitochondrial succinate: ubiquinone reductase, cytochrome c reductase and cytochrome c oxidase from Neurospora crassa using nonionic detergent." *Eur. J. Biochem.* 99:139–149.

Willner, I., V. Heleg-Shabtai, E. Katz, H. K. Rau, and W. Haehnel. 1999. "Integration of a reconstituted de novo synthesized hemoprotein and native metalloproteins with electrode supports for bioelectronic and bioelectrocatalytic applications." *J. Am. Chem. Soc.* 121:6455–6468.

Wilson, G. S. 1983. "Electrochemical studies of porphyrin redox reactions as cytochrome models." *Bioelectrochem. Bioenerg.* 1:172–179.

Wyman, J. 1968. "Regulation in macromolecules as illustrated by haemoglobin." *Quart. Rev. Biophys.* 1:35–81.

Yang, A.-S., M. R. Gunner, R. Sompogna, and B. Honig. 1993. "On the calculation of pKas in proteins." *Prot. Struct. Funct. Genet.* 15:252–265.

You, T., and D. B. Bashford. 1995. "Conformation and hydrogen ion titration of proteins: A continuum electrostatic model with conformational flexibility." *Biophys. J.* 69:1721–1733.

Zauhar, R. Y., and A. Varnek. 1996. "A fast and space-efficient boundary element method for computing electrostatic and hydration effects in large molecules." *J. Comput. Chem.* 17:864–877.

Zheng, Z., and M. R. Gunner. 2009. "Analysis of the electrochemistry of hemes with E(m)s spanning 800 mV." *Proteins* 75:719–734.

Chapter 4

Comparative Study of Native and Synthetic Cytochrome *b*: Modeling, Dynamics, and Bioinformatics Study

Dragan M. Popović*
Department of Chemistry, University of Belgrade, Institute of Chemistry, Technology and Metallurgy, National Institute of the Republic of Serbia, Belgrade, Serbia

Abstract

We defined a procedure to generate the atomic coordinates of an artificial protein (SynCyt*b*), recently synthesized using the template-assembled synthetic strategy. A four-helix bundle, binding two hemes, constitutes the catalytical heart of the photosynthetic cytochrome b_6f (Cytb_6f) complex and the respiratory cytochrome bc_1 (Cytbc_1) membrane-bound protein complex. The model structure mimics the central part of the Cyt*b* subunit and reproduces a native-like function. It consists of a four-helix bundle, two hemes, and a cyclic decapeptide as a template. Since the X-ray structure is not yet solved, the structural elements of SynCyt*b* were assembled from scratch using all known chemical and structural information available, thereby avoiding strain and atomic clashes. A great deal of work was done to generate the coordinates of this protein by a sophisticated stepwise modeling technique. The molecular mechanics was subsequently employed to fully relax the model structure by applying the stepwise constraint energy minimization. To validate the stability and robustness of the computer-generated model structure, SynCyt*b* is tested by monitoring the conformational stability and fluctuations during a long-term molecular dynamics (MD) simulation and comparing the results with values obtained from the crystal structure

* Corresponding Author's Email: dragan.popovic@ihtm.bg.ac.rs.

In: Structure and Function of Cytochromes
Editor: Elaine J. Mathis
ISBN: 979-8-88697-102-6

of native Cyt*b*. The RMSD (root-mean-square deviation) and RMSF (root-mean-square fluctuation) data from the native and artificial proteins MD simulations reveal very similar profiles over time, demonstrating a relative rigidness, stability, and strain-free of the final model structure. The results of the theoretical work presented in this chapter offer an extensive cheminformatics/ bioinformatics analysis to elucidate and compare the structural details and heme conformations of both proteins. Specifically, the details related to redox cofactors and their close surroundings have been examined. We analyzed the relative orientation of axial histidines ligated to heme, the heme conformations, propionates orientations, and orientation of the tilted helical axes in native and SynCyt*b* and compared them with available experimental data. The implemented analysis provides valuable information for further research and facilitates protein design and molecular modeling since the heme redox potential intrinsically and internally depends on the axial ligands' type, orientation, alignment, and basicity. The protein design presented here can be used to generate novel proteins with tailored structural and functional properties.

Keywords: computational chemistry, bioinformatics, cheminformatics, molecular modeling, artificial protein, heme, cytochrome bc_1 complex, cytochrome *b*, molecular mechanics, molecular dynamics

Abbreviations

Cyt*b*	cytochrome *b*
Cytbc_1	cytochrome bc_1
Cytb_6f	cytochrome b_6f
Cytc_1	cytochrome c_1
ET	electron transfer
ETC	electron transport chain
H_h	high potential heme
H_l	low potential heme
ISP	iron-sulfur protein
MD	molecular dynamics
MM	molecular mechanics
PR (PRA, PRD)	heme propionic groups
RMSD	root-mean-square deviation
RMSF	root-mean-square fluctuation
SynCyt*b*	synthetic (artificial) cytochrome *b*.

Introduction

The metabolism of living cells essentially depends on the electron transfer (ET) and proton transfer (PT) reactions, which are involved in numerous anabolic and catabolic processes. The physiological relevance of redox proteins and electron transfer reactions is a topic of growing research interest in enzymology, molecular biology, and bioenergetics. Redox proteins represent a large group of enzymes, and the cofactors or redox-active centers are embedded in their structural domains to perform a broad range of biochemical functions. In particular, the heme-proteins represent a very diverse group involved in the electron transfer (cytochromes), catalysis (catalases, peroxidases), detoxification (cytochrome P450), transport and storage of small molecules (hemoglobin, myoglobin), proton-pumping and energy transduction (cytochrome oxidases, nitric-oxide reductases).

An emerging biomimetic approach to understanding the influence of the protein environment on the heme properties and to exploring protein structure–function relationships has grown out of creating artificial proteins from first principles, i.e., *de novo* protein design (Choma et al. 1994; Bryson et al. 1995; Sharp et al. 1998; Rau and Haehnel 1998; Huffman, Rosenblatt, and Suslick 1998; Reedy and Gibney 2004; Lu et al. 2009; Koder et al. 2009; Watkins, Armstrong, and Anderson 2014; Huang, Boyken, and Baker 2016; Lombardi et al. 2019). The *de novo* design of redox proteins is an attractive approach to investigating the factors involved in the ET (Marcus and Sutin 1985; Page et al. 1999) and creating proteins with novel functions (Willner et al. 1999; Gibney et al. 2000; Lu, Berry, and Pfister 2001; Shleev et al. 2005; Albrecht et al. 2005; Willner and Willner 2005; Joh et al. 2014; Yu et al. 2014; Moser et al. 2016; Bialas et al. 2016; Pinter, Koebke, and Pecoraro 2020). Recent developments in peptide synthesis and chemoselective ligation connecting the unprotected peptides (Mutter et al. 1989; Mutter et al. 1996) and the increasing knowledge gained from the crystal structures of native proteins have allowed the design, chemical synthesis, and characterization of artificial proteins (Lombardi et al. 2000; Gibney et al. 2000; Lu, Berry, and Pfister 2001; Schnepf et al. 2001; Huang et al. 2003; Haehnel 2004; Zhuang et al. 2006; Monien et al. 2007; Fry et al. 2013; Lalaurie et al. 2018; Curnow 2019).

Nature has selected some redox-active molecules and assembled them within a protein matrix to facilitate essential processes such as substrate binding, electron transfer, light-induced excitation, energy conversion, charge separation, DNA repair, chemical catalysis, or energy transduction in photosynthesis and respiration. The protein environment adjusts the physical and chemical properties of the cofactors to perform the designated function satisfactorily and with high efficiency. Hence, the native proteins' functions are achieved through the complexity of their structures. The protein design aims to uncover the construction principles of native proteins to simplify the problem and synthesize minimalistic protein structures that assemble working arrays of cofactors and reproduce native-like functions. Artificial proteins, also named "molecular maquettes," exhibit the main properties of their native counterparts concerning the physicochemical features of the cofactors and the influence that the peptide environment exerts on their properties (Robertson et al. 1994; Sakamoto et al. 1999; Rau, DeJonge, and Haehnel 2000; Ghirlanda et al. 2004; Olson, Williams, and Allen 2013; Vitale et al. 2015).

On the other hand, electrostatic interactions play a central role in various biochemical processes. The importance of the electrostatic effects in understanding the behavior of proteins has been recognized in many contexts, including dynamic properties, enzymatic catalysis, protein solvation, ligand binding, docking, protein-protein association, determining the pK_a values of ionizable residues, redox potentials of cofactors, pH-dependent properties, ionic strength-dependent phenomena, and interactions between polar and charged groups and residues (Perutz 1978; Warshel 1981; Warshel, Russel, and Churg 1984; Klapper et al. 1986; Warshel and Aqvist 1991; Gilson 1995; Nakamura 1996; Ripoll et al. 1996; Gunner et al. 2000; Ullmann, Noodleman, and Case 2002; Georgescu, Alexov, and Gunner 2002; Mao, Hauser, and Gunner 2003; Popović and Stuchebrukhov 2004; Fee, Case, and Noodleman 2008; Ma et al. 2008; Zheng and Gunner 2009; Popović and Djordjević 2020; Zhang et al. 2021).

The combination of experimental and theoretical research reveals new insight into the protein structure and function. Especially site-directed mutation experiments and structural investigations by NMR and X-ray crystallography have increased the understanding of protein function. With the advent of site-specific NMR assignment procedures, it has become possible to identify the titration behavior of individual residues. Other methods such as IR spectroscopy, potentiometry, the assessment of proton uptake or release, or the measurement of the pH dependence of free energy changes can describe the titration of the macromolecule as a whole. All these methods provide essential information about proteins. Classical electrostatics has also proved to be a successful quantitative tool for yielding accurate descriptions of dynamics properties, electrostatic potential, and the free solvation energies of organic molecules and proteins (Warshel and Russel 1984; Warshel and

Aqvist 1991; Honig and Nicholls 1995; Sham, Chu, and Warshel 1997; Tashiro and Stuchebrukhov 2005). The calculations of intra- and intermolecular interactions and electrostatic energies are required to correlate the protein structure with its function.

A detailed characterization of the strength and nature of these interactions requires an understanding of the physicochemical properties of molecules in an aqueous solution. Although the solvent effects on solute properties have been understood for many years, methods that translate these principles into the accurate prediction of experimental observables have been developed in the last three decades. The classical electrostatic approach to studying charged and polar macromolecules in an aqueous solution became possible after developing powerful computers and fast numerical and computational methods to solve Newton's equations of motion governing the MD simulations and the Poisson-Boltzmann equations for solute molecules that have complex shapes and charge distributions. Understanding the solvation properties requires a model of the solute, solvent, and their mutual interactions. In the molecular mechanics (MM) approach, the solute and solvent molecules are described by the force-field, where nonbonded van der Waals and electrostatic terms account for the most details of intermolecular interactions. A proper description of solvent effects at the molecular level requires calculating the mutual interactions of a large number of molecules and their averaging over many solvent configurations and the different protonation states of protein (a solute). Therefore for many applications, the explicit treatment of solvent molecules and mobile ions is not feasible. An alternative approach involves continuum or macroscopic models, in which solvent properties are described in terms of average values. These methods can define solute molecules in atomic details while treating the solvent in terms of the average properties.

The main research interest in this chapter and the following companion chapter is focused on the electrostatic interactions and the electron/proton transfer reactions occurring in redox proteins. Therefore, the heme-proteins with histidines ligated to the iron atom, native and synthetic cytochrome *b* (Cyt*b*), are the central subject of our considerations. Section "*Material and Methods*" introduces the fundamental molecular mechanics and molecular dynamics equations used to model and optimize the synthetic Cyt*b* structure. Section "*Respiratory Electron Transport Chain*" provides necessary information about the structure, function, and working environment of the native cytochrome bc_1 complex in the mitochondrial respiratory electron transport chain (ETC), which is relevant for understanding its biochemical role and redox heterogeneity of the two Cyt*b* hemes.

The results of the theoretical work presented in this chapter are divided into two mutually corresponding parts. First, we generated the atomic coordinates of the model structure of an artificial protein from scratch recently synthesized by Rau and Haehnel (1998) to mimic the central part of the native Cyt*b*. The stability of the computer-generated protein system was tested by monitoring the conformational changes and fluctuations during a long-term molecular dynamics (MD) simulation and comparing the results with values obtained from the crystal structure of native Cyt*b*.

Then, an extensive bioinformatics (cheminformatics) analysis has compared the different structural details of the native and synthetic Cyt*b* proteins. Specifically, the details related to the heme cofactors and their close surroundings have been examined. We analyzed the relative orientation of axial histidines ligated to heme, the heme conformations, propionates orientations, and orientation of the tilted helical axes in native and SynCyt*b* and compared them with available experimental data. This topic is valuable for further research in the field and for facilitating the protein design and molecular modeling since the heme redox potential internally depends on the type (Liu et al. 1995), basicity (Nesset et al. 1996), alignment (Safo et al. 1997), and orientation of the axial ligands (Zarić, Popović, and Knapp 2001; Fufezan, Zhang, and Gunner 2008; Berry and Walker 2008). The protein design presented here aims to generate novel proteins with tailored structural and functional properties.

To validate the computer-generated structure of the synthetic protein, characteristic observables have to be calculated and compared with measured quantities. One of these is the redox potential, a key parameter of redox-active proteins. The results of the redox potentials calculation are discussed in the next chapter, "*Comparative Study on Structure-Function Correlation of Native and Synthetic Cytochrome b: Protonation and Redox Properties.*"

Material and Method

Molecular Mechanics Force Field

For the molecular modeling, energy minimizations, and MD simulations, we used the molecular mechanics force field from the program CHARMM

(Brooks et al. 1983), in which an empirical energy function containing energy terms for bonded and nonbonded interactions is employed. The energy function has the form:

$$E = \sum_{bonds} k_b (b - b_0)^2 + \sum_{angles} k_\theta (\theta - \theta_0)^2 + \sum_{dihedrals} k_\phi (1 - \cos(n\phi - \delta)) + \sum_{impropers} k_\varphi (\varphi - \varphi_0)^2 + \sum_{i=1}^{N} \sum_{j>i}^{N} \left(4\varepsilon_{ij} \left[\left(\frac{\sigma_{ij}}{r_{ij}} \right)^{12} - \left(\frac{\sigma_{ij}}{r_{ij}} \right)^{6} \right] + \frac{q_i q_j}{4\pi\varepsilon_0 r_{ij}} \right) \quad (1)$$

where k_b, k_θ, k_ϕ, and k_φ are the bond, bond angle, dihedral angle, and improper angle force constants, respectively; b, θ, ϕ and φ are bond length, bond angle, dihedral (torsion) angle, and improper angle, respectively, with the subscript zero representing the equilibrium values, dependent on the atom types involved in the particular interaction. These terms describe the bonded interactions in a macromolecule. As one can see, the vibrations of the bond length, bond angle, and improper angle are defined by harmonic potentials oscillating around the equilibrium value. The torsion energy term is a four-atom potential, defined by the phase factor δ and multiplicity n. The multiplicity designates the number of minima on the potential energy surface by rotation around the bond between the two central atoms.

Nonbonded energy terms describe all interactions for atoms separated by three or more bonds, the so-called 1–4 energy terms. Hence, the Coulomb and Lennard-Jones 6–12 terms contribute to the external (nonbonded) interactions, where ε_{ij} are the Lennard-Jones well depths and σ_{ij} are the distances at the Lennard-Jones minimum, q_i are the partial atomic charges, ε_0 is the effective dielectric constant, and r_{ij} are the distances between atom pairs i and j. While the Coulomb potential describes the electrostatic interactions, the Lennard-Jones potential defines van der Waals interactions and atomic contacts. The electrostatic attraction resulting from opposite partial atomic charges is stronger and acts at a more extended distance than the van der Waals forces. Therefore, electrostatic interactions play a crucial role in stabilizing macromolecular conformations and governing interactions between proteins and ligands. The parameterization of the energy function was done using experimental data and quantum-chemical *ab-initio* calculations for small molecules.

Besides already explained potentials, the energy function can also include energy terms to constrain specific molecular conformations. The CHARMM force field contains the energy terms applicable for certain purposes as helpful tools for fixing or restricting the changes occurring in a structure during the minimization or MD simulation. The most straightforward constraint is to rigidly fix a group of atoms in their positions by not allowing them to move at all. Alternatively, one can apply several other types of soft constraints. For example, the harmonic constraints are primarily used to avoid large displacements of atoms during the energy minimization while still allowing the structure to relax. The harmonic constraint energy term has the form:

$$E_{harmonic}(\vec{r}) = k_r \left(\vec{r} - \vec{r}_{fix}\right)^2 \tag{2}$$

In order to maintain the particular structural motives, dihedral constraints can be applied to the torsion angles ϕ by using energy terms like:

$$E_{dihedral}(\phi) = k_\phi \left(\phi - \phi_{fix}\right)^2 \tag{3}$$

In the same way, bond angles (θ) can be harmonically constrained. The NOE constraints in the energy function are originally designed to drive the molecular system to fulfill experimentally derived inter-proton distances. In addition, one can use the NOE-like distance constraints to keep nonbonded atom pairs at a specific distance. They have the following form:

$$E_{NOE}(R) = \begin{cases} \frac{k_{min}}{2}\left(R - R_{min}\right)^2 & \text{for } R < R_{min}, \\ 0 & \text{for } R_{min} < R < R_{max}, \\ \frac{k_{max}}{2}\left(R - R_{max}\right)^2 & \text{for } R > R_{max}. \end{cases} \tag{4}$$

The program **C**HARMM (MacKerell et al. 1998) uses the highlighted empirical energy function for energy minimization, molecular dynamics simulation, and vibrational analysis. There are several minimization methods integrated with **C**HARMM, of which we applied two of them – the steepest descent and conjugate gradient algorithms to relax the atomic coordinates and protein structure. They use the first or second derivatives of the potential energy to adjust the atomic coordinates in order to find the energy (primarily local) minimum.

Molecular Dynamics

The dynamic properties of macromolecules can be studied by simulating their trajectories represented by a set of atomic coordinates and velocities over time. The MD simulation is performed using a classical mechanics approach in which Newton's equations of motion for all atoms of the molecular system are numerically integrated:

$$\frac{d^2\vec{r}_i(t)}{dt^2} = m_i^{-1}\vec{F}_i \tag{5}$$

where m_i and $\mathbf{r}_i$ are mass and position vector of atom *i*. The forces $\mathbf{F}_i$ acting on the atoms *i* are related to the derivatives of the potential energy concerning the atomic positions ($\mathbf{r}_i$). Thus, they are defined as the gradient of the potential energy:

$$\vec{F}_i = \frac{-\partial E(\vec{r}_1, \vec{r}_2, \ldots, \vec{r}_N)}{\partial \vec{r}_i} \tag{6}$$

Hence, the forces acting in the system determine the velocities of the atoms for a given period of time. The heating of the system is achieved by assigning random velocities to the atoms according to a Gaussian distribution for a given temperature since the temperature is directly proportional to the mean kinetic energy of the system. Then, Newton's equations of motion are solved to determine changes in atomic coordinates and velocities as a function of time. The program CHARMM uses the Verlet algorithm to perform this numerical integration. Since the motions in the system (protein) are very fast, the time (integration) step in the MD has to be very small/short. In general, the time step must be significantly smaller than the period of the fastest local motion in protein, which is typically a bond stretch involving the H-atoms (less than 10 fs). For most applications, a time step of about 1 fs is appropriate. Due to these limitations, only relatively short trajectories can be simulated for large systems (up to several microseconds), while most biologically essential processes such as enzymatic reactions and protein folding occur on much larger time scales.

An alternative possibility to simulate the protein dynamics offers the Langevin- equation:

$$\frac{d^2\vec{r}_i(t)}{dt^2} = m_i^{-1}\vec{F}_i + m_i^{-1}\vec{R}_i - \beta_i \frac{d\vec{r}_i(t)}{dt} \tag{7}$$

It describes stochastic dynamics, where the molecular system is coupled to a heat bath of temperature T by random force $\mathbf{R}_i$. In addition to the conventional classical equations of motions, there are two additional terms, a friction term $\beta_i \cdot d\mathbf{r}_i(t)/dt$, which takes away energy from the system, while the random force $\mathbf{R}_i$ introduces energy into the system.

Respiratory Electron Transport Chain

This section describes the position and role of native cytochrome *b* (Cyt*b*) in the mitochondrial ETC. In the framework of this study, we are interested in modeling a synthetic protein that mimics the central part of the Cyt*b* structure consisting of a four helix-bundle motif (Rau and Haehnel 1998). The synthetic protein exhibits the electron transfer between its two hemes, similar to the natural Cyt*b* reproducing native-like function (Willner et al. 1999; DeJonge, Rau, and Haehnel 1999; Fahnenschmidt et al. 1999). Secondly, we aim to evaluate the redox potentials of the hemes in the synthetic and native Cyt*b*. Therefore, to better understand the role of the two hemes (high H_h and low H_l potential) in the Cyt*b*, it is helpful to clarify the structure, function, and role of native Cyt*b* in the mitochondrial electron transfer chain.

The cytochrome bc_1 (Cytbc_1) complex is one of the four major respiratory protein-membrane complexes residing in the inner mitochondrial membrane. The Cytbc_1 complex transfers the electrons from ubiquinol to cytochrome *c* and uses the released energy to form an electrochemical gradient across the inner membrane. Cytochrome *b* (Cyt*b*) appears as a membrane-spanning subunit of the mitochondrial Cytbc_1 complex (Xia et al. 1997; Iwata et al. 1998) and photosynthetic cytochrome b_6f (Cytb_6f) (Martinez et al. 1994) protein complexes. The central part of the Cyt*b* subunit is a four-helix bundle with two hemes, each axially ligated by two histidines.

Coupling of Oxidative Phosphorylation to the Electron Transport

Energy conversion in the biosphere occurs mainly through respiration and photosynthesis and represents a flux several orders of magnitude greater than all anthropogenic energy usage. Natural photosynthetic protein complexes capture sunlight to power the energetic catalysis that supports life on Earth. Then again, natural respiratory protein complexes capture electrons from the digestion of food molecules (via NADH and FAD) to form electrochemical proton gradients across organellar membranes conserving the energy for life processes in living cells. The underlying mechanisms involve coupling electron transfer, along a chain of redox or photo-redox enzymes, to proton translocation across an organellar membrane in which those redox components are embedded. That gives rise to a transmembrane electrochemical proton gradient, which can be coupled to energy-consuming processes, including the synthesis of ATP – a principle first proposed by Peter Mitchell in his famous *chemiosmotic hypothesis* (Mitchell 1961).

In 1951 Lehninger (Lehninger, Nelson, and Cox 1997) provided experimental proof that electron transport from NADH to oxygen is the direct source of energy used for the coupled phosphorylation of ADP. In his experiment, the NADH was rapidly oxidized to NAD^+ at the expense of molecular oxygen. Simultaneously, up to three molecules of ATP were formed from ADP and phosphates. The overall reaction for the respiratory-chain phosphorylation can be written as:

$$NADH + H^+ + 3ADP + 3P_i + \frac{1}{2}O_2 \longrightarrow NAD^+ + 4H_2O + 3ATP \tag{8}$$

This reaction consists of an exergonic part (oxidation of NADH by oxygen) coupled with an endergonic part (phosphorylation of ADP). However, the electron transfer process occurs through a multistep pathway that harnesses the liberated free energy to form ATP in living systems. Thus, the complete oxidation of glucose by molecular oxygen is given by two half-redox reactions:

$$C_6H_{12}O_6 + 6H_2O \longrightarrow 6CO_2 + 24H^+ + 24e^- \tag{9}$$

$$6O_2 + 24H^+ + 24e^- \longrightarrow 12H_2O$$

The 12 electron pairs involved in the glucose oxidation are not transferred directly to O_2; instead, they are transferred to the coenzymes NAD^+ and FAD to form 10 NADH + 2 $FADH_2$ in the reactions catalyzed by the glycolytic enzymes. The electrons then pass into the electron-transport chain where, through reoxidation of NADH and $FADH_2$, they participate in the sequential redox reactions of more than 10 redox centers before finally reducing O_2 to H_2O. In this process, protons are simultaneously ejected from the mitochondrion matrix. The free energy stored in the resulting pH gradient drives the synthesis of ATP from ADP and phosphates through oxidative phosphorylation.

The mitochondrion is the site where many essential biochemical processes, such as the tricarboxylic acid cycle reactions, oxidation of the fatty acids, electron transport, and oxidative phosphorylation. It is a site of eukaryotic oxidative metabolism, which is, so to say, the cell's "power plant." Mitochondria contain a smooth outer membrane, an inner membrane invaginated to form folds called cristae, and an internal matrix rich in protein. The two membranes differ in chemical composition, permeability, and enzyme content. The electron transport and oxidative phosphorylation enzymes reside in the inner membrane, while the tricarboxylic acid cycle enzymes are primarily located in the matrix space. The matrix is the inner mitochondrial compartment consisting of a gel-like substance of about 50% water content. It contains remarkably high concentrations of the soluble enzymes of oxidative metabolism and substrates, nucleotide cofactors, and inorganic ions. The matrix also contains DNA, RNA, and ribosomes. The outer mitochondrial membrane permits free diffusion of up to 10 kDa molecules through nonspecific pores. The inner membrane that is considerably richer in proteins and contains the respiratory chain and transport proteins is freely permeable only to O_2, CO_2, and H_2O. The controlled impermeability of the inner mitochondrial membrane to most ions, metabolites, and low molecular mass compounds permits the generation of ionic gradients across this barrier and results in the compartmentalization of metabolic functions between the cytosol and mitochondria.

Three sites in the respiratory chain have been identified as furnishing the energy for the phosphorylation process: site I, between NAD and ubiquinone; site II, between Cyt*b* and Cyt*c; and* site III, between the binuclear center heme a_3/Cu_B and oxygen. The phosphate and ADP are required components for maximal electron transport rates in intact mitochondria.

Electron Transport

In 1961 Mitchell (Mitchell 1961) proposed the chemiosmotic hypothesis. It postulates that the free energy of electron transport is used to pump the protons from the mitochondrial matrix (a region of low proton concentration and negative electrical potential) to the intermembrane space (an area of high proton concentration and positive electrical potential in contact with cytosol), creating an electrochemical proton gradient across the inner mitochondrial membrane. The energy stored by the proton gradient is used successively for ATP synthesis. The proton pumping across the inner mitochondrial membrane is an endergonic process. The required energy is gained by coupling with the electron-transport chain in which electrons pass through four protein complexes containing several redox centers with progressively greater electron affinity (increasing standard redox potentials). Such an electron transport is thermodynamically favorable. The significant overall free energy change is broken into three smaller energy contributions coupled with ATP synthesis. Complexes I, III, and IV are sites that furnish energy for phosphorylation.

The Sequence of Electron Transport

The free energy necessary to generate ATP is extracted from the oxidation of NADH and $FADH_2$ by the ETC. It consists of four protein complexes and a single-chain protein of cytochrome *c*, through which electrons pass from lower to higher standard redox potentials. Electrons are carried from complex I and II to complex III by the membrane (lipid) soluble coenzyme Q (CoQ or ubiquinone) and between complexes III and IV by the peripheral membrane protein cytochrome *c*. The overall reactions catalyzed by the corresponding protein complexes in the mitochondrial ETC are as follows:

$$\mathrm{NADH} + \mathrm{CoQ(ox)} \xrightarrow{\text{complex I}} \mathrm{NAD^+} + \mathrm{CoQ(red)};$$
$$(\Delta E^0 = 0.360\ \mathrm{V},\ \Delta G^0 = -70\ \mathrm{kJ/mol}) \qquad (10)$$

$$\mathrm{FADH_2} + \mathrm{CoQ(ox)} \xrightarrow{\text{complex II}} \mathrm{FAD} + \mathrm{CoQ(red)};$$
$$(\Delta E^0 = 0.015\ \mathrm{V},\ \Delta G^0 = -2.9\ \mathrm{kJ/mol})$$

$$\mathrm{CoQ(red)} + \mathrm{Cyt}\ c\mathrm{(ox)} \xrightarrow{\text{complex III}} \mathrm{CoQ(ox)} + \mathrm{Cyt}\ c\mathrm{(red)};$$
$$(\Delta E^0 = 0.190\ \mathrm{V},\ \Delta G^0 = -37\ \mathrm{kJ/mol})$$

$$\mathrm{Cyt}\ c\mathrm{(red)} + 2\mathrm{H^+} + \frac{1}{2}\mathrm{O_2} \xrightarrow{\text{complex IV}} \mathrm{Cyt}\ c\mathrm{(ox)} + \mathrm{H_2O};$$
$$(\Delta E^0 = 0.580\ \mathrm{V},\ \Delta G^0 = -110\ \mathrm{kJ/mol}).$$

Complex I (NADH-coenzyme Q reductase) catalyzes NADH oxidation by CoQ(ox). It contains one flavin mononucleotide (FMN) and six to seven iron-sulfur clusters as cofactors (Guerrero-Castillo et al. 2017; Couch, Popović, and Stuchebrukhov 2011). Complex II catalyzes the oxidation of $FADH_2$ by CoQ(ox). It consists of the enzyme succinate dehydrogenase and four other small hydrophobic subunits and enables electron transfer from succinate via $FADH_2$ to CoQ (Schilling et al. 2006). Complex III catalyzes the oxidation of CoQ(red) by cytochrome *c*, while complex IV catalyzes the oxidation of cytochrome *c* by molecular O_2 (Wikström, Krab, and Sharma 2018; Brzezinski and Gennis 2008; Popović 2013). Except for the second one, all four processes supply sufficient free energy to drive ATP synthesis, which is an endergonic process with energy $\Delta G^0 = +30.5$ kJ/mol.

The proteins embedded in the inner mitochondrial membrane are organized into the four respiratory complexes. Each complex consists of several associated protein subunits that contain a variety of redox-active prosthetic groups (Table 1). The redox potentials of prosthetic groups increase through the electron-transport chain, making the electron transfer from one to another redox group energetically favorable. The complexes are relatively mobile within the inner mitochondrial membrane and surprisingly do not form any stable, highly organized structures. Moreover, they are not even present in equimolar ratios.

Complex III

Since the focus of our interest is the cytochrome *b* subunit, we provide here some more information about the complex III of the respiratory chain (Figure 1) and its constitutive part Cyt*b*. The central component of the electron-transport chain in mitochondrial and many aerobic or photosynthetic bacteria is the membrane protein complex known as the cytochrome bc_1 complex, or ubiquinol–cytochrome *c* oxidoreductase (E.C.1.10.2.2). This enzyme complex catalyzes the electron transfer from ubiquinol to a soluble cytochrome *c* generating a proton gradient across the inner mitochondrial membrane. Electron transfer is coupled to the translocation of two protons across the inner mitochondrial membrane for each fully oxidized ubiquinol (Mitchell 1976; Crofts 1985; Hinkle et al. 1991). The cytochrome bc_1 complex contains three subunits with active redox centers – cytochrome *b*, cytochrome c_1, and the Rieske protein (iron-sulfur protein, ISP) (Robertson et al. 1993). The mitochondrial system of eukaryotes contains additional subunits not present in bacteria (Crofts 1985; Schaegger et al. 1986).

Table 1. Components of the mitochondrial electron transport chain

Enzyme complex Name [d]		**Mass (kda)**	**Number of subunits**	**Prosthetic groups**	**Binding partners**
I	NADH-Q reductases	950	44-46	FMN Fe-S Mn-complex	NADH [a] Q [b]
II	succinate-Q reductase	127	4-5	FAD Fe-S	succinate [a] Q [b]
III	ubiquinol-cytochrome *c* oxidoreductase	240	11	heme b_H-562 heme b_L-566 2Fe-2S heme c_1	Q [b] Cyt*c* [c]
	cytochrome *c*	13	1	heme *c*	Cytc_1 [c] Cu_A [c]
IV	cytochrome *c* oxidase	160	13	Cu_A heme *a* heme a_3 Cu_B	Cyt*c* [c] O_2 [b]

[a] location of a binding site is on the matrix side of the membrane.
[b] location of a binding site is within the hydrocarbon core (in the membrane).
[c] location of a binding site is on the intermembrane (cytosolic) side of the membrane.
[d] membrane-bound enzyme complexes of the mammalian (*bovine*) mitochondrial ETC (Guerrero-Castillo et al. 2017; Schilling et al. 2006; Wikström, Krab, and Sharma 2018).

The Cyt*bc*$_1$ complex isolated from beef heart consists of 11 different polypeptide chains and has a total molecular mass of 240 kDa (Schaegger et al. 1995), see Table 1. There are four redox centers, *i.e.*, two heme groups b_H and b_L in Cyt*b*, one heme group in Cytc_1, and one iron-sulfur cluster [2Fe-2S] in the Rieske protein. The enzymatic mechanism of the proton translocation coupled to electron transport is a distinctive version of the "protonmotive Q-cycle" first discovered by Peter D. Mitchell. The mechanism also explains the inhibition patterns by the ubiquinone analogs: antimycin, stigmatellin, undecyl-hydroxy-diazo-benzothiazole, myxothiazole and methoxy-acrylo-stilbene, which are unambiguously bound at one of the two catalytic sites in which quinone is processed (Mitchell 1976; Crofts 1985; Wikström and Berden 1972).

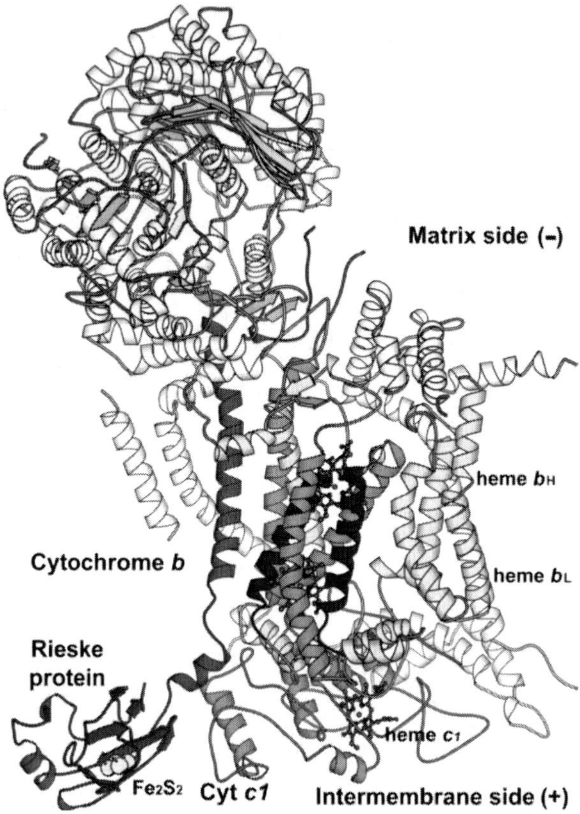

Figure 1. The X-ray structure of the complex III (ubiquinol–cytochrome *c* oxidoreductase or cytochrome bc_1 complex) from the bovine heart was solved by Iwata et al. (1998). The antiparallel four-helix bundle of Cyt*b* is displayed in blue and orange colors, and the rest of the Cyt*b* is shown in yellow. The Rieske iron-sulfur protein (ISP) is depicted in red and cytochrome c_1 in green. All other subunits are shown in gray. A whole mammalian complex III of the respiratory ETC consists of eleven polypeptide chains and contains three hemes and one 2Fe-2S cluster as redox-active groups.

The mitochondrial Cytbc_1 complex performs two functions (bifunctional enzyme). It is a respiratory multienzyme complex that recognizes a mitochondrial targeting presequence (Braun et al. 1992; Braun and Schmitz 1995). The X-ray structures of the complex III reveal two stable steady-state positions for the Rieske iron-sulfur subunit (ISP) (Zhang et al. 1998; Iwata et al. 1998). One location is close to the low potential heme of Cyt*b*, and the other is close enough to Cytc_1. It was suggested that ISP might move around a pivot joint during the catalytic cycle of the functioning bc_1 complex. The reaction mechanism involves the movement of the extrinsic domain of ISP to shuttle electrons from ubiquinol to Cytc_1 (Zhang et al. 1998; Iwata et al. 1998).

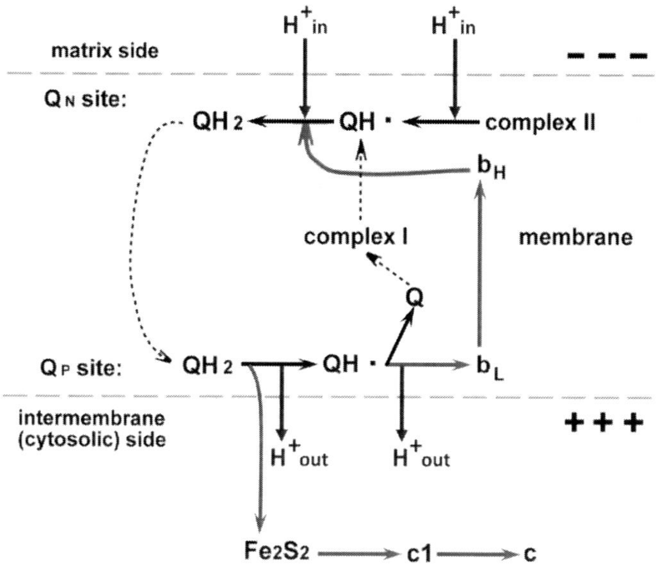

Figure 2. Electron transfer pathway in the complex III. The hemes of Cyt*b* play a central role in the complete oxidation process of semiquinone (QH˙) to quinone (Q), and in that way, the second electron of ubiquinol also gets pushed along the ETC. Red arrows designate the electron transfer, while blue arrows designate the proton transfer reactions of ubiquinones, including a proton uptake (on the *N*-side) and release (on the *P*-side). Since different ubiquinone species come from the so-called Q-pool, the black dashed arrows indicate only a possible pathway of ubiquinones. For more discussion, see the text.

Cyt*b* is a transmembrane protein. Its polypeptide chains from various species have 379 to 385 residues and exhibit considerable sequence homology. Cyt*b* possesses eight transmembrane helices (αA to αH) and four horizontal helices on the intermembrane side (αab, αcd1, αcd2, and αef). All eight stretched transmembrane helices are more than 20 residues long and contain mainly hydrophobic residues. They span the membrane bilayer. The two hemes, b_L and b_H, coordinated to four invariant His residues, are in the center of a four-α-helical bundle formed by helices αA, αB, αC, and αD. Hemes b_L and b_H are close to the intermembrane (cytoplasmic) and matrix side, respectively. There are two possible binding sites, Q_P and Q_N, where semiquinones or the corresponding inhibitors can be bound. The Q_P site is in contact with the membrane's positive (intermembrane) side, while the Q_N site is in contact with the membrane's negative (matrix) side. For example, myxothiazol and antimycin *A* binding sites are observed close to hemes b_L and b_H, respectively. Myxothiazol binds in the semiquinone oxidation (Q_P) site of the Cyt*bc*$_1$ complex, blocking the electron transfer to both the ISP and heme b_L, while antimycin *A* binds in the semiquinone reduction (Q_N) site, preventing the reduction of semiquinone by the b_H heme.

The flow of electrons among the four redox-active prosthetic groups of the ubiquinol-cytochrome *c* reductase complex is explained in Figure 2. First, ubiquinol (QH_2) transfers one of its two high-potential electrons to the Rieske 2Fe-2S cluster. This electron is further shuttled to the cytochrome c_1 heme with higher redox potential. Finally, a peripheral on the matrix side located cytochrome *c* receives that electron and transports the electron from complex III to complex IV. This electron transfer partially oxidizes the ubiquinol (QH_2) to the semiquinone ($QH^{\bullet}$). Since the 2Fe-2S cluster is a one-electron carrier, it cannot accept the second electron from $QH^{\bullet}$ at the same time. Therefore, the electron pathway is bifurcated (Wikström and Berden 1972; Wikström 1973; Brandt 1996). Specifically, the second electron of the $QH^{\bullet}$ species, bound to the semiquinone oxidation (Q_P) site, is transferred to heme b_L. At the same time, the semiquinone ($QH^{\bullet}$) is oxidized to quinone (Q). Then, the electron transfer goes from low (b_L) to high (b_H) potential heme of the Cyt*b*. More details on the bifurcation of the central reaction of the Q-cycle can be found in recent reviews (Brandt and Trumpower 1994; Brandt 1998; Link 1997).

The two cytochrome *b* hemes, although chemically identical, have different electron affinities and redox potentials since they are located in different protein environments. In turn, the high potential (b_H) heme reduces the semiquinone $QH^{\bullet}$ bound to the second binding site (Q_N, close to b_H) to ubiquinol (QH_2). Thus, the net reaction catalyzed by cytochrome *b* is:

$$QH^{\bullet} + QH^{\bullet} \xrightarrow{\text{Cyt } b} QH_2 + Q \tag{41}$$

The cytochrome *b* is, so to say, some kind of the "recycling device" that enables a two-electron carrier (ubiquinol, QH_2) to interact with a one-electron carrier [2Fe-2S] iron-sulfur cluster of the Rieske protein and to pass both of its electrons in the respiratory chain.

Molecular Modeling

Protein Design

The synthetic protein we studied in detail mimics the structure and function of the native Cyt*b*. Both model protein and native Cyt*b*contain two hemes as cofactors, shielded by an antiparallel four-helix bundle. The helices of the synthetic protein are held together by a template. The template is a cyclic decapeptide with an antiparallel β-sheet structure and two β-turns formed by Pro-Gly residues (Rau and Haehnel 1998). The template sequence is given by: *cyclo*-(C A C P G C A C P G). The four-helix bundle is formed by two different types of helices, *F* with 29 residues and *G* with 27 residues. The helix *F* contains phenylalanine at position 15 and the two histidines at positions 8 and 22, while helix *G* contains glycines at positions 8 and 22. Their sequences are closely related to the sequence of helices from *Rhodobacter captulatus* (Hauska et al. 1983; Hauska, Nitschke, and Herrmann 1988; Gennis et al. 1993) and the sequence of the "Maquette" designed by Robertson et al. (1994) to form the four-helix bundle. The underlined amino acid sequences of the native and synthetic four-helix bundles are compared in Figure 3.

A model structure of the SynCyt*b* is displayed in Figure 4. It consists of two pairs of identical helices, F_1, F_2, and G_1, G_2, which form the four-helix bundle. Two hemes are bound to histidine residues of the parallel helices F_1 and F_2, alternating in the bundle with the antiparallel helices G_1 and G_2. The binding helix *F* is modeled with two heme-ligating histidines separated by 13 residues, with central phenylalanine and two arginines at three amino acids away from the histidines. In the shielding helix *G*, two glycines were positioned at the same height as the ligating histidines in helix *F* to provide space for tight packing of the heme edges. To avoid charged end groups and increase the helices' stability, the C-termini are amidated (CO-NH_2), and the N-termini are acetylated (Ac-).

The synthetic four-helix bundle assembly is secured by a template with four cysteines that bind the helices via bromoacetylated groups (BrAc) attached to two N-terminal glycines and two side chains of the C-terminal lysines of the heme-binding *F* and shielding *G* helices, respectively. Hence, the helices are connected by thioether bonds with the cysteines from the template holding the four helices together.

$$\text{template(Cys)-}\mathbf{CH_2-SH} + \mathbf{Br-CH_2\text{-}CO}\text{-NH-helix} \xrightarrow{-\mathbf{HBr}} \text{template(Cys)-}\mathbf{CH_2-S-CH_2\text{-}CO}\text{-NH-helix}$$

N-terminus (at template side) — C-terminus (at open end side)

helix *F*[a]:

1 2 3 4 5 6 7 8 9 0 1 2 3 4 5 6 7 8 9 0 1 2 3 4 5 6 7 8 9

BrAc– G G E L R E L **H** E K L A K Q F E Q L V K L **H** E E R A K K L –NH_2

helix *B*[b, c]:

A M R Y I **H** A N G A S L F F L A V Y I **H** I F R G

helix *D*[b, c]:

R F F S L **H** Y L L P F V I A A L V A I **H** I W A F

Maquette[d]:

C G G G E L W K L **H** E E L L K K F E E L L K L **H** E E R L K K L

N-terminus (at open end side) — C-terminus (at template side)

helix *G*[a]:

1 2 3 4 5 6 7 8 9 0 1 2 3 4 5 6 7 8 9 0 1 2 3 4 5 6 7

Ac– L E E L W K K G E E L A K K L Q E A L E K G K K L A K –NH_2 (K – BrAc)

helix *A*[c]:

W W I W G I V L A F T L V L Q I V T G I V L

helix *C*[c]:

T W I V G M V I Y L L M M G T A F M G Y V L

template:

1	2	3	4	5	6	7	8	9	10
C	A	C	P	G	C	A	C	P	G

Figure 3. Comparison of the amino acid sequences of the four-helix bundles from native Cyt*b*, designed SynCyt*b*, and molecular "Maquette." The amino acid sequences of the two heme-binding helices *B* and *D* and that of the two shielding helices *A* and *C* have been adapted to design the hydrophobic surface of the helices *F* and *G* in the SynCyt*b* (Rau & Haehnel 1998).[a] For comparison, the sequences of the helices *A*, *B*, *C*, and *D* from *Rhodobacter captulatus* (Hauska et al. 1988;[b] Gennis et al. 1993)[c] with the conserved residues marked in bold are shown in addition to the sequence of the *Maquette* designed by Robertson et al. (1994)[d]. The bromoacetyl moieties (BrAc-), attached to the helices *F* and *G*, enable their binding to Cys residues of the template and ensure their antiparallel orientation. Acetyl (Ac-) and amido (-$CONH_2$) blocking groups remove the charges from the ends of the helices stabilizing the entire synthetic structure. His residues (in bold) in the helix type *F* are ligated to the heme iron atoms. The underlined residues are conserved in sequences of Cyt*b* from different species.

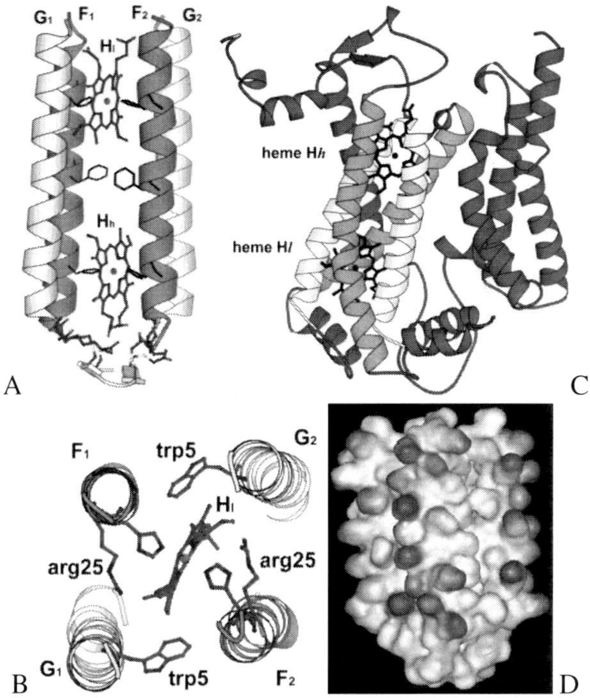

Figure 4. Structures of synthetic and native cytochrome *b*. (A) Side view of the SynCyt*b* structure. The helices of types *F* and *G* are shown in orange and gray. The high potential heme H_h situated in the lower part of the four-helix bundle is shielded from the solvent by a template consisting of an antiparallel cyclic β-strand, which is displayed together with the side chains involved in the four thioether bonds. The low potential heme H_l is located at the open end of the bundle. The hemes are displayed with the axially ligated histidines, His8($F_{1,2}$) and His22($F_{1,2}$). Two central phenylalanines, Phe15(*F*), are also shown. (B) Top view of the SynCyt*b* for the fully relaxed structure. The helices of the bundle are slightly distorted and tilted. For clarity, the lower part involving the template and the high potential heme is not shown. The low potential heme H_l is displayed together with the axially ligated histidines His22(F_1) and His22(F_2) and the two arginines Arg25($F_{1,2}$), forming salt bridges with the PR groups. Furthermore, the two tryptophans Trp5(G_1) and Trp5(G_2) shielding the low potential heme at the top of the bundle from the solvent are shown. (C) The crystal structure of the Cyt*b* subunit of the Cytbc_1 complex from the bovine heart (Iwata et al., 1998). The four-helix bundle of the native Cyt*b*, embedding the two hemes, is displayed in the same orientation as SynCyt*b*, using the same color code. The high potential heme H_h is in the upper part, and the low potential heme H_l is in the lower part of the structure. (D) Electrostatic potential mapped on the surface of SynCyt*b* displays the position of the charged groups and ionic bridges between positively charged lysines (blue) and negatively charged glutamates (red).

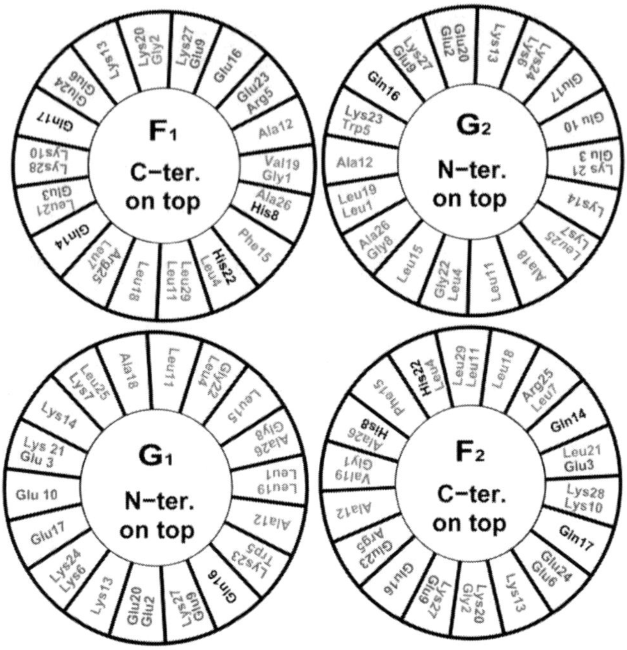

Figure 5. Helical wheel projection of the synthetic four-helix bundle viewed from the top. The hydrophobic residues are given in green, neutral hydrophilic residues by black symbols, while positively and negatively charged residues by blue and red symbols, respectively. The helices of types *F* and *G* correspond to those denoted *H1* and *H2* by Rau et al. (1998).

The hemes are situated in the hydrophobic binding core of the four-helix bundle and axially coordinated by histidines belonging to two oppositely placed *F* helices. The heme planes are in parallel orientation. Both hemes' propionic (PR) groups point to the nearest helix ends as in the native Cyt*b*. The arginine residues at positions 5 and 25 of helix *F* provide positive charges near the position of the PR groups of the hemes, which enables the formation of salt bridges between them. There is a phenylalanine residue at position 15 in the center of helix *F* to isolate the two hemes and fill the hydrophobic space between them. In native structures, helices end up near the edge of the heme groups, and the binding pocket is shielded against the hydrophilic phase with loops connecting helices on both sides of the membrane. In the model structure, the helices are longer, and the shielding of the hydrophobic interior may be provided by tryptophans in position 5 of the helices *G* at one end and by the template at the other end SynCyt*b* molecule.

Except for the four histidines, the synthetic protein conserves the interior packing of the hemes by hydrophobic residues shown by green symbols in Figure 5. The formation of the four-helix bundle is significantly driven by the resulting hydrophobic interactions, thus facilitating the association of the four helices into the helix bundle. The residues on the helix bundle's outer surface are almost all hydrophilic to increase solubility and possibly stabilize the helices by ionic bridges between lysines (in blue) and glutamates (in red) at the solvent-exposed surface. As a result, one obtains a water-soluble model protein, where the antiparallel helices shield the heme-binding pocket; see the helical wheel representation of SynCyt*b* in Figure 5.

Generation of Atomic Coordinates

A crystal structure of the SynCyt*b* consisting of 2218 atoms is not available yet. Nevertheless, we were able to construct atomic coordinates of this model protein since many structural details are very well defined. The generation of the synthetic protein coordinates by advanced modeling techniques and constrained energy minimization was done using the program CHARMM (Brooks et al. 1983; MacKerell et al. 1998). The procedure that we used allows us to repeat individual modeling steps in order to optimize the resulting structure.

To generate atomic coordinates of a stable model structure of the SynCyt*b* required many attempts. We discarded all structures if we observed significant conformational changes after applying energy minimization to relax strain from the model structure. We varied the modeling steps iteratively to avoid strain in the model structure before we even applied steps of structural relaxation. We found that stepwise constrained energy minimization is crucial to finally getting a reliable and stable model structure. If one directly performs the energy minimization without any constraints or the MD simulation on a molecular structure under strain, the well-prepared structural details could get lost, or even worse, the structure may literally explode. Therefore, we used different types of constraints during the various steps of the structural relaxation to stabilize specific features of the molecular structure. Here, we only briefly overview the applied modeling and minimization steps. The code in brackets refers to the corresponding steps described in more detail in the original publication (Popović et al. 2001).

In the first seven modeling steps, we generated the atomic coordinates of the four helices (A1, A2), the two hemes (A4, A5), and the cyclic decapeptide (A6, A7) and assembled them to form the complete model of the SynCyt*b* as faithful as possible without using structural relaxation. The four axially ligated histidines' imidazole rings were appropriately reoriented by resetting the torsion angles and patched to the hemes (A3). The cyclic decapeptide was patched to the four helices via thioether bonds between four cysteines of the decapeptide and acetylated N-terminal Gly1 of the helices *F* or acetylated side chains of the C-terminal Lys27 of the helices *G* (A7).

After the model building, the artificial protein structure is still under strain and has packing problems, particularly the side chains. The strain and van der Waals clashes were removed by applying constrained energy minimization using the force field of the CHARMM22 parameter set (MacKerell et al. 1998). At the beginning of the structural relaxation, we applied several harmonic constraints to the heme-imidazole torsion angles to keep the imidazole planes orthogonal to the heme plane. These constraints were left until the last steps (B10) of minimization. First, the side chains of the four histidines were energy minimized with appropriate constraints (B1). Then, the conformations of the arginines involved in salt bridges with heme PR groups were adjusted by constraint energy minimization (B2). Next, we minimized the side-chain conformations of the two phenylalanines dividing the hydrophobic core of the four-helix bundle (B3). Now all side chains were relaxed by energy minimization except the atoms of arginines, phenylalanines, and cysteines (B4). Namely, we wanted to keep them in their present conformation and to adjust the positions of all other side chains. After that, we could relax all side chains together (B5). Then the cyclic decapeptide was energy minimized (B6, B7). In the following minimization steps also, the hemes were relaxed. First, we simultaneously relaxed the conformations of the propionic acids and arginines, forming salt bridges with each other (B8). Then, we relaxed the SynCyt*b* by gradually reducing the constraints applied to the peptide backbone atoms (B9). Finally, the resulting model structure was energy minimized without any constraints (B10). The entire modeling procedure to construct the SynCyt*b* protein coordinates is described in the original paper by Popović et al. (2001).

Preparation of the Native Cytb Subunit for MD Simulation

We used the crystal structure of the mitochondrial Cytbc_1 complex from the bovine heart, which was solved at 100 K and pH = 6.6 at a resolution of 3.0 Å (Iwata et al., 1998; PDB code: 1BE3). Only the Cyt*b* subunit (chain C in PDB entry) containing two hemes was considered for the MD simulation. During the MD simulation, the titratable groups were considered to be in their standard protonation state, as they are at pH 7. Hydrogen atoms were added to the X-ray structure with the HBUILD command of CHARMM22, leading to an all-atom representation. The total number of atoms of the Cyt*b* subunit, including all hydrogens and two hemes, is 6242. Subsequently, the hydrogen atom positions were energy minimized with the CHARMM22 force field, keeping all non-hydrogen atoms fixed. To provide identical starting conditions for the MD simulation between the native and artificial Cyt, the structure of the native Cyt*b* was relaxed similarly. First, all side-chain atoms were energetically minimized, while both hemes and backbone atoms were fixed. Then, the whole structure of the native Cyt*b* was energetically minimized without applying constraints.

Molecular Dynamics Simulation

To test the stability and lack of strain in the resulting model structure, we performed the MD simulations of the SynCyt*b* and for comparison also for the native Cyt*b* subunit of the bovine heart (Iwata et al. 1998). We ran a Langevin dynamics simulation with a friction coefficient $\beta = 5.0$ ps^{-1} in a vacuum at a temperature of $T = 300$ K, reached after 150 ps for the SynCyt*b* and 100 ps for the native Cyt*b*. The friction mechanism was applied to only the C_α atoms to avoid fast heating and strong coupling between the molecular system and a heat bath. The length of covalent bonds involving hydrogen atoms was constrained with the SHAKE algorithm (Ryckaert, Ciccotti, and Berendsen 1977) so that the time propagation could be done in 1-fs steps. Coulomb and Lennard-Jones (van der Waals) interactions of each atom were calculated with the group switch cut-off scheme (Brooks et al. 1983). The full atom pair interactions are applied at distances less than 9.0 Å. From 9.0 to 10.0 Å, the potential gradually diminishes, and at distances larger than 10.0 Å, atom pair interactions vanish. As a cut buffer radius, we used the value of $r_{cutbuff} = 12.0$ Å. The value of the dielectric constant was $\varepsilon = 1$ everywhere. The trajectories have a length of 10 ns. Atomic coordinates were stored, every 100 fs yielding 100,000 coordinate frames. We considered the last 8 ns of the trajectories for computing averages.

Results and Discussion

Molecular Modeling and Bioinformatics Study

Classical electrostatics (molecular mechanics, dynamics, and modeling) has proved to be a successful quantitative tool for accurately describing the dynamic properties, interaction energies, electrostatic potential, and solvation of organic molecules and proteins (Perutz 1978; Warshel 1981; Warshel and Russel 1984; Warshel, Russel, and Churg 1984; Warshel and Aqvist 1991; Honig and Nicholls 1995; Gilson 1995; Nakamura 1996; Mao, Hauser, and Gunner 2003; Popović and Stuchebrukhov 2004; Ma et al. 2008; Zheng and Gunner 2009; Zhang et al. 2021).

Bioinformatics is a relatively new scientific subdiscipline that uses computer technology to organize and analyze complex biological data, especially in the field of genetics and genomics. However, in a broader sense, bioinformatics is used to collect, store, analyze and disseminate biological data and information, including applications in biological research to analyze or predict the composition or structural details of biomolecules (proteins, nucleic acids, or other macromolecules) and model biological systems. Nowadays, bioinformatics comprises many tools and techniques essential for all research areas in the biological sciences. Thus, bioinformatics (cheminformatics) has also proved valuable for obtaining relevant information on different subjects (Sousa et al., 2016; Carvalho et al., 2019). The different types of cheminformatics, bioinformatics, and PDB data mining analyses have been performed in the past, for example, from studying the cation-π interactions in proteins (Zarić, Popović, and Knapp 2000) to assisting the drug design and discovery of new medicaments (Willett 2011), from searching whether histidines prefer a particular conformation in cytochromes (Fufezan, Zhang, and Gunner 2008) to examine the factors determining the orientation of coordinated histidines in heme proteins (Zarić, Popović, and Knapp 2001) or to facilitating the *de novo* modeling of the synthetic protein structures with desired functional properties (Popović et al. 2001).

Structural Relaxation of the SynCytb

We built up the model structure of SynCyt*b* and carefully relaxed it in 13 discrete minimization steps by gradually reducing the constraints, as mentioned in the method section, and in detail explained in Popović et al. (2001). We monitored the RMSD for different structural parts to see how subsequent modeling and relaxation steps improved the model structure. These data are collected in Table 2, which provides RMSD between the final completely relaxed model structure (B10) and the different partially relaxed structures (X). The RMSD between the initial non-relaxed model structure (A7) and the final completely minimized model structure is 1.5 Å, a typical value obtained by energy minimization of the crystal structure of a protein in a vacuum. It already shows a good quality of our initial model structure (A7).

The RMSDs gradually decrease through individual minimization steps. From Table 2, one can also see that the template atoms were not moved before relaxation step B4, backbone atoms of the helix bundle were not before step B6 and atoms of the hemes were not before step B8. In step B9, the harmonic constraints applied on the backbone atoms were gradually reduced to avoid large conformational changes that can destroy carefully modeled structural details. Hence, Table 2 illustrates how the model structure is improved, and RMSDs become smaller, going from one to the next step during the energy minimization procedure.

Table 2. RMS deviation between structures of the different modeling steps and the final structure of the SynCyt*b*

X – B10[a]	**all atoms**	**backbone atoms**	**side chains**	**4-helix bundle**	**template atoms**	**heme H_l**	**heme H_h**
X = A7	1.536	0.877	1.718	1.555	1.073	0.545	0.874
X = B1	1.527	0.877	1.707	1.547	1.073	0.545	0.874
X = B2	1.483	0.877	1.651	1.496	1.073	0.545	0.874
X = B3	1.477	0.877	1.644	1.489	1.073	0.545	0.874
X = B4	1.067	0.877	1.128	1.042	0.969	0.545	0.874
X = B5	1.050	0.877	1.108	1.023	0.950	0.545	0.874
X = B6	1.023	0.816	1.090	1.023	0.564	0.545	0.874
X = B7	0.994	0.790	1.057	1.002	0.436	0.545	0.874
X = B8	0.988	0.790	1.049	1.000	0.436	0.480	0.831
X = $B9_1$	0.949	0.784	1.004	0.965	0.436	0.440	0.571
X = $B9_2$	0.889	0.742	0.937	0.900	0.430	0.405	0.571
X = $B9_3$	0.691	0.614	0.716	0.691	0.422	0.365	0.553
X = $B9_4$	0.478	0.412	0.499	0.472	0.396	0.322	0.409
X = $B9_5$	0.177	0.173	0.179	0.163	0.356	0.121	0.138

[a] The RMS deviations (in Å) are calculated by superimposing the final model structure (B10) after all steps of energy minimization with the structures obtained after individual steps of structural relaxation (X = A7, ..., $B9_5$) using the Kabsch algorithm (Kabsch 1976, 1978). The steps of structural relaxation are explained in the text. Step B9 consists of five individual steps of energy minimization using constraining forces of decreasing strength.

Starting from the ideal planar conformation of the hemes, the heme conformations are changed through the energy minimization steps, and Table 3 shows these changes by the RMSD. The heme atoms were not moved before step B8, where only the atoms of the PR groups were relaxed. After reducing the constraints applied to the two hemes, their conformations became slightly different through the process of structural relaxation. The RMSDs between all atoms of the two hemes are 1.0 Å after the last step of minimization and 1.2 Å in the average structure of MD simulation. Of course, the RMSDs of porphyrin rings are much smaller (0.3 Å).

Table 3. RMS deviation between the two hemes in SynCyt*b* for the different modeling steps

modeling step [a]	all atoms [b]	porphyrin ring
B8	0.380	0.000
$B9_1$	0.790	0.222
$B9_2$	0.801	0.213
$B9_3$	0.784	0.203
$B9_4$	0.954	0.320
$B9_5$	1.021	0.285
B10	1.010	0.386
average from MD	1.239	0.321

[a] The RMS (in Å) deviations were calculated using the Kabsch algorithm (Kabsch 1976, 1978) by superimposing the structure of the actual modeling step with the initial structure before modeling step B8. [b] All atoms, including the atoms of the porphyrin and propionate groups.

Salt Bridges in the SynCytb

We assumed that an acidic and basic residue pair forms a strong salt bridge if the corresponding oxygen-hydrogen distance is less than 4 Å. This is a reasonable estimation since the salt bridge interactions are of electrostatic nature, to be precise, the long-range interactions between the two oppositely charged residues. All four PR groups from the two hemes form strong salt bridges inside the hydrophobic binding pocket with four arginines. Thereby, the arginines Arg5(F_1), Arg5(F_2), Arg25(F_1), and Arg25(F_2) form salt bridges with PRA(H_h), PRD(H_h), PRA(H_l), and PRD(H_l), respectively. The corresponding O–H distances between the PR groups and the arginines are 2.4Å, 1.6Å, 2.1Å, and 1.5Å.

From the many possibilities to form salt bridges between glutamates and lysines on the outer helix surfaces, our model of the SynCyt*b* exposes only four strong salt bridges. The following residue pairs form these ionic bridges: Glu16(F_2)–Lys13(F_2), Glu24(F_2)–Lys28(F_2), Glu20(G_2)–Lys24(G_2), and Glu10(G_1)–Lys14(G_1). Their corresponding oxygen-hydrogen distances are 2.9 Å, 1.7 Å, 2.4 Å, and 3.8 Å, respectively. This is astonishing since the CHARMM22 energy function used to model the atomic coordinates was applied under vacuum conditions, where electrostatic interactions involving salt bridges on a protein surface should be overestimated. Therefore, one could expect many more potential strong salt bridges. There are two reasonable explanations why only four of them appear in our model structure. First, almost each Lys or Glu has two possible partners for making the salt bridge, which can cause the mutual distances between each acid-base pair to be slightly larger after the energy minimization as a compromise between two possibilities. Secondly, the side chain of Lys is longer than the side chain of Glu, and when they are on the top of each other, Lys side chain has to adopt probably energetically less favorable bent conformation, to be able to form a salt bridge with Glu. In addition, the experiments show that the solvent-exposed salt bridges are generally only marginally stable since these interactions are screened very efficiently in the high dielectric medium of solvent (water) and are strongly affected by the mobile counterions in aqueous solution.

Comparison of Heme Conformations in Synthetic and Native Cytb

The heme conformation, the type and orientation of axial ligands, and their interactions with the heme and the protein environment are essential for the redox potential values in the heme-proteins (Kennedy et al. 2001; Zarić, Popović, and Knapp 2001; Fufezan, Zhang, and Gunner 2008; Popović and Djordjević 2020). Therefore, in the following, we compare the structural differences, mainly related to the histidine–heme conformations in the X-ray structure of the native Cyt*b* (Iwata et al. 1998) and the SynCyt*b* obtained by modeling procedures. The corresponding structural data are collected in Table 4. In the native Cyt*b*, the Nε2–Fe bond lengths differ for the crystal structure, the force-field energy minimized structure, and the MD simulation's average structure. These bond lengths are between 1.98–2.08 Å, close to the ideal value of 2.05 Å supposed in the CHARMM22 force field. In the SynCyt*b*, the bond lengths are slightly larger varying between 2.10 Å and 2.23 Å. They remain somewhat larger even after the MD simulation. This may hint at some steric interactions in the SynCyt*b*, which prevent the helices of type *F* from coming closer to the heme.

The Nε2–Fe–Nε2' bond angles involving the imidazole nitrogen atoms bound to the heme iron exhibit deviations of about ±10° from its ideal value of 180°, except for the angle involving His83 and His182 of the native Cyt*b* with a deviation of almost 30°. However, after energy minimization, that deviation diminishes. The idealized histidine-heme coordination assumes the orthogonality of the imidazole ring and the heme plane. Hence, the angle between the normal of the heme plane and the normal of the axially ligated imidazole plane should be about 90°. The most significant deviation of 21.5° from 90° can be observed for His83 in the crystal structure of the native Cyt*b*, which decreases after energy minimization. Deviations of these angles are generally less pronounced for the SynCyt*b*, probably because this structure is relaxed by energy minimization more extensively. The angle between the heme plane normal and the vector in the imidazole plane pointing from Nε2 to the midpoint of Cγ and Nδ1 characterizes the inclination of the imidazole plane to the heme plane in a direction orthogonal to the imidazole plane normal. Due to steric and electrostatic effects, which enforce perpendicular orientation of axial His to heme plane, this angle should take values near 0° or 180°, depending on whether the imidazole plane is above or below the heme plane.

The torsion angles α, β, and γ refer to the orientation of the imidazole ring relative to the PR groups of heme, concerning the histidine backbone, and the position of the histidine backbone relative to the heme, respectively (Zarić, Popović, and Knapp 2001); see Figure 1 (Chapter "*Factors Determining the Orientation of Ligated Histidines*") for details. For idealized imidazole-heme geometry, the three torsion angles are related to each other ($\gamma = \alpha + \beta$), as was pointed out earlier. The deviations from this relation encountered here are only ±5° (see Table 4). Analyzing the histidine-heme conformations in the model protein, one can see that all angles α are in the allowed regime of about ±90°. In the case of the histidines ligated to the heme H_l, the backbone–heme orientation prevents the favorable orientation of the imidazole rings towards the PR groups, while that is not the case with the histidines of the heme H_h, which roughly point toward the corresponding PR group that is at the same side of the heme plane as the imidazole ring.

The Relative Orientation of the Hemes and Tilted Helices

The angles between the vectors CHA→CHC belonging to different hemes can measure the relative orientation of the two hemes. The vectors CHB→CHD can also be used, giving a different perspective; see Figure 1 in Chapter "*Factors Determining the Orientation of Ligated Histidines*" for a hint. These angles are close to zero for the SynCyt*b*, indicating that the heme planes are parallel and the PR groups point practically in opposite directions (see Table 5). It reflects the regularity in the structure of SynCyt*b*, where the helix axes remain not so much tilted even after MD simulation.

Table 4. Conformation of the hemes in the native and synthetic Cyt*b*

	Fe–Nε2 distance (Å)	**Nε2-Fe-Nε2' angle** [c]	**heme and His normal angle** [d]	**heme normal Nε2–middle angle** [e]	**torsion angle α** [f]	**torsion angle β** [f]	**torsion angle γ** [f]
native Cyt*b* [a]							
His83	2.08	152.1	68.5	155.8	136.1	-17.4	120.2
	2.06	177.0	82.8	171.6	161.5	-42.3	110.6
	2.03	171.9	95.8	163.0	166.2	-67.0	103.8
His182	2.06		107.1	17.6	112.8	65.6	-168.4
	2.07		83.7	6.4	63.4	106.4	-180.0
	1.98		87.2	9.7	68.9	118.0	-168.0
His97	2.08	171.1	81.5	167.6	5.1	-37.8	-37.1
	2.00	167.8	87.7	177.3	49.5	-83.2	-44.3
	2.07	169.0	108.9	173.6	60.0	-85.0	-39.6
	2.06		84.9	9.6	-87.3	72.6	-6.6
His196	1.99		92.4	7.8	-69.7	62.2	5.9
	2.07		73.2	22.7	-34.2	60.9	34.1
SynCyt*b* [b]							
His22(F_1)	2.22	172.1	94.6	161.4	105.2	-101.4	-12.7
	2.11	171.1	91.4	156.6	87.3	-59.9	21.1
His22(F_2)	2.23		78.5	21.3	-120.2	124.1	12.7
	2.21		77.9	4.6	-109.3	93.7	16.6
His8(F_1)	2.17	174.1	107.3	159.2	-45.8	-86.5	138.9
	2.19	171.6	108.6	163.8	-46.0	-71.0	-121.0
His8(F_2)	2.12		91.2	19.5	86.3	94.0	179.4
	2.10		84.2	13.8	82.5	75.0	168.4

[a] The top values are from the crystal structure of the native Cyt*b*; the middle values are from the energy minimized X-ray structure; the bottom values are averages from the MD simulations. [b] Top values are from the minimized structure of the SynCyt*b*; bottom values are averages from MD simulations. [c] The bond angle Nε2–Fe–Nε2' refers to pairs of histidines axially ligated to the same heme. For the native Cyt*b*, the first triplet of values corresponds to the pair His83–His182 and the second triplet of values to His97–His196. For the SynCyt*b*, the first doublet of values corresponds to His22(F_1)–His22(F_2) and the second doublet of values to His8(F_1)–His8(F_2). Footnotes *a* and *b* are applicable, respectively. [d] The vectors Fe–NA, Fe–NB of heme, and the heme normal form a right-handed coordinate system. The same is true for the vectors Cγ–Cδ and Cγ–Nδ1 of the histidine and the imidazole ring normal. [e] The inclination of the axially ligated imidazole ring relative to the heme plane is measured by the angle of the heme plane normal with the vector going through atom Nε2 and the midpoint of the Cγ–Nδ1 bond. [f] For a definition of the torsion angles α, β, and γ, see Chapter "*Factors Determining the Orientation of Ligated Histidines.*"

Table 5. The relative orientation of hemes in native and synthetic Cyt*b*

angles[a]	CHC(H_l)-CHA(H_l) with CHA(H_h)-CHC(H_h)	CHB(H_l)-CHD(H_l) with CHD(H_h)-CHB(H_h)
native Cyt*b*[b]	38.2	41.0
	36.8	36.9
	33.7	35.1
SynCyt*b*[c]	5.0	3.8
	8.9	-14.8

[a] Angles are given in degrees. [b] Top values are from the crystal structure of the native Cyt*b*; middle values are from the energy minimized X-ray structure; bottom values are averages from MD simulations. [c] Top values are from the fully minimized structure of theSynCyt*b*; bottom values are averages from MD simulations.

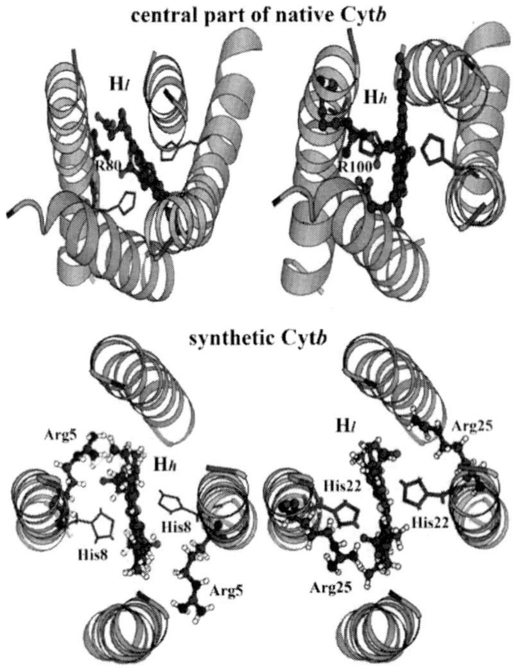

Figure 6. Comparison between the relative orientation of the two hemes and the tilted helices of the four-helix bundle in native and synthetic Cyt*b*. The high (H_h) and low (H_l) potential hemes, coordinated histidines, and arginines (Arg = R) forming the salt bridges with the heme propionates are indicated for native Cyt*b* and SynCyt*b*.

In native Cyt*b*, the helices are, to a great extent, tilted and even bent due to the presence of proline residues in the sequence of the helices αC (Pro134) and αD (Pro186) and glycine in the helix αB (Gly86). As a result, the relative orientation of the hemes is also less regular. The comparison between the relative orientation of the two hemes and the comparison between the tilted helical axes in native and synthetic Cyt*b* are depicted in Figure 6.

Figure 6 illustrates and reflects the following findings:

a) the heme planes are nearly parallel in SynCyt*b* with propionate groups pointing in opposite directions;
b) the angle between heme planes in native Cyt*b* is about 40 degrees in both directions as defined by the abovementioned vectors;
c) the propionic groups are also directed away from each other in native Cyt*b*;
d) the parallel orientation of the two hemes in SynCyt*b* is partially due to the formation of the two salt bridges (Arg···PR) per heme in agreement with Berry and Walker (2008); therefore, the presence of only one Arg per heme propionates is in part responsible for an irregular relative orientation of the Cyt*b* hemes;
e) due to the presence of a cyclic template, the order of H_l and H_h hemes is inversed in SynCyt*b* considering the amino acid sequence;
f) the helical axes in the native and synthetic Cyt*b* are considerably different;
g) while the angle between helices *F*, containing axial histidines, is negligible, the angle between helices *G* is somewhat larger (5-8 degrees) in SynCyt*b*; and
h) the helices of the native four-helix bundle are entirely irregular, largely tilted, and bent due to primarily Pro and Gly residues in the helical sequence and the hydrophobic interactions that drive the protein folding and aggregation/association of the transmembrane helices in native Cyt*b*.

Experimental Data

The binding state of hemes incorporated into the model protein was recently investigated by the EPR and ENDOR spectroscopy (Fahnenschmidt et al. 1999), as well as by Mössbauer and EPR spectroscopy (Iakovleva et al. 2002). The EPR spectra show the occurrence of two types of hemes. The Fe^{3+} low-spin signal refers to the *bis*-histidine ligated heme with approximately parallel histidine planes, corresponding to the low potential heme near the open end of the four-helix bundle. The second contribution corresponds to a highly anisotropic low-spin Fe^{3+} (HALS) species. It is characteristic of *bis*-histidine ligation with tilted or twisted imidazole planes and can be assigned to the high potential heme on the template side. Also, the distance of axial histidines from the metal center and their mutual orientation were studied by ENDOR spectroscopy. However, this method allows conclusions based on comparing the related situations in native proteins or model complexes, where the imidazole orientation and corresponding distances are known. Our model of the SynCyt*b* roughly agrees with these experimental results. The histidine orientation is approximately parallel for the H_l heme (16°), while it is more perpendicular for the H_h heme (48°). However, both hemes possess axial histidines with partially tilted planes relative to the heme plane, such that the corresponding distances Fe–Hε1 and Fe–Hδ2 are not equal. This asymmetry is also not unusual for the native Cyt*b* structures [PDB code: 1BCC and 1BE3, (Xia et al. 1997; Iwata et al. 1998)]. The positions and orientations of the axially ligated histidines in the SynCyt*b* are obtained by energy minimization using the CHARMM22 force field. We did not apply any additional energy term, which would constrain the Fe–Hε1 and Fe–Hδ2 distances and enforce symmetric orientation.

Molecular Dynamics Study

Stability of the Synthetic and Native Cytb

The robustness of the model Cyt*b* structure is tested by monitoring the conformational stability during a long-term molecular dynamics (MD) simulation and comparing the results with the MD simulation of the native Cyt*b* from the bovine Cytbc_1 complex, where a crystal structure (Iwata et al. 1998) is available. The dynamics simulations were done in a vacuum. The trajectories have a length of 10 ns. The conditions of the MD simulations are given in the section "*Molecular Modeling*." Figure 7 displays the time dependence of the RMS deviations for the X-ray structure of the native Cyt*b* and the fully minimized structure of the SynCyt*b*. Two pairs of lines are depicted in the figure. The upper curves display the RMSD of all atoms for the native and synthetic protein structure. The lower pair refers only to the backbone atoms of the four-helix bundle and template. After 2 ns, the RMS deviations of the backbone atoms reach a value of 0.85 Å and practically do not increase anymore. The RMSDs of all atoms reach a plateau value of 1.45 Å for the SynCyt*b* and 1.34 Å for the native Cyt*b*. The RMSD from the MD simulations for native and synthetic protein are very similar, demonstrating that the SynCyt*b* modeled structure is relatively stable, rigid, and strain-free.

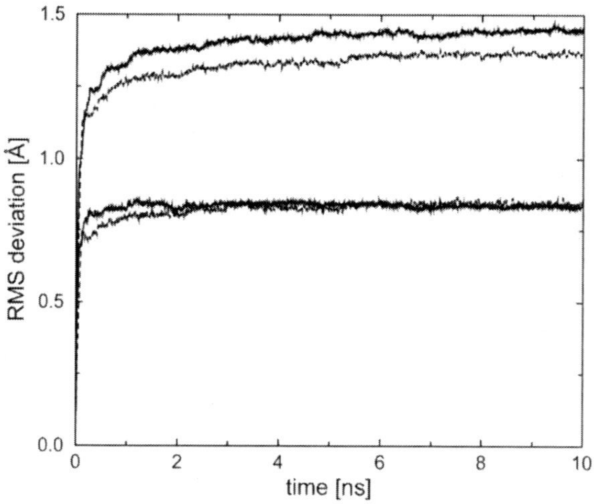

Figure 7. Time evolution of the RMS deviations for the synthetic and native Cyt*b*. To demonstrate the stability of the synthetic and native Cyt*b* structures, the RMS deviations are displayed as a function of time for the 10 ns MD simulations. The RMSD is given for the backbone atoms (lower pair of curves) and all atoms, including the hemes (upper pair of curves). The curves with the slightly smaller values are the data from the native Cyt*b*. For clarity, we removed short-time fluctuations in the RMS deviations and displayed averages over 20 ps corresponding to 200 sets of coordinates. For more details, see the text.

Obtained RMS deviations are typical for MD simulations under vacuum conditions applied to crystal structures of relatively stable proteins. Hence, in that respect, the SynCyt*b* model structure may be as good as a crystal structure. Furthermore, the average structure and the structure of the SynCyt*b* after 10 ns of MD simulation remain all structural details as a model structure. For instance, the helices essentially retain their secondary structure, in agreement with the circular dichroism measurements that estimate the helicities of 81% for the whole structure (Rau and Haehnel 1998). The value based on the modeling is 86%. We observed that the values of the template torsion angles exhibit relatively large fluctuations, but still, we could distinguish two β-strands and two β-turns. However, variations from the ideal conformation angles of cyclic peptides by as much as 30° are also common for native proteins (Voet and Voet 1995).

RMS Fluctuations of the Synthetic and Native Cytb

All these observations favor the relative stability of the SynCyt*b*. By comparing the RMS fluctuations (RMSF) of individual Cα atoms obtained from the MD simulations of the native Cyt*b* with the corresponding values derived from the B-factors of the crystal structure, the values from the MD simulations are generally smaller (Figure 8A). This result also indicates that the structure of the native Cyt*b* is relatively rigid and robust.

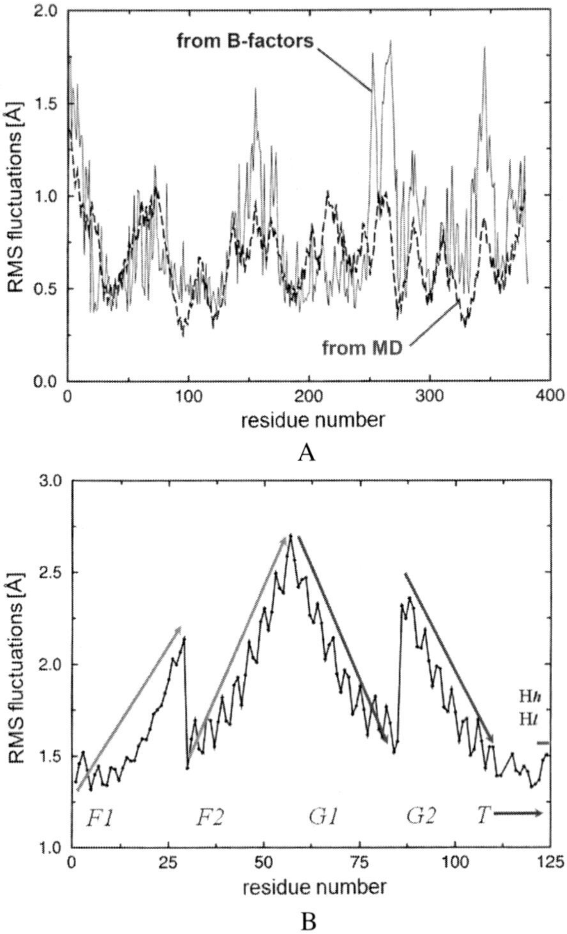

Figure 8. The RMS fluctuations of the native and synthetic Cyt*b*. (A) The RMS fluctuations of Cα atoms were calculated from the crystal structure B-factors and obtained from the MD simulation of native Cyt*b*. (B) The isotropic RMS fluctuations of Cα atoms were obtained from the MD simulation of the SynCyt*b* structure. The orange (helices *F*) and blue (helices *G*) arrows show the orientation of helical axes and demonstrate the antiparallel structure of the four-helix bundle. *T* stands for the template, H_h and H_l for high and low potential hemes. For more details, see the text.

In synthetic protein, the most significant RMS fluctuations of Cα atoms display the residues at the open end of the four-helix bundle (Figure 8B). The arrows indicate the orientation of the helices, where helices *F* have the C-terminus on the top and helices *G* have the N-terminus on the top. The average RMSFs obtained from the MD simulations of native Cyt*b* and SynCyt*b* are around 0.75 Å and 1.85 Å, respectively. Somewhat smaller RMSF values of native Cyt*b* are expected due to the additional supporting transmembrane helices, which stabilize the native structure. In contrast, SynCyt*b* is a simple water-soluble four-helix bundle structure with much more mobility of its atoms due to thermal fluctuations. Nevertheless, this is all expected, and many residues closer to the template exhibit the RMSFs smaller than 1.5 Å. The SynCyt*b* structure is stabilized by linkage to the cyclic template and bonds to the two heme cofactors.

Conclusion

Model building from scratch of an artificial protein with two hemes, recently synthesized, was successfully employed. Its structure consists of a four helix-bundle and a cyclic decapeptide serving as a template and mimics the central part of the native cytochrome *b*. Since there is no X-ray structure available, the structural elements of the artificial protein were carefully assembled from scratch using all known structural information available and avoiding strain and atomic clashes as much as possible. The generation of atomic coordinates by modeling techniques, energy minimization, and molecular dynamics was done using the program CHARMM22 (Brooks et al. 1983). The molecular dynamics (MD) simulations were performed to test the stability and validate the computer-generated SynCyt*b* structure by comparing the RMSD of atomic coordinates with the native counterpart. We used the native Cyt*b* from the bovine heart, where an X-ray structure of 3.0 Å resolution is available from the PDB as a 1BE3 entry (Iwata et al. 1998). The MD simulations of the SynCyt*b* show RMSD (1.45 Å for all atoms), which are typical for very rigid proteins with the known crystal structure. The MD simulations suggest that the assembled model is stable and strain-free. Since the RMSDs for the artificial protein are as small as for the native Cyt*b*, we concluded that the structural stability and robustness of the synthetic Cyt*b* are similar to the native Cyt*b*.

It is supposed that for native proteins, a three-dimensional structure prediction does not generally work since the association of different parts of the structure, sequentially connected by turns in a single polypeptide chain, includes many possibilities of the interfacial packing, which are often difficult to predict (Betz et al. 1996; Betz, Liebman, and DeGrado 1997). However, forming small peptides with the defined secondary structure (helices) and their binding to predetermined positions of a cyclic template offers the possibility of overcoming the unsolved protein folding problem. It opened up the possibility of successfully generating the minimalistic synthetic protein structures. More specifically, the SynCyt*b* structure consists of the four helices, a cyclic template, and two hemes. The template is a cyclic decapeptide with a uniquely defined antiparallel β-strand-β-turn structure. A clear separation between the helical and linker parts enables the helices to maintain their secondary structure. Our findings agree with the experimental data suggesting a high α-helical content in the SynCyt*b* structure (>80%). In addition, the relative position and orientation of the helices to each other are well-defined by the template with its linker elements. There is a specific distribution of hydrophobic and hydrophilic residues along the helical axes, which defines the mutual orientation of the helices.

The hydrophobic faces of helices are oriented toward the center of the four-helix bundle creating the heme-binding pocket, while the hydrophilic side chains are solvent-exposed. The sequences of the two pairs of identical helices *F* and *G* are specially adopted to host the two hemes. Moreover, orienting the hemes along the helix axis facilitates the formation of ionic bridges between the heme propionic groups and four arginines. In the shielding helices *G*, two glycines were positioned at the same height as the ligating histidines in the binding helices *F* to accommodate the bulky edges of the two hemes. All of that makes the SynCyt*b* a structurally well-defined model system to generate its atomic coordinates from scratch by modeling and energy minimization procedure.

We have demonstrated that a careful, stepwise modeling procedure can generate the reliable atomic coordinates of an artificial protein. However, to test whether the approach really works, it would be helpful to solve its X-ray structure. The modeling protocol that we used here can provide a guideline on how to generate the atomic coordinates for similar artificial proteins. Continuation of this work on similar systems is a promising task.

Acknowledgments

This work has been financially supported by the Ministry of Education, Science and Technological Development of the Republic of Serbia (MPNTR): Grant No. 451-03-68/2022-14/200026 and project OI/127035. This article/publication is based upon work from COST Action CA21101, supported by COST (European Cooperation in Science and Technology).

References

Albrecht, T., W. Li, J. Ulstrup, W. Haehnel, and P. Hildebrandt. 2005. "Electrochemical and spectroscopic investigations of immobilized de novo designed heme proteins on metal electrodes." *ChemPhysChem* 6:961–970.

Berry, E. A., and F. A. Walker. 2008. "Bis-histidine-coordinated hemes in four-helix bundles: how the geometry of the bundle controls the axial imidazole plane orientations in transmembrane cytochromes of mitochondrial complexes II and III and related proteins." *J. Biol. Inorg. Chem.* 13:481–498.

Betz, S. F., P. A. Liebman, and W. F. DeGrado. 1997. "De novo design of native proteins: Characterization of proteins intended to fold into antiparallel, rop-like, four-helix bundles." *Biochemistry* 36:2450–2458.

Betz, S. F., J. L. Marmorino, A. J. Saunders, D. F. Doyle, G. B. Young, and G. J. Pielak. 1996. "Unusual effects of an engineered disulfide on global and local protein stability." *Biochemistry* 35:7422–7428.

Bialas, C., L. E. Jarocha, K. B. Henbest, T. M. Zollitsch, G. Kodali, C. R. Timmel, S. R. Mackenzie, P. L. Dutton, C. C. Moser, and P. J. Hore. 2016. "Engineering an Artificial Flavoprotein Magnetosensor." *J. Am. Chem. Soc.* 138:16584–16587.

Brandt, U. 1996. "Bifurcated ubihydroquinone oxidation in cytochrome bc1 complex by proton-gated charge transfer." *FEBS Letters* 387:1–6.

Brandt, U. 1998. "The chemistry and mechanics of ubihydroquinone oxidation at center P (Qo) of the cytochrome bc1 complex." *Biochim. Biophys. Acta-Bioenerg.* 1365:261–268.

Brandt, U., and B. Trumpower. 1994. "The proton motive Q cycle in mitochondria." *Crit. Rev. Biochem. Mol. Biol.* 29:165–197.

Braun, H.-P., M. Emmermann, V. Kruft, and U. K. Schmitz. 1992. "The general mitochondrial processing peptidase from potato is an integral part of cytochrome c reductase of the respiratory chain." *EMBO J.* 11:3219–3227.

Braun, H.-P., and U. K. Schmitz. 1995. "Are the core proteins of the mitochondrial bc1 complex evolutionary relics of a processing protease." *Trends Biochem. Sci.* 20:171–175.

Brooks, B. R., R. E. Bruccoleri, D. J. Olafson, B. D. States, S. Swaminathan, and M. Karplus. 1983. "CHARMM: A program for macromolecular energy, minimization, and dynamics calculations." *J. Comp. Chem.* 4:187–217.

Bryson, J. W., S. F. Betz, H. S. Lu, D. J. Suich, H. X. Zhou, K. T. O'Neil, and W. F. DeGrado. 1995. "Protein design: a hierarchic approach." *Science* 270:935–941.

Brzezinski, P., and R. B. Gennis. 2008. "Cytochrome c oxidase: exciting progress and remaining mysteries." *J. Bioenerg. Biomembr.* 40:521–531.

Carvalho, L. M. de, G. Borelli, A. P. Camargo, M. A. de Assis, S. M. F. de Ferraz, M. B. Fiamenghi, J. José, L. S. Mofatto, S. T. Nagamatsu, G. F. Persinoti, N. V. Silva, A. A. Vasconcelos, G. A. G. Pereira, and M. F. Carazzolle. 2019. "Bioinformatics applied to biotechnology: A review towards bioenergy research." *Biomass and Bioenergy* 123:195–224.

Choma, C. T., J. D. Lear, M. J. Nelson, P. L. Dutton, D. E. Robertson, and W. F. DeGrado. 1994. "Design of a heme-binding four-helix bundle." *J. Am. Chem. Soc.* 116:856–865.

Couch, V., D. Popović, and A. Stuchebrukhov. 2011. "Redox-Coupled Protonation of Respiratory Complex I: The Hydrophilic Domain." *Biophys. J.* 101 (2):431–438.

Crofts, A. R. 1985. *The Enzymes of Biological Membranes*, edited by A. N. Martonosi, 347-382. New York: (Plenum, New York, 1985).

Curnow, P. 2019. "Designing minimalist membrane proteins." *Biochem Soc Trans.* 47:1233–1245.

DeJonge, N., H. K. Rau, and W. Haehnel. 1999. "Light-induced electron transfer in synthetic metalloproteins." *Z. Phys. Chem.* 213:175–180.

Fahnenschmidt, M., H. K. Rau, R. Bittl, W. Haehnel, and W. Lubitz. 1999. "Characterization of a de novo designed heme protein by EPR and ENDOR spectroscopy." *Chem. Eur. J.* 5:2327–2334.

Fee, J. A., D. A. Case, and L. Noodleman. 2008. "Toward a chemical mechanism of proton pumping by the B-type cytochrome c oxidases: application of density functional theory to cytochrome ba(3) of Thermus thermophilus." *J. Am. Chem. Soc.* 130:15002–15021.

Fry, H. C., A. Lehmann, L. E. Sinks, I. Asselberghs, A. Tronin, V. Krishnan, J. K. Blasie, K. Clays, W. F. Degrado, J. G. Saven, and M. J. Therien. 2013. "Computational de novo design and characterization of a protein that selectively binds a highly hyperpolarizable abiological chromophore." *J. Am. Chem. Soc.* 135:13914–13926.

Fufezan, C., J. Zhang, and M. R. Gunner. 2008. "Ligand preference and orientation in b- and c-type heme-binding proteins." *Proteins* 73:690–704.

Gennis, R. B., B. Barquera, B. Hacker, S. R. Van Doren, S. Arnaud, A. R. Crofts, E. Davidson, K. A. Gray, and F. Daldal. 1993. "The bc1 complexes of Rhodobacter sphaeroides and Rhodobacter capsulatus." *J. Bioenerg. Biomembr.* 25:195–209.

Georgescu, R. E., E. G. Alexov, and M. R. Gunner. 2002. "Combining conformational flexibility and continuum electrostatics for calculating pKa's in proteins." *Biophys. J.* 83:1731–1748.

Ghirlanda, G., A. Osyczka, W. Liu, M. Antolovich, K. M. Smith, P. Leslie Dutton, A. J. Wand, and W. F. DeGrado. 2004. "De Novo Design of a D2-Symmetrical Protein that Reproduces the Diheme Four-Helix Bundle in Cytochrome bc1." *J. Am. Chem. Soc.* 126:8141–8147.

Gibney, B. R., Y. Isogai, F. Rabanal, K. S. Reddy, A. M. Grosset, C. C. Moser, and P. L. Dutton. 2000. "Self-assembly of heme a and heme b in a designed four-helix bundle: Implications for a cytochrome c oxidase maquette." *Biochemistry* 39:11041–11049.

Gilson, M. K. 1995. "Theory of electrostatic interactions in macromolecules." *Curr. Opin. Struct. Biol.* 5:216–223.

Guerrero-Castillo, S., F. Baertling, D. Kownatzki, H. J. Wessels, S. Arnold, U. Brandt, and L. Nijtmans. 2017. "The assembly pathway of mitochondrial respiratory chain complex I." *Cell Metabolism* 25:128–139.

Gunner, M. R., E. Alexov, E. Tores, and S. Lipovaca. 2000. "The importance of the protein in controlling the electrochemistry of heme metalloproteins: methods of calculation and analysis." *J. Biol. Inorg. Chem.* 2:126–134.

Haehnel, W. 2004. "Chemical synthesis of TASP arrays and their application in protein design." *Molecular Diversity* 8:219–229.

Hauska, G., E. Hurt, N. Gabellini, and W. Lockau. 1983. *Biochim. Biophys. Acta* 726:97–133.

Hauska, G., W. Nitschke, and R. G. Herrmann. 1988. "Amino acid identities in the three redox center-carrying polypeptides of cytochrome bc1/b6f complexes." *J. Bioenerg. Biomembr.* 20:211–228.

Hinkle, P. C., M. A. Kumar, A. Resetar, and D. L. Harris. 1991. "Mechanistic stoichiometry of mitochondrial oxidative phosphorylation." *Biochemistry* 30:3576–3582.

Honig, B., and A. Nicholls. 1995. "Classical electrostatics in biology and chemistry." *Science* 268:1144–1149.

Huang, P. S., S. E. Boyken, and D. Baker. 2016. "The coming of age of de novo protein design." *Nature* 537:320-327.

Huang, S. S., B. R. Gibney, S. E. Stayrook, P. L. Dutton, and M. Lewis. 2003. "X-ray Structure of a Maquette Scaffold." *J. Mol. Biol.* 326:1219–1225.

Huffman, D. L., M. M. Rosenblatt, and K. S. Suslick. 1998. "Synthetic heme-peptide complexes." *J. Am. Chem. Soc.* 120:6183–6184.

Iakovleva, O., M. Reiner, H. Rau, W. Haehnel, and F. Parak. 2002. "Mössbauer and EPR study of a cytochrome b model." *Phys. Chem. Chem. Phys.* 4:655–660.

Iwata, S., J. W. Lee, K. Okada, J. K. Lee, M. Iwata, B. Rasmussen, T. A. Link, S. Ramaswamy, and B. K. Jap. 1998. "Complete structure of the 11-subunit bovine mitochondrial cytochrome bc1 complex." *Science* 281:64–71.

Joh, N. H., T. Wang, M. P. Bhate, R. Acharya, Y. Wu, M. Grabe, M. Hong, G. Grigoryan, and W. F. DeGrado. 2014. "De novo design of a transmembrane Zn2+-transporting four-helix bundle." *Science* 346:1520–1524.

Kabsch, W. 1976. "A solution for the best rotation to relate two sets of vectors." *Acta Cryst. A* 32:922–923.

Kabsch, W. 1978. "A discussion of the solution for the best rotation to relate two sets of vectors." *Acta Cryst. A* 34:827–828.

Kennedy, M. L., S. Silchenko, N. Houndonougbo, B. R. Gibney, P. Leslie Dutton, K. R. Rodgers, and D. R. Benson. 2001. "Model Hemoprotein Reduction Potentials: The Effects of Histidine-to-Iron Coordination Equilibrium." *J. Am. Chem. Soc.* 123:4635–4636.

Klapper, I., R. Fine, K. A. Sharp, and B. H. Honig. 1986. "Focusing of electric fields in the active site of Cu-Zn superoxide dismutase: Effects of ionic strength and amino-acid modification." *Prot. Struct. Funct. Genet.* 1:47–59.

Koder, R. L., J. L. Ross Anderson, L. A. Solomon, K. S. Reddy, C. C. Moser, and P. L. Dutton. 2009. "Design and engineering of an O2 transport protein." *Nature* 458:305–309.

Lalaurie, C. J., V. Dufour, A. Meletiou, S. Ratcliffe, A. Harland, Wilson O., C. Vamasiri, D. K. Shoemark, C. Williams, C. J. Arthur, R. B. Sessions, M. P. Crump, J. L. R. Anderson, and P. Curnow. 2018. "The de novo design of a biocompatible and functional integral membrane protein using minimal sequence complexity." *Sci. Rep.* 8:14564.

Lehninger, A. L., D. L. Nelson, and M. M. Cox. 1997. *Principles of Biochemistry*. New York: Worth Publishers.

Link, T. A. 1997. "The role of the 'Rieske' iron sulfur protein in the hydroquinone oxidation (Qp) site of the cytochrome bc1 complex - The 'proton-gated affinity change' mechanism." *FEBS Letters* 412:257–264.

Liu, G., W. Shao, S. Zhu, and W. Tang. 1995. "Effects of axial ligand replacement on the redox potential of cytochrome c." *J. Inorg. Biochem.* 60:123–131.

Lombardi, A., F. Pirro, O. Maglio, M. Chino, and W. F. DeGrado. 2019. "De Novo Design of Four-Helix Bundle Metalloproteins: One Scaffold, Diverse Reactivities." *Acc. Chem. Res.* 52:1148–1159.

Lombardi, A., C. M. Summa, S. Geremia, L. Randaccio, V. Pavone, and W. F. DeGrado. 2000. "Inaugural article: Retrostructural analysis of metalloproteins: Application to the design of a minimal model for diiron proteins." *Proc. Natl. Acad. Sci. USA* 97:6298–6305.

Lu, Y., S. M. Berry, and T. D. Pfister. 2001. "Engineering novel metalloproteins: design of metal-binding sites into native protein scaffolds." *Chem. Rev.* 101:3047–3080.

Lu, Y., N. Yeung, N. Sieracki, and N. M. Marshall. 2009. "Design of functional metalloproteins." *Nature* 460:855–862.

Ma, D., N. R. Brandon, T. Cui, V. Bondarenko, C. Canlas, J. S. Johansson, P. Tang, and Y. Xu. 2008. "Four-alpha-helix bundle with designed anesthetic binding pockets. Part I: structural and dynamical analyses." *Biophys. J.* 94:4454–4463.

MacKerell, Jr., A. D., D. Bashford, M. Bellot, Jr. Dunbrack, R. L., J. D. Evanseck, M. J. Field, S. Fischer, J. Gao, H. Guo, S. Ha, D. Joseph-McCarthy, L. Kuchnir, K. Kuczera, F. T. K. Lau, C. Mattos, S. Michnick, T. Ngo, D. T. Nguyen, B. Prodholm, III Reiher, W. E., B. Roux, M. Schlenkrich, J. C. Smith, R. Stote, J. Straub, M. Watanabe, J. Wiórkiewicz-Kuczera, D. Yin, and M. Karplus. 1998. "All-atom empirical potential for molecular modeling and dynamics studies of proteins." *J. Phys. Chem.* 102:3586–3616.

Mao, J., K. Hauser, and M. R. Gunner. 2003. "Multi-conformation continuum electrostatic analysis of heme electrochemistry in cytochromes with different structures." *Biochemistry* 42:829–840.

Marcus, R. A., and N. Sutin. 1985. "Electron transfers in chemistry and biology." *Biochim. Biophys. Acta* 811:265–322.

Martinez, S. E., D. Huang, A. Szczepaniak, W. A. Cramer, and J. L. Smith. 1994. "Crystal structure of chloroplast cytochrome f reveals a novel cytochrome fold and unexpected heme ligation." *Structure* 2:95–105.

Mitchell, P. 1961. "Coupling of phosphorylation to electron transfer by a chemiosmotic type of mechanism." *Nature* 191:144–148.

Mitchell, P. 1976. "Possible molecular mechanisms of the protonmotive function of cytochrome systems." *J. Theor. Biol.* 62:327–367.

Monien, B. H., F. Drepper, M. Sommerhalter, W. Lubitz, and W. Haehnel. 2007. "Detection of Heme Oxygenase Activity in a Library of Four-helix Bundle Proteins: Towards the de Novo Synthesis of Functional Heme Proteins." *Journal of Molecular Biology* 371:739–753.

Moser, C. C., M. M. Sheehan, N. M. Ennist, G. Kodali, C. Bialas, M. T. Englander, B. M. Discher, and P. L. Dutton. 2016. "De Novo Construction of Redox Active Proteins." In *Methods in Enzymology*, 365-388.

Mutter, M., P. Dumy, P. Garrouste, C. Lehmann, M. Mathieu, C. Peggion, S. Peluso, A. Razaname, and G. Tuchscherer. 1996. "Template assembled synthetic proteins (TASP) as functional mimetics of proteins." *Angew. Chem. Int. Ed. Engl.* 35 (13-14):1482–1485.

Mutter, M., R. Hersperger, K. Gubernator, and K. Muller. 1989. "The construction of new proteins: V. A template-assembled synthetic protein (TASP) containing both a 4-helix bundle and β-barrel-like structure." *Proteins Struct. Funct. Bioinform.* 5 (1):13–21.

Nakamura, H. 1996. "Roles of electrostatic interaction in proteins." *Quart. Rev. Biophys.* 29:1–90.

Nesset, M. J. M., N. V. Shokhirev, P. D. Enemark, S. E. Jacobson, and F. A. Walker. 1996. "Models of the Cytochromes. Redox Properties and Thermodynamic Stabilities of Complexes of "Hindered" Iron(III) and Iron(II) Tetraphenylporphyrinates with Substituted Pyridines and Imidazoles." *Inorg. Chem.* 35:5188–5200.

Olson, T. L., J. C. Williams, and J. P. Allen. 2013. "Influence of protein interactions on oxidation/reduction midpoint potentials of cofactors in natural and de novo metalloproteins." *Biochim. Biophys. Acta* 1827:914–922.

Page, C. C., C. C. Moser, X. X. Chen, and P. L. Dutton. 1999. "Natural engineering principles of electron tunneling in biological oxidation-reduction." *Nature* 402:47–52.

Perutz, M. F. 1978. "Electrostatic effects in proteins." *Science* 201:1187–1191.

Pinter, T. B. J., K. J. Koebke, and V. L. Pecoraro. 2020. "Catalysis and Electron Transfer in De Novo Designed Helical Scaffolds." *Angew. Chem. Int. Ed. Engl.* 59:7678-7699.

Popović, D. M. 2013. "Current advances in research of cytochrome c oxidase." *Amino Acids* 45:1073–1087.

Popović, D. M., and I. S. Djordjević. 2020. "Catalytic center of cytochrome c oxidase: Effects of protein environment on pKa values of CuB histidine ligands." *Journal of Serbian Chemical Society* 85:1429–1444.

Popović, D. M., and A. A. Stuchebrukhov. 2004. "Electrostatic study of proton pumping mechanism of bovine heart cytochrome c oxidase." *J. Am. Chem. Soc.* 126:1858–1871.

Popović, D. M., S. D. Zarić, B. Rabenstein, and E. W. Knapp. 2001. "Artificial cytochrome b: Computer modeling and evaluation of redox potentials." *J. Am. Chem. Soc.* 123:6040–6053.

Rau, H. K., N. DeJonge, and W. Haehnel. 2000. "Combinatorial synthesis of four-helix bundle hemoproteins for tuning of cofactor properties." *Angew. Chem. Int. Ed.* 39:250–253.

Rau, H. K., and W. Haehnel. 1998. "Design, synthesis, and properties of a novel cytochrome b model." *J. Am. Chem. Soc.* 120:468–476.

Reedy, C. J., and B. R. Gibney. 2004. "Heme protein assemblies." *Chem. Rev.* 104:617–649.

Ripoll, D. R., Y. N. Vorobjev, A. Liwo, J. A. Vila, and H. A. Scheraga. 1996. "Coupling between folding and ionization equilibria: Effects of pH on the conformational preferences of polypeptides." *J. Mol. Biol.* 264:770–783.

Robertson, D. E., H. Ding, P. R. Chelminski, C. Slaughter, J. Hsu, C. Moomaw, M. Tokito, F. Daldal, and P. L. Dutton. 1993. "Hydroubiquinone-cytochrome c2 oxidoreductase from Rhodobacter capsulatus: Definition of a minimal, functional isolated preparation." *Biochemistry* 32:1310–1313.

Robertson, D. E., R. S. Farid, C. C. Moser, J-L. Urbauer, S. E. Mulholland, R. Pidikiti, J. D. Lear, A. J. Wand, W. F. DeGrado, and P. L. Dutton. 1994. "Design and synthesis of multi-haem proteins." *Nature* 368:425–432.

Ryckaert, J., G. Ciccotti, and H. Berendsen. 1977. "Numerical integration of the Cartesian equations of motion of a system with constraints." *J. Comp. Phys.* 23:327–341.

Safo, M. K., M. J. M. Nesset, F. A. Walker, P. G. Debrunner, and W. R. Scheidt. 1997. "Models of the Cytochromes. Axial Ligand Orientation and Complex Stability in Iron(II) Porphyrinates: The Case of the Noninteracting d(π) Orbitals." *J. Am. Chem. Soc.* 119:9438–9448.

Sakamoto, S., I. Obataya, A. Ueno, and H. Mihara. 1999. "Effects of amino acids substitution of hydrophobic residues on haem-binding properties of designed two-α-helix peptides." *J. Chem. Soc. Perkin Trans.* 2:2059–2069.

Schaegger, H., U. Brandt, S. Gencic, and G. von Jagow. 1995. "Ubiquinol-cytochrome-c reductase from human and bovine mitochondria." *Methods Enzymol.* 260:82–96.

Schaegger, H., T. A. Link, W. D. Engel, and G. von Jagow. 1986. "Isolation of the eleven protein subunits of the bc1 complex from beef heart." *Methods Enzymol.* 126:224–237.

Schilling, B., J. Murray, C. B. Yoo, R. H. Row, M. P. Cusack, R. A. Capaldi, and B. W. Gibson. 2006. "Proteomic analysis of succinate dehydrogenase and ubiquinol-cytochrome c reductase (Complex II and III) isolated by immunoprecipitation from bovine and mouse heart mitochondria." *Biochim. Biophys. Acta* 1762:213–222.

Schnepf, R., P. Hörth, E. Bill, K. Wieghardt, P. Hildebrandt, and W. Haehnel. 2001. "De novo Design and Characterization of Copper Centers in Synthetic Four-Helix Bundle Proteins." *J. Am. Chem. Soc.* 123:2186–2195.

Sham, Y. Y., Z. T. Chu, and A. Warshel. 1997. "Consistent calculation of pKa's of ionizable residues in proteins: Semi-microscopic and microscopic approaches." *J. Phys. Chem. B* 101:4458–4472.

Sharp, R. E., C. C. Moser, F. Rabanal, and P. L. Dutton. 1998. "Design, synthesis, and characterization of a photoactivatable flavocytochrome molecular maquette." *Proc. Natl. Acad. Sci. USA* 95:10465–10470.

Shleev, S., J. Tkac, A. Christenson, T. Ruzgas, A. I. Yaropolov, J. W. Whittaker, and L. Gorton. 2005. "Direct Electron Transfer between Copper-Containing Proteins and Electrodes." *Biosens. Bioelectron.* 20:2517−2554.

Sousa, S. A., J. H. Leitão, R. C. Martins, J. M. Sanches, J. S. Suri, and A. Giorgetti. 2016. "Bioinformatics Applications in Life Sciences and Technologies." *BioMed Research International* 2016: Article ID 3603827.

Tashiro, M., and A. A. Stuchebrukhov. 2005. "Thermodynamic properties of internal water molecules in the hydrophobic cavity around the catalytic center of cytochrome c oxidase." *J. Phys. Chem. B* 109 (2):1015–1022.

Ullmann, G. M., L. Noodleman, and D. A. Case. 2002. "Density functional calculation of pKa values and redox potentials in the bovine Rieske iron-sulfur protein." *J. Biol. Inorg. Chem.* 7:632–639.

Vitale, R., L. Lista, C. Cerrone, G. Caserta, M. Chino, O. Maglio, F. Nastri, V. Pavone, and A. Lombardi. 2015. "An artificial heme-enzyme with enhanced catalytic activity: Evolution, functional screening and structural characterization." *Org. Biomol. Chem.* 13 (17):4859–4868.

Voet, D., and J. G. Voet. 1995. *Biochemistry*. II Edition ed. New York: John Wiley & Sons, Inc.

Warshel, A. 1981. "Electrostatic basis of structure-function correlation in proteins." *Acc. Chem. Res.* 14:284–290.

Warshel, A., and J. Aqvist. 1991. "Electrostatic energy and macromolecular function." *Ann. Rev. Biophys. Biophys. Chem.* 20:267–298.

Warshel, A., and S. T. Russel. 1984. "Calculations of electrostatic interaction in biological systems and in solution." *Quart. Rev. Biophys.* 17:283–422.

Warshel, A., S. T. Russel, and A. K. Churg. 1984. "Macroscopic models for studies of electrostatic interactions in proteins: Limitations and applicability." *Proc. Natl. Acad. Sci. USA* 81:4785–4789.

Watkins, D. W., C. T. Armstrong, and J. L. R. Anderson. 2014. "De novo protein components for oxidoreductase assembly and biological integration." *Current Opinion in Chemical Biology* 19:90–98.

Wikström, M. K., and J. A. Berden. 1972. "Oxidoreduction of cytochrome b in the presence of antimycin." *Biochim. Biophys. Acta* 283:403–420.

Wikström, M. K. F. 1973. "The different cytochrome b components in the respiratory chain of animal mitochondria and their role in electron transport and energy conservation." *Biochim. Biophys. Acta* 301:155–193.

Wikström, M., K. Krab, and V. Sharma. 2018. "Oxygen activation and energy conservation by cytochrome c oxidase." *Chem. Rev.* 118:2469–2490.

Willett, P. 2011. "Chemoinformatics: a history." *Wiley Interdisciplinary Reviews: Computational Molecular Science* 1:46–56.

Willner, B., and I. Willner. 2005. "Reconstituted Redox Enzymes on Electrodes: From Fundamental Understanding of Electron Transfer at Functionalized Electrode Interfaces to Biosensor and Biofuel Cell Applications." In *Bioelectronics – from Theory to Applications*, edited by I. Willner and E. Katz, 35–97. Weinheim, Germany: Wiley-VCH.

Willner, I., V. Heleg-Shabtai, E. Katz, H. K. Rau, and W. Haehnel. 1999. "Integration of a reconstituted de novo synthesized hemoprotein and native metalloproteins with electrode supports for bioelectronic and bioelectrocatalytic applications." *J. Am. Chem. Soc.* 121:6455–6468.

Xia, D., C. Yu, H. Kim, J. Xia, A. M. Kachurin, L. Zhang, L. Yu, and J. Deisenhofer. 1997. "Crystal structure of the cytochrome bc1 complex from bovine heart mitochondria." *Science* 277:60–66.

Yu, F., V. M. Cangelosi, M. L. Zastrow, M. Tegoni, J. S. Plegaria, A. G. Tebo, C. S. Mosly, L. Ruckthong, H. Qayyum, and V. L. Pecoraro. 2014. "Protein Design: Toward Functional Metalloenzymes." *Chem. Rev.* 114:3495–3578.

Zarić, S. D., D. M. Popović, and E. W. Knapp. 2000. "Metal ligand aromatic cation-π interactions in metalloproteins: Ligands coordinated to metal interact with aromatic residues." *Chem. Eur. J.* 6:3935–3942.

Zarić, S. D., D. M. Popović, and E. W. Knapp. 2001. "Factors determining the orientation of axially coordinated imidazoles in heme proteins." *Biochemistry* 40:7914–7928.

Zhang, S., Y. Wang, L. Han, X. Fu, S. Wang, W. Li, and W. Han. 2021. "Targeting N-terminal human maltase glucoamylase to unravel possible inhibitors using molecular docking, molecular dynamics simulations, and adaptive steered molecular dynamics simulations." *Front. Chem.* 9:711242.

Zhang, Z., L. Huang, V. M. Shulmeister, Y.-I. Chi, K. K. Kim, L.-W. Hung, A. R. Crofts, A. Berry, and S. H. Kim. 1998. "Electron transfer by domain movement in cytochrome bc1." *Nature* 392:677–684.

Zheng, Z., and M. R. Gunner. 2009. "Analysis of the electrochemistry of hemes with E(m)s spanning 800 mV." *Proteins* 75:719–734.

Zhuang, J., J. H. Amoroso, R. Kinloch, J. H. Dawson, M. J. Baldwin, and B. R. Gibney. 2006. "Evaluation of electron-withdrawing group effects on heme binding in designed proteins: implications for heme a in cytochrome c oxidase." *Inorg. Chem.* 45:4685-4694.

Chapter 5

Redox Properties of Small Synthetic Hemoproteins and Native Enzymes for Application in Bionanotechnology and Bioelectronic Devices

Dragan M. Popović*

Department of Chemistry, University of Belgrade, Institute of Chemistry, Technology and Metallurgy, National Institute of the Republic of Serbia, Belgrade, Serbia

Abstract

The field of bionanotechnology has recently demonstrated a broad spectrum of applications with nano–bio materials, and these novel hybrid systems have found extensive use for sensing, conductivity, catalysis, energy harvesting, storage or transduction, and the design of new "smart" devices. Specifically, the integrated biomolecule–nanoparticle systems assembled on highly conductive electrodes have been successfully exploited in designing enzyme-based bioelectrochemical devices. The *de novo* protein systems studied here can find use in bionanotechnology and bioelectronics as the central components of biosensors. Hence, understanding their redox properties is essential for fine-tuning the function of such biosensors. In this chapter, a molecular modeling approach has been employed by combining molecular mechanics, molecular dynamics, and quantum mechanics to generate the atomic coordinates of 25 small synthetic mono-heme proteins with a four-helix bundle motif that resembles the catalytic site of native cytochrome *b*. The protonation pattern of proteins and the redox potentials of the heme cofactor are evaluated from the electrostatic energies by solving the

* Corresponding Author's Email: dragan.popovic@ihtm.bg.ac.rs.

In: Structure and Function of Cytochromes
Editor: Elaine J. Mathis
ISBN: 979-8-88697-102-6

Poisson-Boltzmann equation. The factors of protein environment and effects of mutations determining the shift in the redox potential are analyzed and elucidated. The computed redox potentials generally agree well with available experimental measurements (RMSD $\leq$ 15 mV). The predictions of redox potential are made for the structures with E_m values outside of the range used for a redox indicator. The factors responsible for the shift in the redox potential are analyzed in great detail, and the concept presented here may pave the way for the design of proteins with desired properties customized to specific functions while providing new valuable information in both fundamental and applied sciences. In this chapter, some examples of synthetic and native cytochromes used as central components of biosensors in bioelectronic/biocatalytic applications are also discussed.

Keywords: computational biochemistry, materials science, protein design, molecular modeling, synthetic proteins, heme, redox midpoint potential, protein-cofactor interactions, biosensors

Abbreviations

Cyt*b*	cytochrome *b*
Cytbc_1	cytochrome bc_1
Cytb_6f	cytochrome b_6f
Cyt*c*	cytochrome *c*
C*c*O	cytochrome *c* oxidase
E_m	redox midpoint potential
HC1	heme conformation toward a template
HC2	heme conformation toward an open end
MD	molecular dynamics
PB	Poisson-Boltzmann
PRA, PRD	heme propionic groups
QM	quantum-mechanics
RMSD	root-mean-square deviation
Syn-*m*-HP	synthetic mono-heme protein

Introduction

Nanoparticles provide essential building blocks for the construction of functional materials owing to their unique physicochemical, optical, electronic, magnetic, and catalytic properties. The variety of small size (1-100 nm) core materials and adjustable surface properties make them attractive in the design of highly efficient catalysts and sensors (Feynman 1999; Nölting 2005; Gao et al. 2011). Generally, nanoparticles are composed of a metallic or semiconductive core stabilized by a self-assembled organic shell, which can further bind functional molecular units at the nanoparticle periphery (Kim et al. 2014; Heuer-Jungemann et al. 2019). As such, nanoparticle surfaces can be covered with a wide variety of chemical moieties, including molecular receptors that bind analytes and generate a detectable signal for sensing applications (Service 2002). Moreover, nanoparticles can be used as supramolecular building blocks for the construction of nanocomposite assemblies[1] for bioanalytical and bioelectronic applications, as well as for the fabrication of biosensor devices, such as, for example, enzyme-, immuno-, and DNA-sensors (Drummond, Hill, and Barton 2003; Ricci et al. 2007; Wang and Wang 2008; Valenti et al. 2015; Crivianu-Gaita and Thompson 2016; Pellerano and Morris 2021; Uygun and Atay 2021; Zhang et al. 2021).

State-of-the-art methods in bionanotechnology provide new research tools to manipulate nanoparticles' surfaces and to create materials with controlled optical, electronic, magnetic, and catalytic properties (Gazit 2007). The comparable size of biomolecules and nanoparticles suggests that their synergy might yield novel hybrid materials that combine the unique functions of native biomolecules or their synthetic counterparts with the properties of nanoparticles. The field of bionanotechnology has recently demonstrated a broad spectrum of applications, including biomolecule–nanoparticle materials; for some recent reviews and applications on this topic, see the following references (Katz and Willner 2004; Shleev et al. 2005; Leger and Bertrand 2008; Cracknell, Vincent, and Armstrong 2008; Rajendran, Udenigwe, and Yada 2016; Mi et al. 2021; Lopandić et al. 2021; Mukherjee and Mukherjee 2021; Zhang et al. 2021). These novel hybrid systems greatly benefit from the high selectivity, sensitivity, and efficiency of biomolecular units and have found extensive use for sensing, conductivity/circuitry, energy

---

[1] Nanocomposite assemblies are particularly valuable in improving solubility, sensibility, conductivity, signal or energy transduction, and catalysis.

harvesting, storage or transduction, catalysis, and design of new nano-devices (Katz and Willner 2004).

The integrated biomolecule–nanoparticle systems assembled on highly conductive electrodes have been successfully exploited in the design of enzyme-based electrochemical devices such as biosensors, biofuel cells, bioelectrocatalytical, and bioelectronic nano-devices (Willner and Willner 2005; Lisdat et al. 2009; Noll and Noll 2011; Bănică 2012; Zhang et al. 2021). Some of the examples include the assembly of the integrated bioelectrocatalytical Cyt *c*/Cyt oxidase–electrode for the reduction of O_2 (Friedrich et al. 2008); an integrated lactate dehydrogenase (LDH) electrode for the bioelectrocatalyzed oxidation of lactate to pyruvate (Katz, Heleg-Shabtai, Bardea, et al. 1998); cellobiose dehydrogenase (CDH) modified electrodes for detection of carbohydrates, quinones, and catecholamines (Ludwig et al. 2013); the bioelectrochemical sensors for glucose based on glucose oxidase (Qiu et al. 2012) or based on pyrroloquinoline quinine (PQQ)-dependent glucose dehydrogenase (Sarauli et al. 2014), and other enzyme-electrode sensors. Soon after the first synthetically designed enzymes and proteins were created, they were successfully integrated into the hybrid systems for bioelectronic applications (Katz and Willner 2004; Katz, Heleg-Shabtai, Bardea, et al. 1998; Noll and Noll 2011). For example, the synthetic Cyt*b* analogs were electronically active when bound to a gold surface (Katz, Heleg-Shabtai, Willner, et al. 1998) and exhibited light-induced electron transfer when bound to a metalloprotein (Rau, DeJonge, and Haehnel 1998; DeJonge, Rau, and Haehnel 1999).

Design of Novel Proteins

The natural redox-active proteins are essential for biological systems by performing various vital functions in the living cells. Nature has evolved diverse approaches to fine-tune the structure, function, and physicochemical properties of native redox- and heme-proteins. The *de novo* design and engineering of artificial enzymes, which can mimic native counterparts, represent a significant objective in material sciences, intending to obtain new chemical entities for specific applications in analytical electrochemistry, medicine, pharmacy, food industry, ecology, bioelectronics, and bionanotechnology (Reedy and Gibney 2004; Wang and Wang 2008; Lu et al. 2009; Koder et al. 2009; Watkins, Armstrong, and Anderson 2014; Salvati,

Stellacci, and Krol 2015; Huang, Boyken, and Baker 2016; Martin-Banderas et al. 2016; Alexi et al. 2021; Mi et al. 2021). This objective is sometimes hampered by the large size and complexity of native enzymes. Per se, synthetic proteins designed with minimalistic structural requirements can address this issue (Gibney et al. 2000; Lu, Berry, and Pfister 2001; Curnow 2019). There has been a growing interest in manipulating biomolecular structures on length scales, constructing the so-called "molecular maquettes," and thus, approaching the molecular dimensions (Skalicky et al. 1999; Gibney et al. 2000; Moser et al. 2016). The translation and application of the *de novo* engineering principles into biology have enabled the design, synthesis, and use of artificial proteins for biosensing applications and multifunctional purposes (Yu et al. 2014). In contrast to standard native biomolecules, *de novo* designed synthetic polypeptides are generally smaller, less expensive, and more stable, which offers them new perspectives in the advanced fields of science, engineering, and bionanotechnology.

Recent results demonstrate that small synthetic metalloproteins are able to retain a high enzyme activity of the metal cofactors. Moreover, these results anticipate the possibility of designing new catalytic biosensors customized to specific functions (Lombardi et al. 2000; Lu, Berry, and Pfister 2001; Lu et al. 2013; Vitale et al. 2015; Pinter, Koebke, and Pecoraro 2020). Furthermore, synthetic redox-proteins with specifically modeled electrochemical and oxido-reduction properties have been implemented in bioelectronic devices for amperometric biosensing (biosensors) and biofuel cell applications, as widely discussed in the literature (Katz, Heleg-Shabtai, Willner, et al. 1998; Willner et al. 1999; Willner and Willner 2005; Noll and Noll 2011; Bialas et al. 2016).

In general, the aim of *de novo* protein design can be shortly described in the following: a) to uncover the construction principles of native proteins; b) to simplify the native structure by applying the so-called *minimalistic approach*; c) to create only the catalytically active pocket that resembles one of the native proteins; d) synthetic proteins should exhibit the key physicochemical properties of native counterparts; e) to understand the influence of the protein environment and dielectric surrounding on the properties of cofactors; f) to investigate the factors involved in the proton and electron transfer reactions, and g) to create structures with novel functions. The minimalistic modeling approach of synthetic enzymes provides an ideal groundwork for examining the physicochemical properties and biochemical processes in a "simplified state" without the influence and interference of unnecessary structural details. The construction of artificial proteins with heme cofactors has opened up new prospects to investigate the effects of protein interactions with flexibility not available in native proteins (Ghosh and Pecoraro 2005). On the other hand, this approach enables the creation of new chemical and structural entities with substantially altered properties and possibly novel functions (Schnepf et al. 2001; Fry et al. 2013; Joh et al. 2014; Mahajan and Bhattacharjya 2014; Vitale et al. 2015; Bialas et al. 2016; Lalaurie et al. 2018; Curnow 2019; Lombardi et al. 2019).

De novo proteins with tailored structural and functional properties can be designed by site-directed mutagenesis starting from native proteins. That could be done by either screening large libraries of polypeptides to find a protein with desired properties or by chemical synthesis with libraries of short linear and cyclic peptides using a modular building principle (Mutter et al. 1989; Jung 1996). Much effort has been directed to developing the techniques to synthesize polypeptides from scratch and assemble them to create different types of synthetic (artificial) proteins with wide variability in biomimetic chemistry (Bryson et al. 1995; Gibney et al. 2000; Di Costanzo et al. 2001; Reedy and Gibney 2004; Li et al. 2006; Monien et al. 2007; Watkins, Armstrong, and Anderson 2014; Mahajan and Bhattacharjya 2014; Joh et al. 2014). The motivation for these studies was to obtain a deeper understanding of the protein folding and building principles used by nature (DeGrado et al. 1999; Cui et al. 2008; Arai et al. 2012; Moser et al. 2016), to explore the coupling of electron and proton transfer reactions (Willner et al. 1999; Gibney et al. 2000; Shleev et al. 2005; Albrecht et al. 2005; Willner and Willner 2005; Pinter, Koebke, and Pecoraro 2020), and to study the properties of cofactors within the protein matrix (Rau and Haehnel 1998; Popović et al. 2001; Ghosh and Pecoraro 2005; Fry et al. 2013; Bialas et al. 2016). Another focus was to create synthetic redox-active proteins and to examine the interactions of the polypeptide environment with heme as a cofactor (Robertson et al. 1994; Sakamoto et al. 1999; Rau, DeJonge, and Haehnel 2000; Ghirlanda et al. 2004; Olson, Williams, and Allen 2013; Vitale et al. 2015).

The architecture of two pairs of antiparallel α–helices covalently attached to a cyclic peptide is a building principle to generate synthetic proteins with a four-helix bundle structure (Rau, DeJonge, and Haehnel 1998, 2000; Schnepf et al. 2001; Haehnel 2004). The concept of template-assembled synthetic proteins (TASP) was successfully applied to construct models of cytochrome *b* (Cyt*b*) with one (Rau, DeJonge, and Haehnel 2000) or two hemes (Rau and Haehnel 1998). The synthetic Cyt*b* analogs were electronically active when bound to a gold surface (Katz, Heleg-Shabtai, Bardea, et al. 1998) and exhibited light-induced electron transfer when bound to a metalloprotein (Rau, DeJonge, and Haehnel 1998).

The rational design of artificial proteins requires knowledge of protein structures. So far, only a few artificial four-helix bundle proteins have available an NMR (Hill and DeGrado 1998; Skalicky et al. 1999; Ma et al. 2008) or X-ray (Huang et al. 2003; Geremia et al. 2005) resolved structure. However, recently, atomic coordinates for a synthetic Cyt*b* with two hemes (Rau and Haehnel 1998) were generated from scratch through advanced modeling techniques (Popović et al. 2001). This is a reasonable model for the Cyt*b* subunit of the membrane-spanning photosynthetic Cytb_6f and the mitochondrial Cytbc_1 protein complexes. In previous molecular dynamics (MD) simulation studies, root-mean-square deviations (RMSD) obtained for the atomic coordinates of a native Cyt*b* crystal structure and the modeled synthetic Cyt*b* protein showed small and similar values, hence, demonstrating the robustness of the model structure (Popović et al. 2001; Iwata et al. 1998).

Using the model structures and solving the Poisson-Boltzmann (PB) equation, the hemes' redox potentials[2] were computed here by evaluating the electrostatic energy difference between oxidized and reduced states averaged over all possible charge states of titratable residues. The computational procedure and the charge models that we used were validated by considering the Cyt*b* subunit from the crystal structure of the native Cytbc_1 complex (Iwata et al. 1998). The experimentally measured and calculated redox potentials of the two hemes agree well within 20 mV. The generated model structure of the synthetic Cyt*b* was validated by comparing computed and measured redox potentials of the two hemes yielding deviations of less than 15 mV (Popović et al. 2001). In an extensive computational/theoretical study on various native heme-proteins, the redox midpoint potentials (E_m) were evaluated, showing particularly good agreement with experiments for the *bis*-histidine ligated *b*-type hemes (Zheng and Gunner 2009). On the other hand, Lu and coworkers have designed the water-soluble proteins in the same scaffold with the same surface properties, covering a redox potential range from -1 to +1 V by mutating only five residues and using two metal ions (Hosseinzadeh et al. 2016). Spectroscopic methods have revealed the key geometric parameters essential for the high E_m, while the gained theoretical knowledge was used to predict their E_m values.

Success in modeling the structure of a synthetic protein−cofactor complex opens up the possibility of investigating how the protein environment and specific amino acid mutations tune the redox potential of cofactors. Suitable

[2] All redox potentials, i.e., electrochemical midpoint potentials are given [in mV] *vs.* standard hydrogen electrode (SHE).

model systems to probe are the synthetic mono-heme proteins (Syn-*m*-HP) with a single heme as a cofactor, where the redox potential of 400 different mutants can be measured with ease (Rau, DeJonge, and Haehnel 2000).[3]

Biosensors and Bioelectronic Devices Based on Native and Artificial Cytochromes

Advances in nanotechnology, including biomaterials use, have provided new research tools, hybrid materials, structures, and systems for biological and medical research with a wide variety of applications. In particular, the integration of biomaterials distinguished by unique biorecognition, catalytic, and inhibition properties with nanomaterials characterized by unique optoelectronic, magnetic, and catalytic properties results in the creation of novel bio–nano hybrid systems. Therefore, the new hybrid materials possess synergetic properties and functions. In this section, the emphasis is directed to using of biomolecule–nanoparticle assemblies on an Au-electrode surface for bioanalytical applications and the fabrication of bioelectronic devices. In this sense, advances in the field involving natural and synthetic cytochromes and heme-proteins are especially highlighted here.

Enzyme Electrodes and Electrochemical Detection

The electrochemical and optical methods are highly sensitive, and therefore, they are the most commonly employed techniques for signal detection in sensors and biosensors. There are four main types of electrochemical methods: voltammetry, amperometry, electrochemical impedance spectroscopy, and potentiometry. Electrochemical biosensors draw much research attention because of their promising applications in pharmacy, medicine, food industry, ecology, bioelectronics, and bionanotechnology (Leger and Bertrand 2008). Furthermore, the advantages of biosensors lie in low-cost, commercial, rapid, accurate, precise, compact, and convenient methods for quantitative detection of the analyte of interest. The key physicochemical features of electrochemical biosensors include their accuracy, high selectivity, sensitivity, rapid and continuous measurement, fast response time, and fewer amounts of reagents

[3] The numerical values of heme E_m potentials for Syn-*m*-HPs are kindly provided by Prof. Haehnel's research group.

required for calibration (Kalantar-Zadeh 2013). In addition, biosensors are used to measure the concentration of non-polar molecules, which other conventional devices cannot readily determine. The new sensor technology has found its usage in medical diagnostics, cancer detection, and therapy (Peng, Tisch, and Haick 2009; Wang et al. 2016).

Nanoparticles for bioelectronic/biocatalytic applications and biofuel cells are composed of a metallic (semiconductive) core stabilized by a self-assembled organic shell, which binds the functional biomolecular units as a monolayer at the nanoparticle periphery. The biomolecules operate as receptors and bind the specific analyte, generating a detectable signal for sensing applications (Service 2002; Kim et al. 2014; Heuer-Jungemann et al. 2019). The convenient setup exploited in the design of many enzyme-based biosensors renders the integration of reconstituted *de novo* synthetic proteins or native metalloenzymes on highly conductive electrodes. Electrical contact between redox-active proteins and the electrode surface is of fundamental importance and can be achieved in several ways. Modifications of redox enzymes by covalently bonded electroactive relay groups (Degani and Heller 1988; Schuhmann 1995), immobilization of the biocatalysts in redox-functionalized polymers (Willner et al. 1992), physical adsorption, matrix entrapment, or membrane entrapment provide the most common means to establish electrical linkage between the redox sites of proteins and the electrode surfaces (Heller 1990; Riklin et al. 1995). Enzyme electrodes can be organized by cross-linking the cofactor–biocatalyst affinity complexes with conductive support (Heleg-Shabtai, Katz, and Willner 1997). The resulting electrically connected enzyme electrodes are then used as biosensors.

The analyte reacts at the surface of the sensing electrode (modified working electrode) involving either the oxidation or reduction mechanism (voltammetric and amperometric methods). The oxidation and reduction reaction rates are controlled and measured by a potentiostat connected to electrodes submerged in an electrolyte, Figure 1. This reaction is catalyzed by the biomaterial monolayer on an electrode specifically selected or designed for the target analyte. The current generated, in the process, is converted into an electrical signal that can be easily amplified, processed, and displayed by modern electrical devices. In sensing applications, depending on the electrochemical method applied for detection, the analyte generates directly (via electroactive species) or indirectly (via a biorecognition event or mediated enzyme electrodes) an electrical signal proportional to its concentration. The analyte concentration is then obtained by analyzing the intensity of the electrical signal.

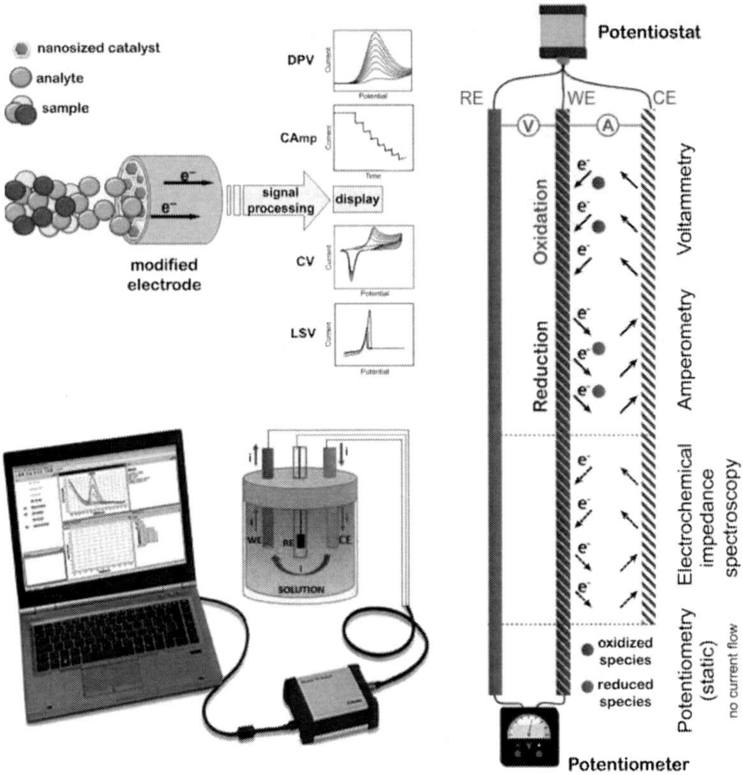

Figure 1. Depiction of electrochemical detection and experimental setup used for electrochemical analysis. The general principle of electrochemical detection is illustrated in the left upper part. In electrochemical sensing, the analyte generates directly (via electroactive species) or indirectly (via a biorecognition event or mediated enzyme electrodes) an electrical signal proportional to its concentration. Electrochemical analysis (the right and left down part) consists of various techniques that use electrical stimulation to analyze the chemical reactivity of a sample surface or a solution. There are four main electrochemical methods: voltammetry, amperometry, electrochemical impedance spectroscopy, and potentiometry. Thereby, various electroanalytical techniques are available and applicable for electroanalysis: constant potential amperometry (CPA), stepped potential chronoamperometry (CA), cyclic voltammetry (CV), differential pulse voltammetry (DPV), linear sweep voltammetry (LSV), square wave voltammetry (SWV), stripping voltammetry (SV), electrochemical impedance spectroscopy (EIS), and potentiometry.

The experimental setup used for electrochemical analysis consists of a conventional three-electrode cell, including the working electrode (WE), reference electrode (RE), counter electrode (CE), potentiostat, and personal computer, Figure 1. The working electrode is the electrode at which the reaction of interest occurs, such as an integrated biomolecule–nanoparticle system assembled on an Au-electrode; a saturated calomel (SCE) or Ag/AgCl electrode is typically used as a reference electrode; an inert conducting material (Pt) is used as a counter electrode. During experiments, charge flow occurs between the working electrode and the counter electrode while the potential of the working electrode is measured with respect to the reference electrode.

The working electrodes for electrochemical sensors commonly include inert metals (Au, Ag, Pt, Pd), semiconducting metal/metal oxides, carbon-based materials, nanosized alloys, binary nanocomposites, conducting polymers, and fabricated electrodes. As a result, the modified electrodes show enhanced sensitivity and response time and improved electrical conductivity and catalytic efficiency due to the reduced overpotentials and faster electron transfer kinetics (Dincer et al. 2019).

Bifunctional Synthetic Redox-Active Protein Electrode

This section discusses some examples of native and artificial cytochromes[4] used as central components of biosensors in bioelectronic/biocatalytic applications. The importance of several natural proteins and synthetic redox systems related to biosensors, with the relevant experimental measurements that show their analytical potential, has been recently demonstrated and reported in the literature (Katz, Heleg-Shabtai, Bardea, et al. 1998; Willner and Willner 2005; Katz, Heleg-Shabtai, Willner, et al. 1998; Willner et al. 1999; Katz and Willner 2004; Friedrich et al. 2008).

Katz et al. (1998) have reported on the assembly of an antiparallel four-helix bundle synthetic protein as a monolayer on an Au-electrode surface. The surface concentration of the helix-bundles corresponds to a densely packed monolayer arrangement. The synthetic analog of Cyt*b* protein consists of the heme-binding helices, type *B*, with two ligating His residues in the amino acid

[4] The discussion is limited to synthetic hemoproteins and native cytochromes only, although there are numerous native proteins, enzyme systems, and engineered models for biosensor applications.

sequence and the heme-protecting helices, type *A*, in the antiparallel orientation. A stepwise procedure for the reconstitution of two Fe(III)–protoporphyrin IX moieties into the synthetic protein is described in the original paper (Willner et al. 1999). The resulting reconstituted protein–cofactor affinity complex reveals electrical contact of the synthetic redox protein with the conductive support. The electrical connection is achieved by merging the terminal Cys residues (by –SH groups) in the helices type *B* with the *N*-succinimidyl-3-maleimidopropionyl (Mp) moieties attached to the Au-surface. Vectorial electron transfer proceeds in the aligned redox-active monolayer assembly, leading the synthetic hemoprotein to act as a rectifier, Figure 2A. More specifically, the electrons go one at a time from Au-electrode, first, to the low potential heme, followed by an electron jump to the high potential heme, while the back electron transfer to the electrode is delayed. It is also demonstrated that the synthetic hemoprotein functions as an electron transfer mediator for the native redox protein, nitrate reductase (NR), and that the bioelectrocatalyzed reduction of nitrate to nitrite is activated, Figure 2B.

Electrical activity and kinetics of the electron transfers between Au-electrode and the two heme cofactors in the enzyme electrode were monitored by cyclic and differential pulse voltammetry, chronoamperometry, and coulometric analysis. Previously, the position of the low and high potential hemes in the helix-bundle and their redox potentials were studied by potentiometric titration but did not provide a definite answer to this issue (Rau and Haehnel 1998).

The reduction potentials of the heme centers are determined here as -182 mV and -116 mV (SHE), being separated by +66 mV. Based on the obtained results, the accurate positioning of the low potential heme (closer to the metal surface and the open end of the helix-bundle) and high potential heme (more distant from the electrode and close to the template) has been assigned.

The electron transfer rates were estimated and analyzed from chronoamperometric experiments. Vectorial electron transfer proceeds in the system due to the appropriate ordering of the redox potentials of the heme centers. The rapid electron transfer first occurs to the heme positioned close to the electrode. Second electron transfer from the low to high potential heme is relatively fast ($\tau = 25$ ms). However, the back electron transfer to the electrode ($\tau = 1$ s) is hindered since the primary heme site, close to the electrode, acts as a barrier to the oxidation process. Specifically, the primary heme site cannot oxidize the secondary heme site due to its low E_m potential.

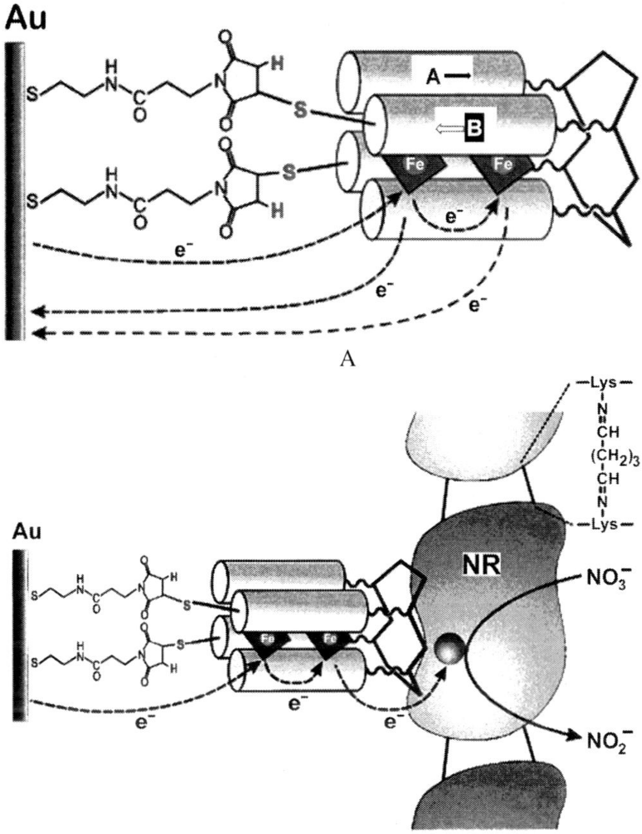

Figure 2. Integration of a reconstituted *de novo* synthesized hemoprotein and native metalloproteins with electrode supports for bioelectronic and bioelectrocatalytic applications. A) Assembly of the monolayered Fe(III)–protoporphyrin IX reconstituted *de novo* synthetic protein on an Au electrode. B) Assembly of the integrated monolayered electrode composed of cross-linked nitrate reductase (NR) as a biocatalyst and the Fe(III)–protoporphyrin IX reconstituted synthetic protein.

Interaction of the synthetic hemoprotein monolayer on the electrode with the native cytochrome-dependent nitrate reductase (NR) results in the formation of a complex (Figure 2B). The resulting affinity complex is cross-linked with glutaric dialdehyde to produce a stable and integrated bioelectrocatalytically active electrode. The cathodic current observed in the presence of added nitrate (NO_3^-) indicated the bioelectrocatalyzed reduction of NO_3^- at the redox potential of the synthetic hemoprotein. The process was monitored by cyclic voltammetry and amperometric measurements. However, the bioelectrocatalyzed NO_3^- reduction does not occur in the absence of NR or after direct coupling of the native biocatalyst (NR) to the maleimide-based monolayer on the electrode. Thus, the synthetic hemoprotein at the electrode enables the mediated electron transfer to NR, which triggers the bioelectrocatalyzed reduction of NO_3^- to NO_2^-. In conclusion, Katz et al. (1998) have presented the bifunctional synthetic redox-active protein electrode as a bioelectronic rectifier and as a base interface to organize integrated enzyme electrodes for biosensors and bioelectronic applications.

Bioelectrocatalytical Cytochrome *c*/Cytochrome *c* Oxidase Electrode

A good example, which includes the integration of natural cytochromes in enzyme-based electrodes and biosensors, is reported on the assembly of the integrated bioelectrocatalytical cytochrome *c*/cytochrome *c* oxidase (Cyt*c*/C*c*O) electrode for the reduction of O_2 (Friedrich et al. 2008). Cytochrome *c* oxidase (C*c*O) from *Paracoccus denitrificans* was immobilized in a fixed orientation using a His-tag attached to subunit I on a gold film and reconstituted *in situ* into a protein-tethered bilayer lipid membrane.

In this orientation, the cytochrome *c* (Cyt*c*) binding site is directed away from the electrode pointing to the outer side of the bilayer lipid membrane (BLM), see Figure 3. Catalytic activity was monitored by impedance spectroscopy (IS) and cyclic voltammetry (CV). Cathodic and anodic currents of C*c*O with Cyt*c* added to the bulk solution were shown to increase under aerobic compared to anaerobic conditions. Thus, C*c*O can be activated by reduced Cyt*c* only under aerobic conditions. Catalytic activity is considered here in terms of a continuous electrochemical oxidation/reduction process of the C*c*O/Cyt*c* complex in the presence of oxygen. The electron hopping mechanism through the redox sites of the C*c*O protein complex associated with Cyt*c* bound to the C*c*O enzyme electrode has been suggested in this study. In this mechanism, electrons jump along the chain of redox sites inside the protein, with the final redox site exchanging electrons with the electrode. By contrast, in the non-hopping model, the substrate (Cyt*c*) would exchange electrons directly with the electrode. That is not a very likely scenario considering the presence of a lipid membrane, which prevents Cyt*c* from directly accessing the electrode. Simulations supporting this hypothesis are included in the study.

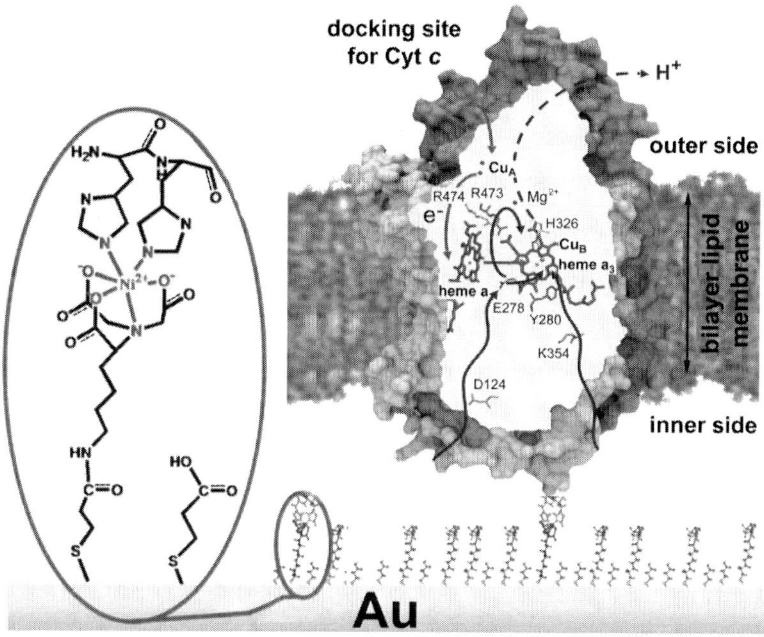

Figure 3. Schematic representation of cytochrome *c* oxidase embedded into a bilayer lipid membrane. The protein is attached to a gold film on the NTA-Ni^{2+} complex-modified surface by a His-tag attached to subunit I. The lipid bilayer is assembled around the protein by *in situ* dialysis of micelles made from detergent-soluble lipid.

C*c*O is a remarkably intricate enzymatic system that undergoes several oxidation intermediates (R, A, P, F, H, O, E) and many transient states in the catalytic cycle. The catalytic mechanism of C*c*O includes the coupling of electron and proton transfer reactions, reduction of O_2 to H_2O, and proton pumping against the membrane potential. Having a cross-linked enzyme electrode shown in Figure 3 and including Cyt*c* and O_2 in the bulk phase makes the system even more complex for understanding.

The CV experiments of the reduced Cyt*c* on a C*c*O-containing-BLM electrode under aerobic and anaerobic conditions give rise to cathodic and anodic currents depending on the potential. The amplification is considered here as a feature of a catalytic current controlled by the turnover of the C*c*O enzyme. Two processes proceed simultaneously:

1. the electrochemical process of Cyt*c* without oxygen, $\mathrm{Cytc}_{oxd} + e^{-} \rightleftarrows \mathrm{Cytc}_{red}$ and

2. in the presence of O_2 and with Cytc_{red} present in the bulk phase, the enzymatic turnover occurs,

$$4Cyt\,c_{red} + O_2 + 8H^+ \rightleftarrows 4Cyt\,c_{oxd} + 2H_2O + 4H^+_{pumped}\,.$$

The Cytc_{oxd} formed here is reduced repeatedly, resulting in an amplification of the cathodic current. In contrast, the enzymatic reaction is expected to proceed via P(peroxy)/F(ferryl-oxo) states of the C*c*O. The P/F-state, on the other hand, gets anodically oxidized, restoring the oxidized O-state; $CcO\,(P/F) \rightleftarrows CcO\,(O) + O_2 + e^-$. Therefore, the O-state in C*c*O may be reduced repeatedly to regenerate the P/F state, which may again be oxidized, amplifying the anodic current.

In conclusion, in the presence of oxygen and Cytc_{red} in the bulk solution, the substrates of both the cathodic and the anodic processes, Cytc_{oxd} and CcO(P/F), respectively, are restored by the enzymatic reaction, giving rise to increased (catalytic) currents in both directions. Friedrich et al. (2008) have suggested the electron transfer pathways through the enzyme in the case of an anodic current flowing through the electrode with the electrons traveling in the direction opposite from the electrode. However, in the case of a cathodic current flowing through the electrode, the electrons migrate from the metal electrode.

Biosensors and Biomolecular Recognition

A biosensor is an analytical detection system consisting of a biological component combined with a transducer to measure a biochemical quantity. The development of biosensors has started with the invention of enzyme electrodes by Leland C. Clark in 1962 (Clark and Lyons 1962). Biosensors are operated based on the principle of signal transduction. Their components include a biorecognition element, a biotransducer, and an electronic system composed of a display, processor, and amplifier. Bioreceptors are recognition elements that retain a high binding affinity toward a specific analyte. Recognition elements can be natural or artificially bioinspired molecules derived synthetically using biological principles (Zourob 2010; Piletsky and Whitcombe 2013).

Biomolecular recognition is primarily achieved by noncovalent interactions, including hydrogen-bonding, electrostatic, van der Waals forces, and hydrophobic interactions. In the presence of a target molecule, recognition elements either undergo a bio/chemical reaction, producing a measurable signal directly, or they are tagged with signaling molecules, such as enzymes, for signal transduction. Natural recognition elements are native bioreceptor molecules, such as nucleic acids, proteins, enzymes, antibodies, membranes, organelles, cells, and tissues. Nowadays, genetically modified multifunctional molecules based on natural biomolecules and synthetic peptides with receptor properties are increasingly implemented in biosensors (Weber and Fussenegger 2012). Hence, each biosensor typically includes two main components: a bioelement, such as an enzyme, antibody, or cell receptor, and a sensing element or a transducer. These two elements are combined and integrated through several methods such as covalent bonding, matrix entrapment, physical adsorption, and membrane entrapment (Katz and Willner 2004).

The biorecognition element, essentially a bioreceptor, is allowed to interact with a specific analyte. The transducer measures this interaction and generates an output signal. The intensity of the signal output is proportional to the concentration of the analyte. The signal is then amplified and processed by the electronic system. One type of biosensor, specifically discussed here, is the electrochemical biosensor (Dincer et al. 2019). The principle behind the operation of these sensors involves the generation of ions by various chemical events that change the electrical properties of the analyte solution. The concentration of the analyte is then measured concerning this change.

In a biosensor, the bioreceptor is designed to interact with the specific analyte of interest, producing an effect measurable by the transducer. High selectivity for the analyte among a matrix of other chemical or biological components is a crucial requirement of the bioreceptor (so-called "matrix effect"). While the used biomolecular types can vary widely, biosensors can be classified according to common types of bioreceptor interactions involving: antibody/antigen, enzymes/ligands, nucleic acids/DNA, cellular structures/cells, or biomimetic materials (Crivianu-Gaita and Thompson 2016).

The novel bio-nano devices provide the opportunity for an accurate, reliable, and user-friendly automation analysis, and therefore, they are an emerging alternative to the conventional analytical techniques, which require time-consuming procedures and expensive instruments. Finally, in conclusion, the importance of developing the hybrid catalytically active and highly selective bio-nano electrodes and biosensor devices, applicable for quality and safety assessment of food, environmental monitoring, wastewater treatment, toxin detection, pharmaceutical analysis, and clinical diagnostics, has received extensive scientific and research attention nowadays (Wang and Wang 2008; Donzella and Crea 2011; Weber and Fussenegger 2012; Juzgado et al. 2017; Dincer et al. 2019; Alexi et al. 2021; Mukherjee and Mukherjee 2021; Mi et al. 2021; Pellerano and Morris 2021; Zhang et al. 2021).

Material and Method

Studied Models

The synthetic hemoproteins studied here mimic the structure and function of the catalytic part of native Cyt*b*. The Syn-*m*-HPs consist of an antiparallel four-helix bundle formed by two identical pairs of helices *A* with 15 residues and helices *B* with 16 residues, covalently linked to a cyclic decapeptide serving as a template (Figure 4). Two histidines axially ligated to heme are positioned in the helix pair of type *B*. Therefore, helices type *B* are the heme-binding, while helices type *A* are the heme-protecting and shielding. The assembly of four antiparallel helices provides a heme-binding site placed in the hydrophobic interior.

The variation of mutated amino acids is restricted to hydrophobic side-chains of different sizes (Gly, Ala, Val, Leu, Ile, Phe), two polar (Tyr, Gln), and one basic amino acid residue (Arg). Since the synthetic structures are modeled using a minimalistic approach, it is not surprising that their redox properties are altered compared with the native Cyt*b*.

The atomic coordinates of Syn-*m*-HP with the variable sequence part $(AFAAF)_A(QLAL)_B$ (mutant 4) were constructed from scratch with advanced modeling techniques, applied successfully before to model a synthetic Cyt*b* with two hemes (Popović et al. 2001), see Chapter I. Molecular mechanics have been employed for molecular modeling, stepwise energy minimizations and MD simulations by using the program CHARMM with its force-field (MacKerell et al. 1998). Essential for obtaining a robust structure was to avoid strain during the assembly of the components by placing them in agreement with chemical and steric requirements before the structure was carefully relaxed by energy minimization. Moreover, all known structural details from the experimental studies were used (Rau and Haehnel 1998; Katz, Heleg-Shabtai, Bardea et al. 1998; DeJonge, Rau, and Haehnel 1999; Willner et al. 1999; Rau, DeJonge, and Haehnel 2000; Albrecht et al. 2005). For example, the imidazole planes are initially set up to be parallel, with the pseudo-torsion angle $\alpha = +90^{o}$;[5] for a definition of angles α, β, and γ, see Chapter IV. Angles α close to +90° and nearly parallel orientation of the two imidazole planes were deduced from the EPR and ENDOR spectroscopy characterization of the *de novo* designed Syn-*m*-HP hemoproteins (Fahnenschmidt et al. 1999).

helix A:

N-terminus (at open end of Syn-*m*-HP) — C-terminus (at template of Syn-*m*-HP)

	1	2	3	4	5	6	7	8	9	10	11	12	13	14	15	
Ac –	L	E	E	α	α	K	K	α	E	E	α	α	K	K	K	– $CONH_2$
															| Mp	

helix B:

N-terminus (at template of Syn-*m*-HP) — C-terminus (at open end of Syn-*m*-HP)

	1	2	3	4	5	6	7	8	9	10	11	12	13	14	15	16	
Mp –	G	L	E	β	β	L	K	β	**H**	E	E	β	L	K	K	L	– $CONH_2$

template:

1	2	3	4	5	6	7	8	9	10
C	A	**C**	P	G	**C**	A	**C**	P	G

Figure 4. Amino acid sequences of the helical peptides, type *A* and *B*, of the four-helix bundle with a cyclic decapeptide serving as a template. The letters α and β denote the variable residues in sequences of the different mutants. The maleimidopropionyl moieties (Mp-), attached to the helices *A* and *B*, enable their binding to Cys residues (in bold) of the template and ensure their antiparallel orientation. Acetyl (Ac-) and amido (-$CONH_2$) blocking groups remove the charges from the ends of the helices stabilizing the entire synthetic structure. His residue (in bold) in the helix type *B* is ligated to the heme iron.

[5] The orientation of the imidazole ring relative to the heme plane and with respect to the heme propionic groups is measured by the pseudo-torsion angle α (Cε1-Nε2-FE-CHA).

In the MD simulations, the model structure of Syn-*m*-HP exhibited an RMSD profile over time, which is very similar to the RSMD profile obtained for the crystal structure of native Cyt*b*. The other 24 mutant structures were constructed from the relaxed structure – sequence 4. Then, the corresponding amino acid side-chains were replaced, and the resulting structures were relaxed by energy minimizing first the exchanged side-chains alone, followed by optimization of all other side-chains, and finally the minimization of all atoms.

The RMSD values of the atomic coordinates of 24 structures generated by starting from the reference structure 4 do not exceed 0.62 Å for the backbone and 0.89 Å for all atoms except heme and side-chain atoms of residues that differ, but in most cases, the deviations are much smaller (see Results Section). Details on the molecular modeling and MD simulation procedure were extensively discussed previously by Popović et al. (2001).

Redox Potential Calculations

The electrostatic calculations have been performed using the MEAD program suite (Bashford 1997). Usage of the program is described in the following references (Bashford and Karplus 1990; Bashford and Gerwert 1992; Bashford et al. 1993; Bashford and Karplus 1991), while the fundamental concepts of the protein electrostatics are repeatedly discussed in the literature (Warshel, Russel, and Churg 1984; Honig and Nicholls 1995; Warshel, Papazyan, and Muegge 1997; Warshel and Papazyan 1998; Sharp and Honig 1990). The electrostatic energies of the protein-water system were evaluated by solving the PB equation:

$$\nabla\left[\varepsilon(\vec{r})\,\nabla\phi(\vec{r})\right] = -4\pi\rho(\vec{r}) + \kappa^2(\vec{r})\phi(\vec{r}) \tag{1}$$

with the charge distribution $\rho(\vec{r})$, the ionic strength of 100 mM determining the value of the Debye length $1/\kappa$, a dielectric constant of 80 for the solvent and 4 within the protein to obtain the electrostatic potential $\phi(\vec{r})$. A suitable choice for the dielectric constant of protein was broadly discussed in the literature (Gilson and Honig 1986; Honig and Nicholls 1995; Simonson and Brooks 1996; Schutz and Warshel 2001).

The Gibbs free energy of a specific protonation/redox state of the enzyme $(\vec{n})$ is expressed as:

$$G(\vec{n})=\sum_{\mu=1}^{P} q_{\mu}^{(n)} RT\ln 10\left(pH-pK_{a,\mu}^{intr}\right)-\sum_{\eta=1}^{O} q_{\eta}^{(n)} F\left(E_{ext}-E_{intr,\eta}^{0}\right)+\frac{1}{2}\sum_{\mu=1}^{N}\sum_{\nu=1}^{N} q_{\mu}^{(n)} q_{\nu}^{(n)} W_{\mu\nu} \quad (2)$$

where $(\vec{n})$ is the N-dimensional protonation/redox state vector ($N= P + O$; the total number of protonatable sites plus redox-active groups). The charge state of group μ for the charge pattern $(\vec{n})$ is characterized by the formal charge $q_{\mu}^{(n)}$, which adopts the values of −1 (deprotonated state) and 0 (protonated state) for acids, and 0 (deprotonated or reduced state) and +1 (protonated or oxidized state) for bases and heme as redox center, respectively. E_{ext} is the external redox potential of solvent, T is the absolute temperature, R is the universal gas constant, and F is the Faraday constant. The first sum runs over all P protonatable sites, the second sum runs over all O redox groups, and the third term is associated with the charge-charge interactions between all N ionizable groups in the system. The intrinsic quantities, $E_{intr,\eta}^{0}$ and $pK_{a,\mu}^{intr}$, are defined as E_m or pK_a of the site in protein when all other sites are in their neutral state, i.e., the reference states (Bashford and Gerwert 1992).

This method aims to reproduce the shift in E_m (pK_a) associated with transferring a model compound representing the redox (protonatable) group with a known solution $E_{model,\eta}^{0}$ ($pK_{a,\mu}^{model}$) value from the aqueous phase to the protein environment. The *bis*-histidine/imidazole heme model compound in an aqueous solution is used with the corresponding experimental E_{model}^{0} value of -220 mV for heme *b*-type (Moore and Pettigrew 1990; Wilson 1983). The redox midpoint potential of heme at pH 7 ($E_{m,7}$) is calculated simultaneously in a self-consistent way, as the thermodynamic average over the charge states of all ionizable groups of the synthetic mono-heme protein, as explained in detail, for example, in the following references (Ullmann and Knapp 1999; Baptista, Martel, and Soares 1999; Gunner et al. 2000; Ullmann 2000; Popović and Stuchebrukhov 2004; Song, Mao, and Gunner 2006; Popović and Stuchebrukhov 2012; Soares and Baptista 2012).

In the present calculations, the independent titratable groups are the *bis*-imidazole porphyrin, Arg, Glu, Lys, Tyr, and the two heme propionates (PRA and PRD). The *bis*-imidazole porphyrin system and the heme propionic groups are treated as the independent variably charged groups that interact only electrostatically. For the propionates, we used the same charges as for the side-chain of glutamate, taken from the CHARMM22 parameter set (MacKerell et al. 1998).

Since the *N*- and *C*-terminal groups are neutralized by blocking groups, i.e., acetylated (Ac-) and amidated (-$CONH_2$), respectively (Figure 4), they are not available for titration. Atomic partial charges are taken from the CHARMM22 force field (MacKerell et al. 1998) except for the heme moiety, where the charges are computed quantum-mechanically (QM). Here, we applied the DFT calculations to optimize the *bis*-imidazole porphyrin structure (QM system) and obtain the partial atomic charges for both reduced and oxidized states. Both states were optimized with the 6–31+G* basis set, while the electrostatic potential in the vicinity of the heme moiety was calculated in vacuum at the B3LYP/6–311+G* level by using the GAUSSIAN 09 program package (Frisch et al. 2016). The atomic partial charges for the QM system are obtained from electrostatic potentials by using the restraint electrostatic potential (RESP) method developed by Kollman and coworkers (Bayly et al. 1993; Cornell et al. 1993). To reduce the number of independent charge parameters in the RESP procedure, the atomic partial charges of the pyrrole rings of porphyrin were symmetrized as described by Popović et al. (2001). More details of the computational procedure and a list of the used atomic charges are given in reference (Popović et al. 2001).

The protein charges and atomic radii that we use in this application have been used successfully before to calculate the pK_a values of titratable groups, redox potentials of cofactors, and bioenergetics of electron and proton transfer reactions in different enzymes and proteins (Popović et al. 2001; Popović et al. 2002; Popović and Stuchebrukhov 2004, 2005, 2006; Quenneville, Popović, and Stuchebrukhov 2006; Popović et al. 2010; Popović and Stuchebrukhov 2012; Couch, Popović, and Stuchebrukhov 2011; Vagedes et al. 2000; Rabenstein and Knapp 2001).

Results and Discussion

The axially coordinated ligands (their type/identity (Liu et al. 1995), orientation (Zarić, Popović, and Knapp 2001; Fufezan, Zhang, and Gunner 2008), alignment (Safo et al. 1997), and basicity (Nesset et al. 1996)), protein environment (Gunner et al. 2000; Popović and Djordjević 2020), and the inhomogeneous dielectric surrounding around the heme cofactor (Popović et al. 2001) are essential in tuning the properties of heme to perform in the desired manner. The local protein charges from the polar groups, protein backbone dipoles, and the charged ionizable side-chains (Varadarajan et al. 1989; Ullmann and Knapp 1999), as well as the strong proton coupling to the redox reaction (such as, *e.g.*, the coupling with the protonation state of the heme propionates (Lancaster et al. 1996; Shifman et al. 1998; Gunner and Alexov 2000; Popović 2013)) are all part of the protein environment effects on the E_m value, as discussed in reference (Popović et al. 2001) and the related references therein. For a long time, it has been assumed that heme solvent exposure makes a substantial contribution to the equilibrium redox midpoint potential E_m of native hemoproteins due to the desolvation (energy) penalty of burying the formally charged ferric heme in the low dielectric medium of protein (Gunner and Honig 1991; Tezcan, Winkler, and Gray 1998). The other described effects include a nonplanar distortion and ruffling of the heme plane (Olea, Kuriyan, and Marletta 2010; Neya et al. 2010; MacGowan and Senge 2013), as well as the effect of the imidazole-heme binding affinity, defined by the individual His–Fe(III) coordination equilibrium constants (Kennedy et al. 2001), on the heme E_m potential.

Heme Orientation in the Helix-Bundle

The atomic coordinates of 25 synthetic mono-heme proteins were generated in a similar fashion as described in our previous work related to the synthetic di-heme protein (Popović et al. 2001). The sequences listed in Table 1 were chosen to cover the whole range of measured redox potentials (-89 mV to -148 mV) for a whole family of synthetic mono-heme proteins related by mutations at the different amino acids sites (Rau, DeJonge, and Haehnel 2000). The computer-generated structures of synthetic proteins were carefully chosen to include the most characteristic structural and functional properties to avoid being redundant. Moreover, the different sequences of helices type A_i and B_j are considered, particularly those with interesting structural features and specific amino acid mutations. The sequences of the synthetic proteins considered are numbered from 1 to 25 in order of decreasing measured redox potential values. The complete sequences are:

$L\,E\,E\,\alpha_4\,\alpha_5\,K\,K\,\alpha_8\,E\,E\,\alpha_{11}\,\alpha_{12}\,K\,K\,K$ for helix type *A* and
$G\,L\,E\,\beta_4\,\beta_5\,L\,K\,\beta_8\,H\,E\,E\,\beta_{12}\,L\,K\,K\,L$ for helix type *B*.

Greek letters denote positions with variable amino acids. The variable parts of the sequences in helices *A* and *B* are denoted as $(\alpha_4\,\alpha_5\,\alpha_{11}\,\alpha_{12})_A$ $(\beta_4\,\beta_5\,\beta_8\,\beta_{12})_B$ in Table 1.

We have also considered sequences 24 and 25, where the corresponding redox potential could not be measured since the values were outside the range used for the redox indicator. A significant portion of the mutants considered focuses on the interface between helix pairs A_1/B_1 and A_2/B_2, where α4α11 and β5β12 are varied such that the hydrophobic residues come together except for Arg, see helical wheel projection of the four-helix bundle, Figure 5A. The helical wheel projection of the antiparallel four-helix bundle is a valuable representation to locate the residues on the surface, in the interfacial region, or in the interior of the helix bundle. In the case of model structures 24 and 25, the helices B_1 and B_2 (on β5β12 positions) do not provide sufficient shielding of the heme in the binding pocket, making it more water accessible. As a result, their heme midpoint potentials are the lowest of all structures.

Two possible modes to place the heme in the binding pocket of the four-helix bundle were considered. The propionates are oriented toward the template – HC1 or the open end of the four-helix bundle – HC2. We generated all 25 structures with heme conformation HC1. For sequences 1, 4, 8, 14, 15, 16, and 23, we also generated structures with heme conformation HC2. The structures of Syn-*m*-HP with sequences 4 and 17 and heme conformation HC1 are shown in Figure 5. The free energy difference of the two heme conformations was computed from the electrostatic energies by solving the PB equation averaged over the protonation states of all titratable groups, including the solvation energy calculated by program suite MEAD (Bashford 1997; Bashford and Gerwert 1992), as well as, the non-electrostatic conformation energy and a surface energy term obtained from CHARMM program package (MacKerell et al. 1998). The total free energy differences are given in Table 2, clearly favoring the heme conformation HC1 in the structures 4, 8, 14, and 23 but preferring the heme conformation HC2 in the structures 1 and 15 that contain Arg residue at the α5 position, as well as, in the structure 16 with Arg residue at the α4 position.

Table 1. The 25 considered mutant structures with variable sequence parts

$(\alpha_4\ \alpha_5\ \alpha_8\ \alpha_{11}\ \alpha_{12})_A\ (\beta_4\ \beta_5\ \beta_8\ \beta_{12})_B$									
1	$(LRALL)_A$ $(QVLL)_B$	**2**	$(AFAAF)_A$ $(QVLL)_B$	**3**	$(AFAAF)_A$ $(QVLL)_B$	**4**	$(AFAAF)_A$ $(QLAL)_B$	**5**	$(LLLLL)_A$ $(QVLL)_B$
6	$(LLVLL)_A$ $(QRLA)_B$	**7**	$(VLALL)_A$ $(QVLL)_B$	**8**	$(AFAAF)_A$ $(QLLY)_B$	**9**	$(AFAAF)_A$ $(QAYA)_B$	**10**	$(AFAAF)_A$ $(QLLV)_B$
11	$(LLAVL)_A$ $(QAVA)_B$	**12**	$(LLLLL)_A$ $(QAVA)_B$	**13**	$(LLLLL)_A$ $(QLLV)_B$	**14**	$(AYAAY)_A$ $(QLLV)_B$	**15**	$(LRALL)_A$ $(QAYA)_B$
16	$(RLALL)_A$ $(QLLV)_B$	**17**	$(LLLLL)_A$ $(QRLA)_B$	**18**	$(FAAFA)_A$ $(QLLV)_B$	**19**	$(LAVLL)_A$ $(QAVA)_B$	**20**	$(ALAAL)_A$ $(QLLV)_B$
21	$(ALAAL)_A$ $(QRLL)_B$	**22**	$(LLALL)_A$ $(QAYA)_B$	**23**	$(GLGGL)_A$ $(QAVA)_B$	**24**	$(VLALL)_A$ $(QALA)_B$	**25**	$(ALAAL)_A$ $(QALA)_B$

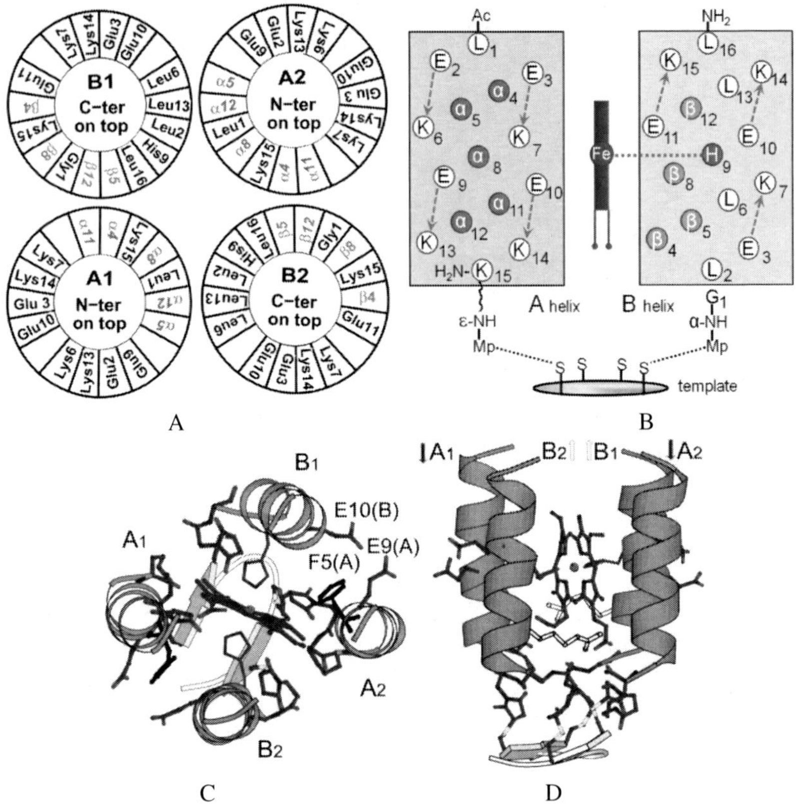

Figure 5. Modeling of the synthetic mono-heme proteins. (A) The topology of the four-helix bundle is shown as the helical wheel projection. (B) The schematic side view – the vertical depiction. By combining both representations, one can locate the residues on the surface, in the interfacial region, or in the interior of the four-helix bundle. (C) The computer-generated structure of Syn-*m*-HP with sequence 4. The helices, *A* and *B,* are displayed as green and orange rubber bands, respectively. The β-stand-β-turn structure of the cyclic decapeptide, serving as a template, is shown by a pair of gray arrows. (D) The structure of Syn-*m*-HP mutant with sequence 17. The two pairs of glutamates E9(A_1)/E10(B_2) and E9(A_2)/E10(B_1), arginines R5(B_1)/R5(B_2), and the *bis*-histidine heme moiety are displayed in atomic detail.

Table 2. Conformational energy difference (in kcal/mol) in the studied structures

Structure 1	**Structure 4**	**Structure 8**	**Structure 14**	**Structure 15**	**Structure 16**	**Structure 23**
+6.1	-5.9	-5.0	-3.4	+7.9	+1.2	-9.7

In many native heme-proteins, the propionic groups tend to get oriented toward the aqueous solution or to be solvent-exposed, which is not frequently the case in the studied synthetic hemoproteins. The propionates' orientation toward the template is often stabilized by the hydrogen bonds with the backbone amido-groups. In contrast, the orientation toward the open end of the helix bundle (HC2) makes the propionates partially deprotonated ($pKa \geq 7$) and only partially solvent-exposed due to the screening and shielding effects of the internal hydrophobic side-chains on the top of the helix-bundle. Three leucines commonly shield each propionic group – Leu1(A_1), Leu13(B_2), and Leu16(B_1) or Leu1(A_2), Leu13(B_1), and Leu16(B_2), unless the structure contains the arginines, which can be then involved in the salt bridges with the PRA and PRD groups.

Protonation State and Effects of Ionizable Groups

The protonation pattern of titratable residues and the redox potentials of the 25 modeled synthetic mono-heme proteins were calculated from the solution of the PB equation. Each Syn-*m*-HP structure possesses 14 glutamates, 14 lysines, two propionic groups, and a heme moiety as titratable sites. Some mutants additionally include two arginines and/or two to four tyrosines. The total charge of Syn-*m*-HPs at pH 7, given in Table 3, indicates nearly the electroneutrality of all studied protein systems. The results of the acid-base protein titration also demonstrate that the isoelectric points (pI) of all studied mutants are similar and only slightly above 7, typically in the range of 7.0 to 7.5.

The neighboring charged residues exhibit a significant effect on the heme E_m value. The negatively charged residues stabilize a formally charged *bis*-imidazole ferric heme decreasing its redox potential (showing a negative shift – greater ease of oxidation). In contrast, the positively charged residues favor the *bis*-imidazole ferrous heme increasing its redox potential (leading to a positive shift – greater ease of reduction).

Generally, lysines and glutamates are all charged except for the two pairs of glutamates, E9(A_2)/E10(B_1) and E9(A_1)/E10(B_2), which are partially protonated due to their mutual proximity and partially hydrophobic neighborhood. Moreover, the residues E9($A_{1,2}$) and E10($B_{1,2}$) are situated at the interface between helices A_1/B_2 and A_2/B_1, roughly at the equal height as the iron atom of heme (Figure 5); therefore, they show the most substantial effect of all ionizable sites on the redox potential. The combined contribution of the four glutamates to the E_m value is typically between -60 to -90 mV (Table 3). Similarly, the side-chains K7(A_1)/β8(B_1) and K7(A_2)/β8(B_2) are positioned on the interface between helices A_1/B_1 and A_2/B_2 at the same height as the iron atom of the porphyrin ring, but the joint contribution of the two Lys is less than +20 mV. Compared to glutamate, lysine possesses a longer side-chain and, therefore, can stick farther from heme in the aqueous phase. The reason for the relatively small contribution of K7(*A*) (less than 10 mV per site on average) *vs.* E9(*A*) or E10(*B*) (roughly 20 mV per site) lies in the fact that the residue K7(*A*) is spatially close to β8(*B*), which is typically a hydrophobic side-chain of Leu, Val, Ala or occasionally a polar Tyr. The hydrophobic side-chains in the neighborhood at position β8(*B*) can shield heme from the influence of the positively charged K7(*A*). In addition, Lys7(*A*) can be engaged in a salt bridge with Glu3(*A*) residue from the same helix *A*.

The individual contribution to the heme E_m potential of any other fully charged ionizable site (glutamates and lysines) is smaller than 5 mV but often only 2-3 mV. That is not surprising, considering the distant positions of their side-chains relative to heme, which stick into the aqueous phase, and the possibility of forming the ionic bridges, $Lys^+ \cdots Glu^-$, stabilizing the helical structure. Nevertheless, we observed that the ionic bridges are relatively weak because of the screening effects of high dielectric of water and the presence of the counterions in an aqueous solution.

The apparent exceptions are the propionic groups sited in the binding pocket if they are in the charged state. The contribution of a single propionate site (fully deprotonated) to the heme E_m potential could be up to -120 mV. In the structures with arginines, in the heme orientation HC1 or the heme orientation HC2, the combined contribution of both propionates is up to -200 mV or -70 mV, respectively.

If the structure contains arginines, the magnitude of contribution largely depends on their conformation, a location in or out of the four-helix bundle, and a position relative to the heme propionic groups, *e.i.*, if they can form the salt bridges or not. The charged and fully protonated arginine contributes to E_m potential up to +90 mV.

Arg at position α5 in helix A_2 (sequence 1, 15) is fully protonated and forms an ionic bridge with the corresponding pair of glutamates E9(A_2)/E10(B_1) that are consequently fully deprotonated. The ionic bridge is partially shorter and stronger between the Glu9 and Arg5 side-chains that come on top of each other, belonging to the same helix A_2. Arg5(A_1), whose side-chain is more located in the hydrophobic interface between helix A_1 and

B_2, is just partially protonated (0.4 – 0.6) and interacts only weakly with the glutamates E9(A_1)/E10(B_2) that are therefore partially deprotonated.

Table 3. Redox potentials of the studied synthetic hemoprotein structures[a]

#	Structure A(i)/B(j)	Exp. value [mV]b	Calc. value [mV]b	Total contribution of four Glu at α9, β10 positions [mV]c	Total charge at pH 7d	RMSD [Å] all atomse
1	A20/B11	-89	-86	-83.8	0.144	0.61
2	A13/B11	-94	-98	-68.7	0.148	0,54
3	A13/B02	-102	-95	-61.3	0.154	0.42
4	A13/B17	-103	-103	-61.5	0.153	0.00
5	A01/B11	-105	-127	-84.2	0.195	0.56
6	A02/B19	-108	-140	-82.7	0.257	0.70
7	A10/B07	-109	-114	-84.9	0.087	0.36
8	A13/B06	-110	-98	-68.9	0.134	0.44
9	A13/B13	-117	-130	-56.5	0.196	0.37
10	A13/B09	-117	-112	-88.2	0.188	0.49
11	A09/B14	-117	-138	-74.4	0.267	0.45
12	A01/B14	-120	-130	-66.4	0.407	0.43
13	A01/B09	-123	-110	-84.1	0.197	0.35
14	A14/B09	-125	-115	-56.5	0.309	0.38
15	A20/B13	-130	-126	-66.4	0.085	0.53
16	A19/B09	-132	-159	-79.4	0.230	0.69
17	A01/B19	-133	-139	-81.2	0.368	0.65
18	A12/B09	-135	-139	-84.7	0.028	0.36
19	A06/B14	-136	-140	-73.6	0.235	0.48
20	A08/B09	-137	-126	-73.1	0.121	0.37
21	A08/B04	-141	-155	-73.5	0.502	0.70
22	A18/B13	-143	-153	-69.3	0.115	0.34
23	A17/B14	-148	-146	-69.4	0.229	0.43
24	A10/B08	n/a	-157	-69.3	0.184	0.39
25	A08/B08	n/a	-166	-68.8	0.108	0.89

[a] The variable parts in the sequences of helix *A* and *B* denoted as $(\alpha_4\alpha_5\alpha_8\alpha_{12})_A$ $(\beta_4\beta_5\beta_8\beta_{12})_B$ are given in Table 1; for the complete sequences, see Figure 4.

[b] The measured (Rau, DeJonge, and Haehnel 2000) and calculated E_m values at pH 7.

[c] The total contribution of the four Glu at α9 and β10 positions to the heme E_m potentials.

[d] The total charge of Syn-*m*-HPs at pH 7.

[e] The RMSD of all atoms relative to the starting model structure 4 (in bold).

If there is a bulky hydrophobic side-chain like Phe or Tyr at position α5, the environment of the four glutamates gets more hydrophobic. Consequently, the sum of protons of these four glutamates rises around unity by about one-fourth of the proton on average. However, the sum of protons remains well below unity with Ala or Leu at position α5. In heme conformation HC1, the propionates are embedded in a hydrophobic pocket and therefore fully protonated (i.e., uncharged and neutral) unless they are involved in salt bridges with Arg at the position β5, as in the structures 6, 17, and 21 where both Arg5(B_1) and Arg5(B_2) form salt bridges with the two propionates. In heme conformation HC2, the propionates are more solvent-exposed and, therefore, mainly deprotonated. The two propionic groups attached to the pyrrole ring A (PRA) and pyrrole ring D (PRD) belonging to the same heme moiety are practically never full deprotonated at pH 7 due to the local proximity and the mutual strong electrostatic coupling, which is in agreement with the findings of the studies on native hemoproteins (Fufezan, Zhang, and Gunner 2008; Zheng and Gunner 2009) or Cyt*bc₁* complex (Popović et al. 2001). However, if the negative charges on the heme propionates are effectively compensated by the positive charges and dipoles of surrounding residues, as this is the case in cytochrome *c* oxidase, then all four propionic groups of two hemes can be fully deprotonated, as found in the recent computational studies (Popović and Stuchebrukhov 2004; Song, Michonova-Alexova, and Gunner 2006).

Furthermore, the helices of the four-helix bundle are slightly tilted up to 5 degrees among themselves in all Syn-*m*-HP structures. In general, the helix-bundles in native proteins can be more significantly tilted and even considerably bent or distorted (Iwata et al. 1998; Popović et al. 2001; Zarić, Popović, and Knapp 2001). As a result, the helices A_1/A_2 or B_1/B_2 and their corresponding residues are not entirely equivalent. Also, the matching inter-residue distances, interaction energies, and the protonation states of identical sites belonging to the same helix type may slightly differ.

Comparison of Calculated vs. Measured E_m

The calculated and measured redox potentials are compared in the correlation diagram, Figure 6, and Table 3. Somewhat larger deviations are only observed for the sequences 6 and 16 that contain arginines but also in the mutants 5 and 11. One reason for these discrepancies is the length of the Arg side-chain, which renders its exact position in the aqueous phase uncertain. Another reason is the protonation state of arginines, which varies depending on the mutual position relative to the other titratable groups, specifically hydrophobic side-chains or acidic and basic residues, including the heme propionic groups.

Exchanging the bulky side-chain Phe5(*A*) by Leu (sequence 2 to 5), the pair of glutamates E9(*A*)/E10(*B*) gets more solvent-exposed and therefore less protonated. The increase in negative charge in the heme neighborhood leads to a downshift of the redox potential. A general observation is that the redox potential moves closer to the solution value of -220 mV if large hydrophobic side-chains inside the heme-binding pocket are exchanged by small side-chains (sequence 11 to 23). That is caused by the cavities created in the neighborhood of the heme, where the dielectric constant becomes equal to 80.

Analyzing the different synthetic hemoproteins, we observe shifts in the redox potential by the following mechanisms. At position α12 near the propionates and α8 at the lateral sides of the heme, hydrophobic residues of different sizes can create cavities inside the heme-binding pocket. Filling the cavities with water would shift the redox potential on average by 10 mV. The glutamic acids, E9(*A*) and E10(*B*), at the helix interface (Figure 5A), are relatively close to the central part of the heme. They are generally negatively charged and yield a downshift of the redox potential by more than 20 mV per Glu. We observe upshifts of the heme redox potential depending on the hydrophobicity, polarity, and size of the residues placed at position α5, which is close to the glutamates E9(*A*) and E10(*B*). Bulky hydrophobic residues like Phe or Tyr at α5 shield the glutamates from the solvent. Consequently, these become partially protonated and increase the heme redox potential by up to 25 mV, counteracting the downshift influence of fully charged glutamates.

In the presence of Arg at α4, α5, and β5 that can form salt bridges with the acidic groups of propionates, E9(*A*) or E10(*B*), these groups remain deprotonated. Nevertheless, the positive charges of the arginines, which are closer to heme than the glutamates, generate a combined shift of the redox potential, which is typically upward, varying from +10 to +35 mV.

Other mutations at α4α11 and β5β12 are located on the interface between helix pairs A_1/B_1 and A_2/B_2 and refer to hydrophobic residues only. In these cases, the redox potentials are shifted upward if the packing of the corresponding side-chains is more dense and downward if the packing of the side-chains is looser, providing space for the solvent molecules.

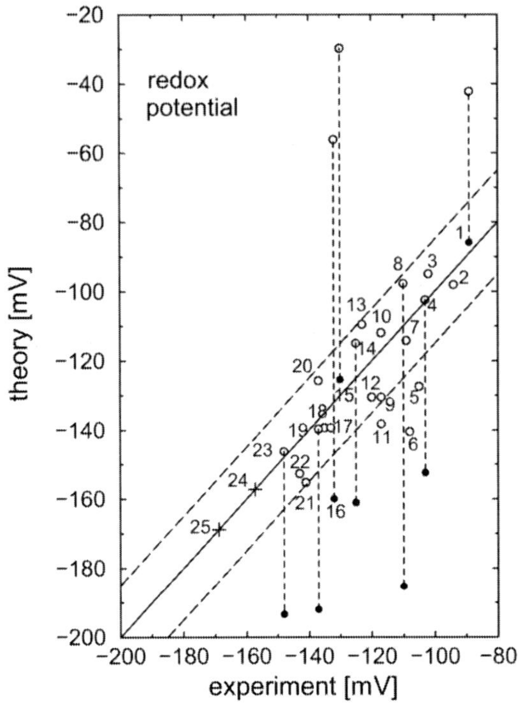

Figure 6. Correlation diagram of measured (Rau, DeJonge, and Haehnel 2000) *vs.* calculated redox potentials of 25 different synthetic mono-heme Cyt*b* proteins. Numbers denote the sequences from 1 to 25 with a falling value of the measured redox potential. The complete sequence is shown in Figure 4, with the variable parts in the sequence of helices *A* and *B* listed in Table 1. White circles refer to heme conformation HC1, black circles to heme orientation HC2 (see text). For the structures of sequences 24 and 25 denoted by "+," no measured redox potentials are available. The dashed diagonals mark an RMSD of 15mV between measured and computed redox potential.

Most of the hydrophilic and some of the hydrophobic residues at β4 (Gln) and β8 (Ala, Val, Leu, Tyr) positions are solvent-exposed and distant from the heme. They all have practically no influence on the heme redox potential.

During the modeling process, we noticed in some structures that the bulky hydrophobic residues at the α8 position might cause the heme ruffling and distortion of the porphyrin plane instead of making heme less water accessible. As a result, the heme ruffling may lead to an opposite effect of the expected one contributing to a decrease in the heme midpoint potential. The details and range of these effects in native heme-proteins are described in the literature (MacGowan and Senge 2013; Olea, Kuriyan, and Marletta 2010). A recent study on the tetra-heme cytochrome subunit of the reaction center has demonstrated that the redox-potential variations due to heme distortion can count up to 55 mV (MacGowan and Senge 2013). We observed that such features could play a specific role only in some of the Syn-*m*-HP structures, although the effect does not exceed 25 mV.

The charges of the template and inner hydrophobic side-chains show only a minor influence on the heme redox potential, but they act as a way to make heme less water accessible. Indeed, their screening and shielding effects might generally lead to a significant increase in the heme redox potential in agreement with the results of calculations obtained for native heme-proteins (Zheng and Gunner 2009). If all charges of the template and its linking parts are set to zero, it will affect the redox potential by only 2–8 mV. However, removing the template from the structure would result in the heme redox potential downshift of 50 to 70 mV. Hence, there are effects of screening the protein charges by a high dielectric of the aqueous phase and shielding the porphyrin and heme propionates from the solvent (aqueous solution), which generally causes a significant increase in the redox E_m potential.

Finally, by comparing experimental and calculated redox potentials, as well as by analyzing the conformational energies of the synthetic mono-heme proteins, we found that in structures 1, 15, and 16, which have arginines at position α4 or α5 close to the open end of the four-helix bundle, the heme is preferentially in HC2 with the propionates oriented toward the open end of the four-helix bundle. A much better agreement for other studied structures is obtained if the propionates of the heme point toward the template (HC1). Interestingly, the calculated E_m values of the heme conformation HC2 are more similar to the one computed, -167 mV (Popović et al. 2001), or measured, -170 mV (Rau and Haehnel 1998) and -182 mV (Willner et al. 1999), for the synthetic di-heme protein with the same heme propionates orientation, *i.e.*, toward the open end of the four-helix bundle. Generally, the heme conformation HC1 generates 40 to 100 mV higher midpoint potential values than the conformation HC2. Moreover, in the conformation HC2 with the heme propionates orientated toward the open end, the E_m potentials approach the solution value of the model compound of -220 mV (Wilson 1983; Moore and Pettigrew 1990). In these structures, depending on the solvent accessibility, the propionates could be more or less deprotonated, which would considerably affect the magnitude of the E_m downshift.

General Findings

In this chapter, we have implemented a molecular modeling approach by combining molecular mechanics with the CHARMM force field, molecular dynamics, and quantum mechanics to generate the structures of synthetic hemoproteins. Furthermore, we applied the continuum electrostatics method with protonation and redox titrations to study the protein−cofactor interactions in high resolution. In particular, the influence of the protein environment and the different mutations on the redox potential of synthetic hemoproteins has been examined. The applied concept seems intriguing and has scientific merit and shows the potential to provide new information relevant to both basic and applied science.

We have also considered sequences 24 and 25, where the corresponding redox potentials could not be measured since the values were outside the range used for the redox indicator. A large portion of the mutants considered focuses on the interface between helix pairs A_1/B_1 and A_2/B_2, where α4α11 and β5β12 are varied such that the hydrophobic residues come together except for arginines, see helical wheel projection of the four-helix bundle, Figure 5A. The helical wheel projection of the antiparallel four-helix bundle is a practical means to locate the residues on the surface, in the interfacial region or in the interior of the helix bundle. In the case of the model structures 24 and 25, the binding-helices B_1 and B_2 on β5β12 positions (Ala5Ala12), along with the corresponding hydrophobic residues (str. 24, Val4Leu11 and str. 25, Ala4Ala11) in the protecting-helices type *A*, do not provide a sufficient shielding of heme in the binding-pocket, making heme more water accessible. As a result, their heme midpoint potentials are the lowest of all structures.

Generally, the total shift in the redox potential relative to the redox potential of a model compound in the aqueous phase is mainly due to the low dielectric medium within the protein, bulky hydrophobic residues, the protein backbone charges and dipoles, as well as the possible salt bridges formed between arginines and heme propionate groups. The differences in the shift of the redox potential between different mutants are due to the altered interactions with the specific hydrophilic and charged side-chains (Glu9($A_{1,2}$), Glu10($B_{1,2}$), and arginines), salt bridges formed with the heme propionates, and particularly, the shielding effects of the hydrophobic residues. In native proteins, many non-bonding interactions of the polypeptide environment with the heme cofactor are formed with side chains of charged and polar amino acids, whereas the backbone groups are not frequently involved. However, the analysis performed in this study shows the strong impact of the backbone charges on the heme redox potential. Specifically, the backbone charges cause a positive shift (+30 to +50 mV) of the E_m potential due to the favorable orientation of –N-H groups toward the heme-binding pocket. Besides, it has been shown here that decreasing heme solvent exposure correlates with increasing its redox potential (greater ease of reduction) due to the desolvation penalty of burying a formally charged ferric heme in a low dielectric environment. The substantial impact of a dielectric medium in the hydrophobic interior of antiparallel helix-bundle is related to decreasing heme solvent exposure; however, this is not the only or decisive effect on the E_m values. We have found that the multiple interactions and the energy contributions are responsible for tuning the heme E_m potential in synthetic mono-heme proteins.

Ultimately, one can ask an interesting simple-minded question. What are the biological relevance and practical usage of *de novo* synthetic hemoproteins? First of all, synthetic hemoproteins are valuable model systems in many experimental and theoretical studies, and their biological significance is reflected through:

1. the model studies of the proton and electron transfer reactions in Syn-*m*-HPs;
2. the assessment of the effects of protein charges, different interaction types, and the local dielectric environment on the redox properties of the heme cofactors;
3. studying the protein structure–function relationship, and
4. understanding the protein–cofactor interactions at the atomic level.

Moreover, these specific protein systems have a practical purpose in bioelectronics (Willner et al. 1999; Katz and Willner 2004; Lisdat et al. 2009; Noll and Noll 2011) and bionanotechnology (Gazit 2007; Leger and Bertrand 2008; Cracknell, Vincent, and Armstrong 2008; Lisdat et al. 2009) where the ability to understand the factors controlling and fine-tuning their redox properties could be essential for the design and functioning of biosensors. The transfer and application of the construction principles from biology to protein engineering have enabled the design, synthesis, and use of artificial proteins as bioreceptors for sensing applications (Mutter et al. 1996; Zourob 2010; Piletsky and Whitcombe 2013). These proteins can be either fully synthetic or semi-synthetic and rationally engineered to improve or substitute natural variants in different applications. In contrast to their native counterparts, artificial bioreceptors usually offer improved stability and high affinity at a lower cost rendering them ideal structural entities for bioelectronic applications and disposable biosensors (Dincer et al. 2019).

Conclusion

Model building from scratch with advanced modeling techniques has been used to generate the atomic coordinates of the synthetic four-helix bundle mono-heme proteins. In this chapter, we have analyzed the impact of the protein environment, including the interactions of protein charges, an inhomogeneous dielectric medium, and the different mutations in the amino acid sequence on the redox properties and E_m potential of the heme cofactor in *de novo* synthetic hemoproteins by computational methods. The study contributes to rationalizing the factors that modulate the redox potential of the heme cofactor in proteins and to understanding protein structure−function relationship and apoprotein−cofactor interactions in atomic detail. It also shows how the synergy of modeling and computation can be used to design synthetic proteins with desired redox properties customized to specific functions. In general, the presented results of the study provide new valuable information relevant to both fundamental and applied sciences. More specifically, the gained knowledge is of broad interest to the protein design field since it can help the creation of new metalloproteins with predetermined redox potentials and, hence, with programmed functions. Furthermore, the applied approach could help, in bionanotechnology, to fabricate new materials and enzyme-based biosensors as novel bioelectronic devices. The elucidation of catalytic properties and customized redox functions of the synthetic proteins still remains an outstanding theoretical, experimental, and practical challenge for future nanoscale bioelectronics.

Acknowledgments

This work has been financially supported by the Ministry of Education, Science and Technological Development of the Republic of Serbia (MPNTR): Grant No. 451-03-68/2022-14/200026 and project OI/127035. This article/publication is based upon work from COST Action CA21101, supported by COST (European Cooperation in Science and Technology).

References

Albrecht, T., W. Li, J. Ulstrup, W. Haehnel, and P. Hildebrandt. 2005. "Electrochemical and spectroscopic investigations of immobilized de novo designed heme proteins on metal electrodes." *ChemPhysChem* 6:961–970.

Alexi, N., H. Jeanette, W. L. Birgitte, N. Lawrence, M. H. Roana, and et al. 2021. "Potential of novel cadaverine biosensor technology to predict shelf life of chilled yellowfin tuna (Thunnus albacares)." *Food Control* 119:107458.

Arai, R., N. Kobayashi, A. Kimura, T. Sato, K. Matsuo, A. F. Wang, J. M. Platt, L. H. Bradley, and M. H. Hecht. 2012. "Domain-swapped dimeric structure of a stable and functional de novo four-helix bundle protein, WA20." *J. Phys. Chem. B* 116:6789–6797.

Bănică, F.-G. 2012. Chemical Sensors and Biosensors: Fundamentals and Applications. Chichester, UK: John Wiley & Sons.

Baptista, A. M., P. J. Martel, and C. M. Soares. 1999. "Simulation of electron-proton coupling with a Monte Carlo method: Application to cytochrome c3 using continuum electrostatics." *Biophys. J.* 76:2978–2998.

Bashford, D. 1997. "An object-oriented programming suite for electrostatic field effects in biological molecules." In *Scientific Computing in Object-Oriented Parallel Environments*, edited by Y. Ishikawa, R. R. Oldehoeft, J. V. W. Reynders and M. Tholburn, 233–240. Berlin, Germany: Springer.

Bashford, D., D. A. Case, C. Dalvit, L. Tennant, and P. E. Wright. 1993. "Electrostatic calculations of side-chain pKa values in myoglobin and comparison with NMR data for histidines." *Biochemistry* 32:8045–8056.

Bashford, D., and K. Gerwert. 1992. "Electrostatic calculations of the pKa values of ionizable groups in bacteriorhodopsin." *J. Mol. Biol.* 224:473–486.

Bashford, D., and M. Karplus. 1990. "pKas of ionizable groups in proteins: Atomic detail from a continuum electrostatic model." *Biochemistry* 29:10219–10225.

Bashford, D., and M. Karplus. 1991. "Multiple-site titration curves of proteins: An analysis of exact and approximate methods for their calculations." *J. Phys. Chem.* 95:9557–9561.

Bayly, C. I., P. Cieplak, W. D. Cornell, and P. A. Kollman. 1993. *J. Phys. Chem.* 97:10269–10280.

Bialas, C., L. E. Jarocha, K. B. Henbest, T. M. Zollitsch, G. Kodali, C. R. Timmel, S. R. Mackenzie, P. L. Dutton, C. C. Moser, and P. J. Hore. 2016. "Engineering an Artificial Flavoprotein Magnetosensor." *J. Am. Chem. Soc.* 138:16584–16587.

Bryson, J. W., S. F. Betz, H. S. Lu, D. J. Suich, H. X. Zhou, K. T. O'Neil, and W. F. DeGrado. 1995. "Protein design: a hierarchic approach." *Science* 270:935–941.

Clark, L. C., and C. Lyons. 1962. "Electrode Systems for Continuous Monitoring in Cardiovascular Surgery." *Annals of the NY Academy of Sciences* 120:29-45.

Cornell, W., P. Cieplak, C. Bayly, and P. Kollman. 1993. *J. Am. Chem. Soc.* 115:9620–9631.

Couch, V., D. Popović, and A. Stuchebrukhov. 2011. "Redox-Coupled Protonation of Respiratory Complex I: The Hydrophilic Domain." *Biophys. J.* 101 (2):431–438.

Cracknell, J. A., K. A. Vincent, and F. A. Armstrong. 2008. "Enzymes as Working or Inspirational Electrocatalysts for Fuel Cells and Electrolysis." *Chem. Rev.* 108:2439−2461.

Crivianu-Gaita, V., and M. Thompson. 2016. "Aptamers, antibody scFv, and antibody Fab' fragments: An overview and comparison of three of the most versatile biosensor biorecognition elements." *Biosens Bioelectron.* 85:32–45.

Cui, T., V. Bondarenko, D. Ma, C. Canlas, N. R. Brandon, J. S. Johansson, Y. Xu, and P. Tang. 2008. "Four-alpha-helix bundle with designed anesthetic binding pockets. Part II: halothane effects on structure and dynamics." *Biophys. J.* 94:4464–4472.

Curnow, P. 2019. "Designing minimalist membrane proteins." *Biochem Soc Trans.* 47:1233–1245.

Degani, Y., and A. Heller. 1988. "Direct electrical communication between chemically modified enzymes and metal electrodes. 2. Methods for bonding electron-transfer relays to glucose oxidase and D-amino-acid oxidase." *J. Am. Chem. Soc.* 110:2615–2620.

DeGrado, W. F., C. M. Summa, V. Pavone, F. Nastri, and A. Lombardi. 1999. "De novo design and structural characterization of proteins and metalloproteins." *Annu. Rev. Biochem.* 68:779–819.

DeJonge, N., H. K. Rau, and W. Haehnel. 1999. "Light-induced electron transfer in synthetic metalloproteins." *Z. Phys. Chem.* 213:175–180.

Di Costanzo, L., H. Wade, S. Geremia, L. Randaccio, V. Pavone, W. F. DeGrado, and A. Lombardi. 2001. "Toward the de novo design of a catalytically active helix bundle: a substrate-accessible carboxylate-bridged dinuclear metal center." *J. Am. Chem. Soc.* 123:12749–12757.

Dincer, C., R. Bruch, E. Costa-Rama, M. T. Fernández-Abedul, A. Merkoçi, A. Manz, G. A. Urban, and F. Güder. 2019. "Disposable Sensors in Diagnostics, Food, and Environmental Monitoring." *Advanced Materials* 31 (30):1806739.

Donzella, V., and F. Crea. 2011. "Optical biosensors to analyze novel biomarkers in oncology." *J. Biophotonics.* 4 (6):442–452.

Drummond, T. G., M. G. Hill, and J. K. Barton. 2003. "Electrochemical DNA sensors." *Nature biotechnology* 21:1192–1199.

Fahnenschmidt, M., H. K. Rau, R. Bittl, W. Haehnel, and W. Lubitz. 1999. "Characterization of a de novo designed heme protein by EPR and ENDOR spectroscopy." *Chem. Eur. J.* 5:2327–2334.

Feynman, R. 1999. "There is Plenty of Room at the Bottom." In *The Pleasure of Finding Things Out*, edited by J. Robbins. Perseus Books.

Friedrich, M. G., M. A. Plum, M. G. Santonicola, V. U. Kirste, W. Knoll, B. Ludwig, and R. L. C. Naumann. 2008. "In Situ Monitoring of the Catalytic Activity of Cytochrome c Oxidase in a Biomimetic Architecture." *Biophysical Journal* 95:1500–1510.

Gaussian 09 Revision A.02. Gaussian, Inc., Wallingford, CT.

Fry, H. C., A. Lehmann, L. E. Sinks, I. Asselberghs, A. Tronin, V. Krishnan, J. K. Blasie, K. Clays, W. F. Degrado, J. G. Saven, and M. J. Therien. 2013. "Computational de novo design and characterization of a protein that selectively binds a highly hyperpolarizable abiological chromophore." *J. Am. Chem. Soc.* 135:13914–13926.

Fufezan, C., J. Zhang, and M. R. Gunner. 2008. "Ligand preference and orientation in b- and c-type heme-binding proteins." *Proteins* 73:690–704.

Gao, W., S. Sattayasamitsathit, J. Orozco, and J. Wang. 2011. "Highly Efficient Catalytic Microengines: Template Electrosynthesis of Polyaniline/Platinum Microtubes." *J. Am. Chem. Soc.* 133:11862–11864.

Gazit, E. 2007. *Plenty of room for biology at the bottom: An introduction to bionanotechnology*. London: Imperial College Press.

Geremia, S., L. Di Costanzo, L. Randaccio, D. E. Engel, A. Lombardi, F. Nastri, and W. F. Degrado. 2005. "Response of a designed metalloprotein to changes in metal ion coordination, exogenous ligands, and active site volume determined by X-ray crystallography." *J. Am. Chem. Soc.* 127:17266–17276.

Ghirlanda, G., A. Osyczka, W. Liu, M. Antolovich, K. M. Smith, P. Leslie Dutton, A. J. Wand, and W. F. DeGrado. 2004. "De Novo Design of a D2-Symmetrical Protein that Reproduces the Diheme Four-Helix Bundle in Cytochrome bc1." *J. Am. Chem. Soc.* 126:8141–8147.

Ghosh, D., and V. L. Pecoraro. 2005. "Probing metal–protein interactions using a de novo design approach." *Curr. Opin. Chem. Biol.* 9:97–103.

Gibney, B. R., Y. Isogai, F. Rabanal, K. S. Reddy, A. M. Grosset, C. C. Moser, and P. L. Dutton. 2000. "Self-assembly of heme a and heme b in a designed four-helix bundle: Implications for a cytochrome c oxidase maquette." *Biochemistry* 39:11041–11049.

Gilson, M. K., and B. Honig. 1986. "The dielectric constant of a folded proteins." *Biopolymers* 25:2097–2119.

Gunner, M. R., and E. Alexov. 2000. "A pragmatic approach to structure based calculation of coupled proton and electron transfer in proteins." *Biochim. Biophys. Acta* 1458:63–87.

Gunner, M. R., E. Alexov, E. Tores, and S. Lipovaca. 2000. "The importance of the potein in controlling the electrochemistry of heme metalloproteins: methods of calculation and analysis." *J. Biol. Inorg. Chem.* 2:126–134.

Gunner, M. R., and B. Honig. 1991. "Electrostatic control of midpoint potentials in the cytochrome subunit of the Rhodopseudomonas viridis reaction center." *Proc. Natl. Acad. Sci. USA* 88:9151–9155.

Haehnel, W. 2004. "Chemical synthesis of TASP arrays and their application in protein design." *Molecular Diversity* 8:219–229.

Heleg-Shabtai, V., E. Katz, and I. Willner. 1997. "Assembly of Microperoxidase-11 and Co(II)-Protoporphyrin IX Reconstituted Myoglobin Monolayers on Au-Electrodes: Integrated Bioelectrocatalytic Interfaces." *J. Am. Chem. Soc.* 119:8121–8122.

Heller, A. 1990. "Electrical wiring of redox enzymes." *Acc. Chem. Res.* 23:128–134.

Heuer-Jungemann, A., N. Feliu, I. Bakaimi, M. Hamaly, A. Alkilany, I. Chakraborty, A. Masood, M. F. Casula, A. Kostopoulou, E. Oh, K. Susumu, M. H. Stewart, I. L. Medintz, E. Stratakis, W. J. Parak, and A. G. Kanaras. 2019. "The Role of Ligands in the Chemical Synthesis and Applications of Inorganic Nanoparticles." *Chem. Rev.* 119:4819–4880.

Hill, R. B., and W. F. DeGrado. 1998. "Solution Structure of Alpha2D, A Native like De Novo Designed Protein." *J. Am. Chem. Soc.* 120:1138–1145.

Honig, B., and A. Nicholls. 1995. "Classical electrostatics in biology and chemistry." *Science* 268:1144–1149.

Hosseinzadeh, P., N. M. Marshall, K. N. Chacón, Y. Yu, M. J. Nilges, S. Y. New, S. A. Tashkov, N. J. Blackburn, and Y. Lu. 2016. "Design of a single protein that spans the entire 2-V range of physiological redox potentials." *Proc. Natl. Acad. Sci. USA* 113:262–267.

Huang, P. S., S. E. Boyken, and D. Baker. 2016. "The coming of age of de novo protein design." *Nature* 537:320-327.

Huang, S. S., B. R. Gibney, S. E. Stayrook, P. L. Dutton, and M. Lewis. 2003. "X-ray Structure of a Maquette Scaffold." *J. Mol. Biol.* 326:1219–1225.

Iwata, S., J. W. Lee, K. Okada, J. K. Lee, M. Iwata, B. Rasmussen, T. A. Link, S. Ramaswamy, and B. K. Jap. 1998. "Complete structure of the 11-subunit bovine mitochondrial cytochrome bc1 complex." *Science* 281:64–71.

Joh, N. H., T. Wang, M. P. Bhate, R. Acharya, Y. Wu, M. Grabe, M. Hong, G. Grigoryan, and W. F. DeGrado. 2014. "De novo design of a transmembrane Zn2+-transporting four-helix bundle." *Science* 346:1520–1524.

Jung, G. 1996. *Combinatorial Peptide and Nonpeptide Libraries: A Handbook*. Edited by G. Jung. Weinheim, Germany: Wiley-VCH.

Juzgado, A., A. Solda, A. Ostric, A. Criado, G. Valenti, S. Rapino, G. Conti, G. Fracasso, F. Paolucci, and M. Prato. 2017. "Highly sensitive electrochemiluminescence detection of a prostate cancer biomarker." *J. Mater. Chem. B.* 5 (32):6681–6687.

Kalantar-Zadeh, K. 2013. *Sensors: An Introductory Course*. Boston, MA, USA: Springer US.

Katz, E., V. Heleg-Shabtai, A. Bardea, I. Willner, H. K. Rau, and W. Haehnel. 1998. "Fully integrated biocatalytic electrodes based on bioaffinity interactions." *Biosensors and Bioelectronics* 13:741–756.

Katz, E., V. Heleg-Shabtai, I. Willner, H. K. Rau, and W. Haehnel. 1998. "Surface Reconstitution of a de novo Synthesized Hemoprotein for Bioelectronic Applications." *Angew. Chem. Int. Ed.* 37:3253–3256.

Katz, E., and I. Willner. 2004. "Integrated Nanoparticle–Biomolecule Hybrid Systems: Synthesis, Properties, and Applications." *Angew. Chem. Int. Ed. Engl.* 43:6042–6108.

Kennedy, M. L., S. Silchenko, N. Houndonougbo, B. R. Gibney, P. Leslie Dutton, K. R. Rodgers, and D. R. Benson. 2001. "Model Hemoprotein Reduction Potentials: The Effects of Histidine-to-Iron Coordination Equilibrium." *J. Am. Chem. Soc.* 123:4635-4636.

Kim, M. R., Z. Xu, G. Chen, and D. Ma. 2014. "Semiconductor and Metallic Core–Shell Nanostructures: Synthesis and Applications in Solar Cells and Catalysis." *Chem. Eur. J.* 20:11256–11275.

Koder, R. L., J. L. Ross Anderson, L. A. Solomon, K. S. Reddy, C. C. Moser, and P. L. Dutton. 2009. "Design and engineering of an O2 transport protein." *Nature* 458:305–309.

Lalaurie, C. J., V. Dufour, A. Meletiou, S. Ratcliffe, A. Harland, Wilson O., C. Vamasiri, D. K. Shoemark, C. Williams, C. J. Arthur, R. B. Sessions, M. P. Crump, J. L. R. Anderson, and P. Curnow. 2018. "The de novo design of a biocompatible and functional integral membrane protein using minimal sequence complexity." *Sci. Rep.* 8:14564.

Lancaster, C. R., H. Michel, B. Honig, and M. R. Gunner. 1996. "Calculated coupling of electron and proton transfer in the photosynthetic reaction center of Rhodopseudomonas viridis." *Biophys. J.* 70:2469–2492.

Leger, C., and P. Bertrand. 2008. "Direct Electrochemistry of Redox Enzymes as a Tool for Mechanistic Studies." *Chem. Rev.* 108:2379−2438.

Li, W.-W., P. Hellwig, M. Ritter, and W. Haehnel. 2006. "De novo design, synthesis, and characterization of quinoproteins." *Chem. Eur. J.* 12:7236–7245.

Lisdat, F., R. Dronov, H. Mohwald, F. W. Scheller, and D. G. Kurth. 2009. "Self-Assembly of Electro-Active Protein Architectures on Electrodes for the Construction of Biomimetic Signal Chains." *Chem. Commun.*:274–283.

Liu, G., W. Shao, S. Zhu, and W. Tang. 1995. "Effects of axial ligand replacement on the redox potential of cytochrome c." *J. Inorg. Biochem.* 60:123–131.

Lombardi, A., F. Pirro, O. Maglio, M. Chino, and W. F. DeGrado. 2019. "De Novo Design of Four-Helix Bundle Metalloproteins: One Scaffold, Diverse Reactivities." *Acc. Chem. Res.* 52:1148–1159.

Lombardi, A., C. M. Summa, S. Geremia, L. Randaccio, V. Pavone, and W. F. DeGrado. 2000. "Inaugural article: Retrostructural analysis of metalloproteins: Application to the design of a minimal model for diiron proteins." *Proc. Natl. Acad. Sci. USA* 97:6298–6305.

Lopandić, Z., L. Dragačević, D. M. Popović, U. Andjelković, R. Minić, and M. D. Gavrović-Jankulović. 2021. "BanLec-eGFP chimera as a tool for evaluation of lectin binding to high-mannose glycans on microorganisms." *Biomolecules* 11:1–13, article number 180.

Lu, Y., S. M. Berry, and T. D. Pfister. 2001. "Engineering novel metalloproteins: design of metal-binding sites into native protein scaffolds." *Chem. Rev.* 101:3047–3080.

Lu, Y., S. Chakraborty, K. D. Miner, T. D. Wilson, A. Mukherjee, Y. Yu, J. Liu, and N. M. Marshall. 2013. "Metalloprotein Design." In *Comprehensive Inorganic Chemistry II (Second Edition): From Elements to Applications*, 565–593.

Lu, Y., N. Yeung, N. Sieracki, and N. M. Marshall. 2009. "Design of functional metalloproteins." *Nature* 460:855–862.

Ludwig, R., R. Ortiz, C. Schulz, W. Harreither, C. Sygmund, and L. Gorton. 2013. "Cellobiose Dehydrogenase Modified Electrodes: Advances by Materials Science and Biochemical Engineering." *Anal. Bioanal. Chem.* 405:3637−3658.

Ma, D., N. R. Brandon, T. Cui, V. Bondarenko, C. Canlas, J. S. Johansson, P. Tang, and Y. Xu. 2008. "Four-alpha-helix bundle with designed anesthetic binding pockets. Part I: structural and dynamical analyses." *Biophys. J.* 94:4454–4463.

MacGowan, S. A., and M. O. Senge. 2013. "Computational Quantification of the Physicochemical Effects of Heme Distortion: Redox Control in the Reaction Center Cytochrome Subunit of Blastochloris viridis." *Inorg. Chem.* 52:1228–1237.

MacKerell, Jr., A. D., D. Bashford, M. Bellot, Jr. Dunbrack, R. L., J. D. Evanseck, M. J. Field, S. Fischer, J. Gao, H. Guo, S. Ha, D. Joseph-McCarthy, L. Kuchnir, K. Kuczera, F. T. K. Lau, C. Mattos, S. Michnick, T. Ngo, D. T. Nguyen, B. Prodholm, III Reiher, W. E., B. Roux, M. Schlenkrich, J. C. Smith, R. Stote, J. Straub, M. Watanabe, J.

Wiórkiewicz-Kuczera, D. Yin, and M. Karplus. 1998. "All-atom empirical potential for molecular modeling and dynamics studies of proteins." *J. Phys. Chem.* 102:3586–3616.

Mahajan, M., and S. Bhattacharjya. 2014. "Designed di-heme binding helical transmembrane protein." *Chembiochem.* 15 (9):1257–1262.

Martin-Banderas, L., M. A. Holgado, M. Duran-Lobato, J. J. Infante, J. Alvarez-Fuentes, and M. Fernandez-Arevalo. 2016. "Role of nanotechnology for enzyme replacement therapy in lysosomal diseases. A focus on Gaucher's disease (Review)." *Current Medicinal Chemistry* 23:929–952.

Mi, F., M. Guan, C. Hu, F. Peng, S. Sun, and X. Wang. 2021. "Application of lectin-based biosensor technology in the detection of foodborne pathogenic bacteria: A review." *Analyst* 146 (2):429–443.

Monien, B. H., F. Drepper, M. Sommerhalter, W. Lubitz, and W. Haehnel. 2007. "Detection of Heme Oxygenase Activity in a Library of Four-helix Bundle Proteins: Towards the de Novo Synthesis of Functional Heme Proteins." *Journal of Molecular Biology* 371:739–753.

Moore, G. R., and G. W. Pettigrew. 1990. "Cytochromes c: Evolutionary, structural and physicochemical aspects." In, 309–362. Berlin, Germany: Springer.

Moser, C. C., M. M. Sheehan, N. M. Ennist, G. Kodali, C. Bialas, M. T. Englander, B. M. Discher, and P. L. Dutton. 2016. "De Novo Construction of Redox Active Proteins." In *Methods in Enzymology*, 365-388.

Mukherjee, S., and N. Mukherjee. 2021. "Current developments in diagnostic biosensor technology: Relevance to therapeutic intervention of infectious and inflammatory diseases of human." In *Modern Techniques in Biosensors*, 1-36. Berlin, Germany: Springer.

Mutter, M., P. Dumy, P. Garrouste, C. Lehmann, M. Mathieu, C. Peggion, S. Peluso, A. Razaname, and G. Tuchscherer. 1996. "Template assembled synthetic proteins (TASP) as functional mimetics of proteins." *Angew. Chem. Int. Ed. Engl.* 35 (13-14):1482–1485.

Mutter, M., R. Hersperger, K. Gubernator, and K. Muller. 1989. "The construction of new proteins: V. A template-assembled synthetic protein (TASP) containing both a 4-helix bundle and β-barrel-like structure." *Proteins Struct. Funct. Bioinform.* 5 (1):13–21.

Nesset, M. J. M., N. V. Shokhirev, P. D. Enemark, S. E. Jacobson, and F. A. Walker. 1996. "Models of the Cytochromes. Redox Properties and Thermodynamic Stabilities of Complexes of "Hindered" Iron(III) and Iron(II) Tetraphenylporphyrinates with Substituted Pyridines and Imidazoles." *Inorg. Chem.* 35:5188–5200.

Neya, S., M. Suzuki, T. Hoshino, H. Ode, K. Imai, T. Komatsu, A. Ikezaki, M. Nakamura, Y. Furutani, and H. Kandori. 2010. "Molecular Insight into Intrinsic Heme Distortion in Ligand Binding in Hemoprotein." *Biochemistry* 49:5642–5650.

Noll, T., and G. Noll. 2011. "Strategies for "Wiring" Redox-Active Proteins to Electrodes and Applications in Biosensors, Biofuel Cells, and Nanotechnology." *Chem. Soc. Rev.* 40:3564−3576.

Nölting, B. 2005. "Biophysical Nanotechnology." In *Methods in Modern Biophysics*. Berlin, Heidelberg, New York: Springer.

Olea, Jr. C., J. Kuriyan, and M. A. Marletta. 2010. "Modulating Heme Redox Potential through Protein-Induced Porphyrin Distortion." *J. Am. Chem. Soc.* 132:12794–12795.

Olson, T. L., J. C. Williams, and J. P. Allen. 2013. "Influence of protein interactions on oxidation/reduction midpoint potentials of cofactors in natural and de novo metalloproteins." *Biochim. Biophys. Acta* 1827:914–922.

Pellerano, M., and M. C. Morris. 2021. "Fluorescent peptide biosensors for probing CDK Kinase activity in cell extracts." In *Cell Cycle Oscillators*, 39-50. Berlin, Germany: Springer.

Peng, G., U. Tisch, and H. Haick. 2009. "Detection of Nonpolar Molecules by Means of Carrier Scattering in Random Networks of Carbon Nanotubes: Toward Diagnosis of Diseases via Breath Samples." *Nano Lett.* 9:1362–1368.

Piletsky, S. A., and M. J. Whitcombe. 2013. *Designing Receptors for the Next Generation of Biosensors*. Berlin, Germany: Springer.

Pinter, T. B. J., K. J. Koebke, and V. L. Pecoraro. 2020. "Catalysis and Electron Transfer in De Novo Designed Helical Scaffolds." *Angew. Chem. Int. Ed. Engl.* 59:7678-7699.

Popović, D. M. 2013. "Current advances in research of cytochrome c oxidase." *Amino Acids* 45:1073–1087.

Popović, D. M., and I. S. Djordjević. 2020. "Catalytic center of cytochrome c oxidase: Effects of protein environment on pKa values of CuB histidine ligands." *Journal of Serbian Chemical Society* 85:1429–1444.

Popović, D. M., I. V. Leontyev, D. G. Beech, and A. A. Stuchebrukhov. 2010. "Similarity of cytochrome c oxidases in different organisms." *Proteins Struct. Funct. Bioinform.* 78:2691–2698.

Popović, D. M., and A. A. Stuchebrukhov. 2004. "Electrostatic study of proton pumping mechanism of bovine heart cytochrome c oxidase." *J. Am. Chem. Soc.* 126:1858–1871.

Popović, D. M., and A. A. Stuchebrukhov. 2005. "Proton exit channels in bovine cytochrome c oxidase." *J. Phys. Chem. B* 109:1999–2006.

Popović, D. M., and A. A. Stuchebrukhov. 2006. "Two conformational states of Glu242 and pKas in bovine cytochrome c oxidase." *Photochem. Photobiol. Sci.* 5:611–620.

Popović, D. M., and A. A. Stuchebrukhov. 2012. "Coupled electron and proton transfer reactions during the O to E transition in bovine cytochrome c oxidase." *Biochim. Biophys. Acta* 1817 (4):506–517.

Popović, D. M., S. D. Zarić, B. Rabenstein, and E. W. Knapp. 2001. "Artificial cytochrome b: Computer modeling and evaluation of redox potentials." *J. Am. Chem. Soc.* 123:6040–6053.

Popović, D. M., A. Zmiric, S. D. Zarić, and E. W. Knapp. 2002. "Energetics of radical transfer in DNA photolyase." *J. Am. Chem. Soc.* 124:3775–3782.

Qiu, C., X. Wang, X. Liu, S. Hou, and H. Ma. 2012. "Direct electrochemistry of glucose oxidase immobilized on nanostructured gold thin films and its application to bioelectrochemical glucose sensor." *Electrochimica Acta* 67:140-146.

Quenneville, J., D. M. Popović, and A. A. Stuchebrukhov. 2006. "Combined DFT and electrostatics study of the proton pumping mechanism in cytochrome c oxidase." *Biochim. Biophys. Acta* 1757:1035–1046.

Rabenstein, B., and E. W. Knapp. 2001. "Calculated pH-dependent population of CO-myoglobin conformers." *Biophys. J.* 80:1141–1150.

Rajendran, S. R. C. K., C. C. Udenigwe, and R. Y. Yada. 2016. "Nanochemistry of Protein-Based Delivery Agents." *Front. Chem.* 4:31.

Rau, H. K., N. DeJonge, and W. Haehnel. 1998. "Modular synthesis of de novo-designed metalloproteins for light-induced electron transfer." *Proc. Natl. Acad. Sci. USA* 95:11526–11531.

Rau, H. K., N. DeJonge, and W. Haehnel. 2000. "Combinatorial synthesis of four-helix bundle hemoproteins for tuning of cofactor properties." *Angew. Chem. Int. Ed.* 39:250–253.

Rau, H. K., and W. Haehnel. 1998. "Design, synthesis, and properties of a novel cytochrome b model." *J. Am. Chem. Soc.* 120:468–476.

Reedy, C. J., and B. R. Gibney. 2004. "Heme protein assemblies." *Chem. Rev.* 104:617–649.

Ricci, F., G. Volpe, L. Micheli, and G. Palleschi. 2007. "A review on novel developments and applications of immunosensors in food analysis." *Analytica chimica acta* 605:111–129.

Riklin, A., E. Katz, I. Willner, A. Stocker, and A. F. Bückmann. 1995. "Improving enzyme–electrode contacts by redox modification of cofactors." *Nature* 376:672–675.

Robertson, D. E., R. S. Farid, C. C. Moser, J-L. Urbauer, S. E. Mulholland, R. Pidikiti, J. D. Lear, A. J. Wand, W. F. DeGrado, and P. L. Dutton. 1994. "Design and synthesis of multi-haem proteins." *Nature* 368:425–432.

Safo, M. K., M. J. M. Nesset, F. A. Walker, P. G. Debrunner, and W. R. Scheidt. 1997. "Models of the Cytochromes. Axial Ligand Orientation and Complex Stability in Iron(II) Porphyrinates: The Case of the Noninteracting d(π) Orbitals." *J. Am. Chem. Soc.* 119:9438–9448.

Sakamoto, S., I. Obataya, A. Ueno, and H. Mihara. 1999. "Effects of amino acids substitution of hydrophobic residues on haem-binding properties of designed two-α-helix peptides." *J. Chem. Soc. Perkin Trans.* 2:2059–2069.

Salvati, E., F. Stellacci, and S. Krol. 2015. "Nanosensors for early cancer detection and for therapeutic drug monitoring." *Nanomedicine (Lond)* 10 (23):3495-3512.

Sarauli, D., K. Peters, C. Xu, B. Schulz, D. Fattakhova-Rohlfing, and F. Lisdat. 2014. "3D-Electrode Architectures for Enhanced Direct Bioelectrocatalysis of Pyrroloquinoline Quinone-Dependent Glucose Dehydrogenase." *ACS Appl. Mater. Interfaces* 6:17887−17893.

Schnepf, R., P. Hörth, E. Bill, K. Wieghardt, P. Hildebrandt, and W. Haehnel. 2001. "De novo Design and Characterization of Copper Centers in Synthetic Four-Helix Bundle Proteins." *J. Am. Chem. Soc.* 123:2186–2195.

Schuhmann, W. 1995. "Electron-transfer pathways in amperometric biosensors. Ferrocene-modified enzymes entrapped in conducting-polymer layers." *Biosens. Bioelectron.* 10:181–193.

Schutz, C. N., and A. Warshel. 2001. "What are the dielectric ''constants'' of proteins and how to validate electrostatic models." *Proteins* 44:400–417.

Service, R. F. 2002. "Nanotechology: Biology offers nanotechs a helping hand." *Science* 298:2322–2323.

Sharp, K., and B. Honig. 1990. "Electrostatic interactions in macromolecules: Theory and applications." *Ann. Rev. Biophys. Biophys. Chem.* 19:301–332.

Shifman, J. M., C. C. Moser, W. A. Kalsbeck, D. F. Bocian, and P. L. Dutton. 1998. *Biochemistry* 37:16815.

Shleev, S., J. Tkac, A. Christenson, T. Ruzgas, A. I. Yaropolov, J. W. Whittaker, and L. Gorton. 2005. "Direct Electron Transfer between Copper-Containing Proteins and Electrodes." *Biosens. Bioelectron.* 20:2517–2554.

Simonson, T., and C. L. Brooks. 1996. "Charge separation and the dielectric constant of proteins: Insights from molecular dynamics." *J. Am. Chem. Soc.* 118:8452–8458.

Skalicky, J. J., B. R. Gibney, F. Rabanal, R. J. Bieber-Urbauer, P. L. Dutton, and A. J. Wand. 1999. "Solution structure of a designed four-α-helix bundle maquette scaffold." *J. Am. Chem. Soc.* 121:4941–4951.

Soares, C. M., and A. M. Baptista. 2012. "Simulation of multihaem cytochromes." *FEBS Lett.* 586:510–518.

Song, Y., J. Mao, and M. R. Gunner. 2006. "Electrostatic environment of hemes in proteins: pKas of hydroxyl ligands." *Biochemistry* 45:7949–7958.

Song, Y., E. Michonova-Alexova, and M. R. Gunner. 2006. "Calculated Proton Uptake on Anaerobic Reduction of Cytochrome c Oxidase: Is the Reaction Electroneutral?" *Biochemistry* 45:7959–7975.

Tezcan, F. A., J. R. Winkler, and H. B. Gray. 1998. "Effects of Ligation and Folding on Reduction Potentials of Heme Proteins." *J. Am. Chem. Soc.* 120:13383–13388.

Ullmann, M. G. 2000. "The coupling of protonation and reduction in proteins with multiple centers: theory, computational method, and application to cytochrome c3." *J. Phys. Chem. B* 104:6293–6301.

Ullmann, M. G., and E. W. Knapp. 1999. "Electrostatic models for computing protonation and redox equilibria in proteins." *Eur. Biophys. J.* 28:533–551.

Uygun, Z. O., and S. Atay. 2021. "Label-free highly sensitive detection of DNA approximate length and concentration by impedimetric CRISPR-dCas9 based biosensor technology." *Bioelectrochemistry* 140:107812.

Vagedes, P., B. Rabenstein, J. Åqvist, J. Marelius, and E. W. Knapp. 2000. "The deacylation step of acetylcholinesterase: Computer simulation studies." *J. Am. Chem. Soc.* 122:12254–12262.

Valenti, G., E. Rampazzo, E. Biavardi, E. Villani, G. Fracasso, M. Marcaccio, F. Bertani, D. Ramarli, E. Dalcanale, F. Paolucci, and L. Prodi. 2015. "An electrochemiluminescence-supramolecular approach to sarcosine detection for early diagnosis of prostate cancer." *Faraday Discuss.* 185:299–309.

Varadarajan, R., T. E. Zewert, H. B. Gray, and S. G. Boxer. 1989. "Effects of buried ionizable amino acids on the reduction potential of recombinant myoglobin." *Science* 243:69–72.

Vitale, R., L. Lista, C. Cerrone, G. Caserta, M. Chino, O. Maglio, F. Nastri, V. Pavone, and A. Lombardi. 2015. "An artificial heme-enzyme with enhanced catalytic activity: Evolution, functional screening and structural characterization." *Org. Biomol. Chem.* 13 (17):4859–4868.

Wang, X. H., and S. Wang. 2008. "Sensors and biosensors for the determination of small molecule biological toxins." *Sensors* 8:6045–6054.

Wang, Z., S. Lee, K. Koo, and K. Kim. 2016. "Nanowire-Based Sensors for Biological and Medical Applications." *IEEE Trans Nanobioscience* 15:186–199.

Warshel, A., and A. Papazyan. 1998. "Electrostatic effects in macromolecules: Fundamental concepts and practical modeling." *Curr. Op. Struct. Biol.* 8:211–217.

Warshel, A., A. Papazyan, and I. Muegge. 1997. "Microscopic and semimicroscopic redox calculations: What can and cannot be learned from continuum models." *J. Biol. Inorg. Chem.* 2:143–152.

Warshel, A., S. T. Russel, and A. K. Churg. 1984. "Macroscopic models for studies of electrostatic interactions in proteins: Limitations and applicability." *Proc. Natl. Acad. Sci. USA* 81:4785–4789.

Watkins, D. W., C. T. Armstrong, and J. L. R. Anderson. 2014. "De novo protein components for oxidoreductase assembly and biological integration." *Current Opinion in Chemical Biology* 19:90–98.

Weber, W., and M. Fussenegger. 2012. "Emerging biomedical applications of synthetic biology." *Nat. Rev. Genet.* 13:21–35.

Willner, B., and I. Willner. 2005. "Reconstituted Redox Enzymes on Electrodes: From Fundamental Understanding of Electron Transfer at Functionalized Electrode Interfaces to Biosensor and Biofuel Cell Applications." In *Bioelectronics – from Theory to Applications*, edited by I. Willner and E. Katz, 35–97. Weinheim, Germany: Wiley-VCH.

Willner, I., V. Heleg-Shabtai, E. Katz, H. K. Rau, and W. Haehnel. 1999. "Integration of a reconstituted de novo synthesized hemoprotein and native metalloproteins with electrode supports for bioelectronic and bioelectrocatalytic applications." *J. Am. Chem. Soc.* 121:6455–6468.

Willner, I., E. Katz, N. Lapidot, and P. Bäuerle. 1992. "Bioelectrocatalysed Reduction of Nitrate Utilizing Polythiophene Bipyridinium Enzyme-Electrodes." *Bioelectrochem. Bioenerg.* 29:29–45.

Wilson, G. S. 1983. "Electrochemical studies of porphyrin redox reactions as cytochrome models." *Bioelectrochem. Bioenerg.* 1:172–179.

Yu, F., V. M. Cangelosi, M. L. Zastrow, M. Tegoni, J. S. Plegaria, A. G. Tebo, C. S. Mocny, L. Ruckthong, H. Qayyum, and V. L. Pecoraro. 2014. "Protein Design: Toward Functional Metalloenzymes." *Chem. Rev.* 114:3495-3578.

Zarić, S. D., D. M. Popović, and E. W. Knapp. 2001. "Factors determining the orientation of axially coordinated imidazoles in heme proteins." *Biochemistry* 40:7914–7928.

Zhang, Y., S. Jingxue, L. Liansheng, and Q. Hong. 2021. "A review of biosensor technology and algorithms for glucose monitoring." *Journal of Diabetes and its Complications* 35 (8):107929.

Zheng, Z., and M. R. Gunner. 2009. "Analysis of the electrochemistry of hemes with E(m)s spanning 800 mV." *Proteins* 75:719–734.

Zourob, M. 2010. *Recognition Receptors in Biosensors*. New York, NY, US: Springer US.

Index

D

E

F

G

I

J

K

L

M

N

O

P

Q

R

S

T

U

V

W

Z